GOD *and the* History of the Universe

GOD
and the
History of the Universe

JARVIS STREETER

Edited by David J. Lull

PICKWICK *Publications* · Eugene, Oregon

GOD AND THE HISTORY OF THE UNIVERSE

Pickwick Publications
An Imprint of Wipf and Stock Publishers
199 W. 8th Ave., Suite 3
Eugene, OR 97401

www.wipfandstock.com

PAPERBACK ISBN: 978-1-4982-3678-2
HARDBACK ISBN: 978-1-4982-3680-5

Cataloguing-in-Publication data:

Streeter, Jarvis.

God and the history of the universe / Jarvis Streeter ; edited and with a foreword by David J. Lull.

xxx + 470 pp. ; 23 cm. Includes bibliographical references and indexes.

ISBN: 978-1-4982-3678-2 (paperback) | 978-1-4982-3680-5 (hardback)

1. Theology. 2. Religion and science. 3. Cosmology. 4. World view. I. Lull, David J. II. Franklin, Craig. III. Thomas, Sam. IV. Tierney, Nathan L. V. Streeter, Susan Gresham. VI. Title.

BL265 S77 2016

Manufactured in the U.S.A. 03/22/2016

For Susan, my loving friend on this surprising
and challenging adventure that is life.

Contents

Who Was Jarvis?

Craig Franklin
San Francisco, CA

I'm guessing not many books about theology are introduced by someone with huge doubts about God and religion. But Jarvis Streeter was the only deeply religious person I ever knew whose beliefs made sense to me, without argument or preaching. It helped that he was a trusted friend since sixth grade. But beyond friendship, he offered compelling facts and logic as a roadmap to belief, minus the detours and dogma that have made people like me skeptical of formal Christianity. "God is love," Jarvis said. "Perfect self-giving Love." No surprise there; that's the essence of Christianity. But Jarvis spent decades studying and cutting away the bogus add-ons, like the concept of eternal Hell. As he once said, "People like a God who sends other people to Hell, but they want mercy and forgiveness for themselves. It's embarrassing." Embarrassing, because it strays so far from original Christian doctrine, and because we tend to twist God to our needs rather than shaping ourselves to God's perfect self-giving Love.

So who was Jarvis?

A funny guy, with Tom Cruise good looks and towering intelligence.

In sixth grade he appeared on Art Linkletter's *Kids Say the Darndest Things*. Clearly, he was destined to be a great comedian. Two years later, the Beatles inspired Jarvis to play the guitar. Clearly, he was destined to be a great rock star. Then there was his winning science project in junior high school, showing how ants left chemical trails when scouting for food. So clearly, was he destined to be a great scientist? In high school, his two-man team from auto shop placed third in the national automobile troubleshooting competition. Was he clearly destined to be a great mechanic? Or did his winning speech for senior class president clearly show he was destined to be a great politician?

Is there such a thing as a "Renaissance Kid"?

Along the way, Jarvis and I played sports and talked about girls. Since my career goal was journalism, I didn't pay much attention to his straight-A grades in chemistry, physics, and advanced mathematics. But one day at lunch, in our senior year of high school, Jarvis pointed to the heavens and declared his true destiny: he was going to be an astronaut.

So I was surprised when he got his college degree in sociology, and *shocked* when he took off for Kenya as a Lutheran missionary. Where did *that* come from? Of course, it had been there all along. Jarvis Streeter was always seeking answers to the biggest questions. Next stops along the way: Lutheran seminary and advanced degrees in theology at Yale and Southern Methodist University. His coursework required tremendous academic rigor, including analysis of Bible texts in their original, ancient languages (Greek and Hebrew).

In the mid-1980s, I spent a bum's weekend with Jarvis at SMU, laughing and drinking. That's when he first mentioned his idea to write a book integrating science and religion. I think my profound philosophical response was, "You're shitting me." No kidding, Jarvis insisted. Science and religion are not mutually exclusive. Neither are intellect and spirituality. You don't have to pick sides.

Thirty years later, and shortly before his death, Jarvis said in a TV interview, "My whole life has been a search for the truth. And my intuition has been that religion and science *both* have important things to tell us about that." While science is the imperfect attempt to understand the natural world, as real as Higgs boson and evolution, religion is the imperfect attempt to understand the spiritual world, as real as consciousness. Where they come together, Jarvis believed, was in our search for meaning in the cosmos.

What were Jarvis's ultimate qualifications for writing this inspiring book? He had the brains to comprehend complex science, the backbone to question the politics and dogma of religion, and the heart to know love.

In the last two years of his life, Jarvis talked openly about God, love, and death in his church and to his theology students. I visited him as often as I could, until he was on his deathbed, foggy and in pain. At some point in my visits, I realized Jarvis had gone from being my life-long friend to being my teacher. I still have huge doubts about God and religion. But thanks to knowing Jarvis, and reading this book, he's got me thinking.

Jarvis Streeter as a Teacher

Sam Thomas
Associate Professor of Religion
California Lutheran University, Thousand Oaks, CA

During the last few years of his long teaching career at California Lutheran University, Jarvis had two statements posted next to his office door. Streeter's Law Number 1: "In the end, most people believe what they want to believe, not what a rational analysis of the facts would lead them to believe." Streeter's Law Number 2: "The less people know about something, the more they typically think they know about it." In many ways these laws—not, of course, as apodictic regulations but as his own empirical observations of a kind of "natural law" of human beings—intersected with the way Jarvis understood and practiced his role as a teacher. In his pedagogy, as in his research and writing—which were rooted in his twin commitments to both science and theology—Jarvis never gave up on the possibility and pursuit of knowledge and truth.

Jarvis had a low tolerance for sloppy or lazy thinking. Yet he also had a high degree of patience and compassion—expressions, no doubt, of his pastoral sensibilities. And as his students and colleagues knew well, he had a sense of humor in the classroom. These, then, were the two poles of his pedagogical practice: relating and giving comfort to his students while also unsettling them intellectually. His students knew they could not pretend to have done the reading; they knew he would, in his Socratic style, find out just how much they had prepared for class, or just how far they could carry their reasoning, or whether they had considered this or that possible objection. But they also knew that he would listen to and care about the stories of their lives.

One semester I was teaching in a room that had a rather thin partition separating it from the next classroom. Jarvis was teaching on the other side, and my students and I would always know when his class was discussing one of his favorite topics. As he lectured or responded to student comments, his voice would gradually get louder and more animated. Sometimes we would simply stop and listen, and enjoy the moment for what it was. At least a couple times during that semester, when he reached a crescendo his students erupted into spontaneous applause, knowing themselves, I suppose, to be in the presence of a master teacher. On one of these occasions, my students and I joined them from the other side of the partition.

Jarvis, I applaud you now: for being such an inspiring teacher, a supportive colleague, and a loyal friend—and for finishing and publishing this book with the help of David Lull and your life partner Susan. With the aid of this book, more people will find their way to thinking clearly and creatively about the intersections between science and theology, and to the possibility of embracing both faith and reason. And if they knew you personally, they will hear in the words of this book the kind and passionate voice of a pastor still tending to his flock.

Teaching "Faith and Reason" with Jarvis

Nathan L. Tierney
Professor and Chair, Philosophy Department
California Lutheran University, Thousand Oaks, CA

I CO-TAUGHT THE PHILOSOPHY/RELIGION Capstone course, "Faith and Reason," with Dr. Streeter for 23 years. It was specifically planned as an interdisciplinary course for seniors in Religion and Philosophy, and covered the long dialogue between faith and reason from the ancient Greeks and Church Fathers to contemporary debates. We both felt strongly about the value of the course, believing it not only encapsulated the mission of a Lutheran university, but raised questions about the nature of human understanding and faith commitments that are a significant part of our current intellectual landscape.

Students often assumed that Dr. Streeter represented faith, while I represented reason, though this of course was not true. Jarvis was as committed to rational examination of religion, as I was to my Christian faith. Nevertheless, the approach of our two disciplines was quite different: we both learned a lot from each other as, I hope, did our students.

Jarvis's particular area of expertise was the religion-science interface, as he had pursued graduate studies in science before going on to his PhD in theology. He was very much opposed to what he called *fundamentalism* in religion, by which he meant an overly literal reading of the Bible, coupled with an ignorance of, or unreflective rejection of, both biblical criticism and countervailing scientific evidence. Normally so cool and tolerant of opposing views, this view would make him quite heated. He was convinced that faith and religion must continually evolve to accommodate themselves to new scientific discoveries. Indeed, his own views evolved during the period that I knew and taught with him. Starting from a process philosophy perspective, he integrated this with Whiteheadian metaphysics and developing scientific cosmologies. Toward the end of his life he became convinced that the touchstone of true religion was love: any religious perspective, such as exclusivism in theology or portrayals of God in warlike or punishing terms, which did not center itself on God's love, was mistaken.

My own area of understanding was the religion-philosophy interface. Jarvis helped me read early church documents and contemporary theologies with hermeneutic rigor, while I expanded his appreciation of philosophical treatments of religion. It made for many interesting conversations, both in the classroom and out! I certainly feel that my understanding of the faith-reason dialogue has been much deepened by what I learned from Jarvis. We did not always agree (our students relishing such moments!) but we always learned something from our disagreements.

Jarvis had a particular way with children. He was the favorite of my daughter from the age of four, who frequently visited our campus. She would always rush to Jarvis's office, often curling up in his bookcase for a nap while Jarvis met with students. Jarvis smiled and introduced her as his special friend.

It was a pleasure to teach with Jarvis. He was unfailingly collegial and open-minded without sacrificing rigor. He was a good friend. We worked smoothly and happily together. He had the intellectual virtue of listening carefully to new ideas, or views that he disagreed with, and treating them fairly and respectfully. He had great compassion for his students; he was always approachable and cared for their spiritual growth as much as their intellectual expansion. As a colleague, he never insisted on his own way, and was always ready to try a new teaching method or new text. I trusted him. He loved to teach, but also managed to find time to write two good books on topics close to his heart. In short, Jarvis was a friend and colleague whose spirit and vocation I greatly miss: we are the poorer for his absence.

Foreword

David J. Lull

Professor Emeritus of New Testament

Wartburg Theological Seminary, Dubuque, IA

In 1982, I was teaching at Yale University Divinity School when Jarvis Streeter asked me to work with him on an independent study of the parables and the historical Jesus. We immediately discovered we shared the same passion for a Christian theology grounded in rigorous, rational thinking about history, science, and Christian doctrines. On my recommendation, he applied for and was accepted in the PhD program in theology at Southern Methodist University, where he studied with Schubert M. Ogden, my theological mentor. In 1987, Jarvis, with his Greek New Testament in hand, came to my hotel room at the annual meeting of the Society of Biblical Literature and American Academy of Religion. (I had just become the SBL executive director.) He wanted to discuss Romans 5, especially verses 12–14.[1] Jarvis had decided that Ernest Becker's book, *The Denial of Death*,[2] would be the centerpiece of his doctoral thesis, which would later become his 2009 book, *Human Nature, Human Evil, and Religion: Ernest Becker and Christian Theology* (Lanham, MD: University Press of America, 2009).

1. "Therefore, just as sin came into the world through one man, and death came through sin, and so death spread to all because all have sinned—sin was indeed in the world before the law, but sin is not reckoned when there is no law. Yet death exercised dominion from Adam to Moses, even over those whose sins were not like the transgression of Adam, who is a type of the one who was to come" (NRSV).

2. This book, copyrighted in 1973 but published posthumously in 1974, received the 1974 Pulitzer Prize for nonfiction. Diagnosed with colon cancer in late 1972, Ernest Becker died in 1974, at age 49. See Ernest Becker Foundation, "A Four-part Sketch of Ernest Becker and His Work."

Just shy of thirty years after the beginning of what would become our lifelong friendship, Jarvis was diagnosed with pancreatic cancer in August 2011. My wife and I moved to southern California in July 2013. Jarvis had almost completed *God and the History of the Universe.* I had offered Jarvis comments on it via email, but now we could discuss it face to face. Late one afternoon, as Jarvis drove his wife, Susan, my wife, Karen, and me to one of his favorite restaurants for dinner, we were casually discussing how close he was to finishing his book, when I asked him whether he had made plans for someone to be sure it would be published, given how little time cancer would allow him to do that. Jarvis answered, "No, I can't burden anyone with that task." I said, "Let me do it." Cancer would allow Jarvis only a few more substantive conversations about his book. He died December 23, 2013, at age 64.

After getting feedback on the unedited manuscript from Craig Franklin, he and I concluded that it needed heavily invasive editing. It would have been more reader-friendly if it were not so overloaded with scientific details, and if it reflected Jarvis's conversational, or even lecture, style. But heavy pruning would have entailed too many guesses about what details were essential for Jarvis's purposes. Besides, that pruning and changing Jarvis's writing style would have been too labor-intensive. More importantly, the result would be a book that would be less *Jarvis's* book. In the end, Susan Streeter and I decided to keep the manuscript pretty much as Jarvis wrote it.

The web address for the Rosetta mission's 2014 landing on a comet is http://www.esa.int/Our_Activities/Space_Science/Rosetta/Highlights/Landing_on_a_comet/.

Every week—almost every day—during the editing of Jarvis's manuscript, news reports about new about scientific discoveries and studies appeared: the Rosetta mission's 2014 landing on a comet (see the QR code above);[3] the New Horizon spacecraft's 2015 photos of Pluto;[4] the Kepler exoplanet explorer's 2015 discovery of Kepler-452b, "the first near-Earth-size planet in the 'habitable zone' around a sun-like star";[5] a September 1, 2015 report of an Iowa Geological Survey team's discovery of a giant sea scorpion that lived about 460 mya;[6] and the September 10,

3. For the use of QR codes in place of images, see p. xvii below.

4. See NASA, "NASA Spacecraft Returns New Images of Pluto En Route to Historic Encounter."

5. See NASA, "NASA's Kepler Mission Discovers Bigger, Older Cousin to Earth"; and NASA, "Finding Another Earth."

6. See Geggel, "'Bizarre,' Human-Size Sea Scorpion Found in Ancient Meteorite Crater"; and Lamsdell, et al. "The Oldest Described Eurypterid."

2015 report by the Rising Star Expedition discovery (dating back to October 2013—March 2014) of a burial site in a South African cave containing bone fossils of a new species of hominid, the *Homo naledi*.[7] Jarvis would have loved these stories.

These new scientific discoveries and studies might make Jarvis's book out of date. In one sense, any book of this kind would be out of date before the ink dries on the first page. That proves what Jarvis says about the ever-changing nature of "the facts" and of scientific theories. In another sense, however, this book is a snapshot both of the state of science at a point in time and of Jarvis's love affair with science. That is a snapshot Susan and I wanted to preserve in this book!

I have intruded my own views sparingly and only in the footnotes, and always with the marker "Editor's note." Some of these address problems that Jarvis acknowledged in our conversations, but which cancer prevented him from solving. Others arose as I verified his footnotes.

One important feature of Jarvis's manuscript that presented a challenge is the inclusion of many color images. Jarvis believed beauty was a characteristic of truth! However, we would have to pay copyright owners of some of the images costly fees or eliminate them. We would have had to delete some others because we could not find or get a response from copyright holders. Printing the rest of the images in color would have raised the price of the book significantly. Printing them in grayscale might have created additional copyright issues and, in any case, would made them less and less effective. Jarvis would not be happy about that! Instead, we are following the suggestion of Susan and Jarvis's friend Tim Hengst to replace the images with grayscale QR codes with embedded websites for the images. Readers with a smartphone and QR code reader can view the images on the Internet. Captions explain the images and provide web addresses for readers without a smartphone. Please note, however, due to the dynamic nature of the Internet/World Wide Web, an image might have migrated to another web address—it might even have disappeared from the Web.

Finally, it is my pleasure to acknowledge the contributions of others to the completion of Jarvis's manuscript and the publication of his book: Ed Smith, a professional book editor and my longtime friend, who came out of retirement to perform the hard work of a copy editor; and K. C. Hanson (editor in chief), Jim Tedrick (managing editor), Matt Wimer (assistant managing editor), and Ian Creeger (typesetter) of Wipf and Stock Publishers.

7. See Berger, "*Homo naledi*"; Howley, "Homo naledi"; and Shreeve, "This Face Changes the Human Story."

Preface

I GREW UP AS a boy who went to church each Sunday. Even before I entered elementary school, science—specifically, all things having to do with dinosaurs and "outer space"—fascinated me. As I grew older, these scientific interests persisted and expanded into other areas. By the time I was in junior high school, I had determined to become a scientist and an astronaut. Never did it occur to me that my religious beliefs and the science I loved were incompatible. Rather, as I learned more about each, I integrated the new knowledge into my existing worldview, gradually altering and maturing it.

Following the landing of men on the Moon in 1969, NASA announced that there would be no more missions to the Moon beyond the handful already scheduled. As a result, there were massive layoffs of scientists and engineers who had worked in the space program. This led me to give up my dream of being an astronaut and, with it, my physics and aerospace-engineering college majors. My immersion in the social upheavals of the late 1960s, and a sense of their importance, also played a part in that decision. I decided to conclude my college career by pursuing studies in the social sciences. Nevertheless, my interest in the natural sciences, like my adherence to the Christian faith, continued.

A few years later, and somewhat to my surprise, I decided to begin graduate work in Christian studies with the aim of becoming a college or university professor. Yet, even then, science—specifically, my study of the archaeology of the ancient Near East during the so-called "biblical era"—was integral to that decision. Two master's degrees focused on biblical studies, church history, and theology led to doctoral studies in systematic and philosophical theology and, thereafter, to my becoming a professor of Christian theology.

My goal as a theologian has always been to articulate an understanding of reality that not only makes sense from a rationally considered Christian perspective but also takes fully into account the understanding of reality described by contemporary science. Thus, in a previous book, *Human Nature, Human Evil, and Religion,* I sought to integrate certain social-scientific perspectives on human nature and behavior with traditional and contemporary Christian understandings of these topics. In this book, I turn my attention to the integration of the reigning theories in the various natural sciences with what I take to be a credible Christian theology, in order to create a worldview that is both scientific and Christian.

In its method and form, this book is different from most works dealing with Christian theology. First, it is unusual in its method in that it attempts to integrate a comprehensive summary of contemporary scientific theories about the origins and development of the universe and a sophisticated Christian understanding of God and God's relationship to the universe into a coherent whole. I take this approach for two reasons. First, I believe that any Christian theology that fails to account for what we know about reality from the whole range of intellectual disciplines runs the risk of making claims that are clearly false in light of our best current knowledge or that lack any grounding in the real world of our everyday existence. This leads to my second, related reason for this approach, which is that failure to tie Christian theology to the world of our lived experience makes Christianity appear irrelevant to people who are not themselves Christian, and less relevant than it might be to some who are.

The second way in which this book deals with theology in a manner different from the norm is in its form. Typically, theologies are organized topically, moving from one traditional Christian doctrine to the next, with the most common organizing principle being the three articles of the Apostles' and Nicene Creeds.[1] Thus, Christian theologies typically first treat doctrines about God generally and God the Father specifically, discussing the person of God (God's attributes and God as Triune) and the works traditionally associated with God generally or God the Father particularly (the doctrines of creation, humankind, sin, and divine providence, which is God's continuing action within the universe). The second-article doctrines come next, including reflection on the person of Jesus the Christ,[2] usually construed as the incarnation of God the Son in the person of Jesus, and on

1. Creeds, coming from the Latin *credo* ("I believe"), are statements of belief. The Apostles' and Nicene Creeds profess faith in the Triune God, or Trinity, the one God being conceived to exist in three "persons"—a complex technical term—and so these creedal confessions begin with God the Father, continue with God the Son, and conclude with God the Holy Spirit.

2. "Christ" is the Greek translation of the Hebrew term "messiah," which means "anointed one." As a title, it refers to the promised righteous king for whom Jews have waited to restore their fortunes and lead them and the world into a proper relationship with God. Thus, Christ/Messiah is a title given Jesus, not his last name.

his saving work. Finally, the third-article doctrines take up the person and work of the Holy Spirit, including discussion of the Church and its sacraments along with thought about the final end of human life and the universe. This is a pattern that stretches back at least to the Protestant Reformation of the sixteenth century and is the one used in most contemporary works on Christian theology.

In this book I proceed quite differently. I present my understanding of Christian theology in conjunction with a narrative of the origin and subsequent history of the universe as understood by contemporary science. Thus, following each chapter or chapters dealing with a particular phase in the history of the universe, I discuss Christian doctrines that are relevant to that discussion and/or are necessary as groundwork for the theological topics addressed in future chapters. Taken together, these sections describe how I, as a theologian, understand the nature and work of God in light of both science and Christian thought, past and present. To my knowledge, no one has previously attempted this kind of historically- and scientifically-oriented theological presentation. Where Christian theologians have sought to integrate science into their thought, it has typically followed the more traditional topical format noted above. It is my hope that this different way of presenting Christian thought will provide the reader with a basic knowledge of aspects of science and theology, and the way they can mutually inform each other to create a credible Christian—or, more generally, theistic[3]—worldview for our time.[4]

I am writing this book primarily for readers who do not have any significant prior knowledge of either science or Christian theology, but who have a reasonably intelligent, open mind and a commitment to think rationally. Dogmatists of either the theistic or the atheistic[5] variety will find little of interest here. However, this book will be of interest and value to those

3. Editor's note: The term *theistic* (and *theism*) is derived from the Greek word for God (*theos*). A *theistic* worldview, also called *theism*, includes belief in the existence of at least one divine being, who plays a role in creating and ordering everything that exists.

4. While I believe that much of the theology presented in this book is also applicable to other theistic religions, especially the Jewish and Islamic traditions, which are historically related to Christianity, I will not address this issue here, because it would divert us from the central and already substantial topic of this book.

5. Editor's note: The term *atheistic* (and *atheism*) has two parts. The first is the initial a, which is the alpha-privative in Greek and Latin grammar, which signifies the negation of the second part of the word, which is derived from the Greek word for God (*theos*). An *atheistic* worldview entails the denial of the existence of even one divine being and, therefore, of any role for a divine being in the origin and ordering of the universe.

who have an interest in science and theology—or both—and would like to learn more about them and see one way to integrate them and to form a comprehensive understanding of reality.

I have made every effort to make the sometimes quite difficult material discussed in this book comprehensible for anyone who is willing to make an effort to understand it. For reasons of brevity and flow, I will rarely discuss how or why scientists have come to hold the theories they do about the natural world or present chronological accounts of the people and research that over time have led to those beliefs. Rather, I have attempted to present only what is important to know for our purposes of recounting the history and nature of the universe (or multiverse,[6] if it does indeed exist). Thus, for example, in recounting the evolution of Earth[7] and its life, I present the currently-accepted dates of the various eras of Earth's history and some life-forms representative of each era, but I do not discuss the theory and practice of geology and paleontology, or the radiometric process[8] that has established the accepted dates.[9]

In some places, for the sake of brevity and flow, I do not go into philosophical debates on certain topics addressed in this book, nor the nuances of these discussions. One example of this concerns the definition of "fact." In the first chapter, I offer a common-sense definition, one supported in the philosophical literature, yet one that is by no means the only possible definition, as the complex history of the debate in philosophy shows (see the QR code below). The same is true regarding my discussion of logic, where I discuss only the use of logic in the analysis of data and in the synthesis of facts to form theories.

To give you an idea of the philosophical complexity of the topic, briefly examine the *Stanford Encyclopedia of Philosophy* **article on facts (http://plato.stanford.edu/entries/facts/).**

6. As we will see, scientists have recently begun theorizing about the possible existence of a multiverse—of other universes that have existed prior to our own, exist alongside our universe today, and/or will come to exist in the future. Some scientists speak of this larger reality, combining both our universe and the multiverse, as the Universe (capitalized).

7. Editor's note: Jarvis capitalized "Earth" when it refers to our planet.

8. Editor's note: "Radiometric dating" employs the measurement of the presence of radioactive elements to determine the age of organic and inorganic materials.

9. Later in this book, I will discuss the scientific, philosophical, and theological examples used briefly here and in what follows, many in greater detail. Below, the section on the structure of the book gives a summary of topics discussed in each chapter.

This narrow focus in my discussions of science and philosophy is also reflected in most of my discussions of Christian theology, such as those concerning the nature of God and God's relation to the universe. Thus, I do not provide a comprehensive account of the variety of theological positions, the theologians who have posited them, nor all the biblical, historical, and theological arguments they have used to support their views. Instead, I lay out only those perspectives on any given doctrine that many, if not most, Christians today have come to hold, some of the reasons why I agree or disagree with the traditional views, and my own take on how the doctrine in question ought to be understood. If you are interested in pursuing further the scientific, philosophical, and/or theological topics addressed in this book, the sources listed in the bibliography should be of assistance.

I have also attempted to present only the scientific information I deem necessary to gain a basic understanding of the history of the universe, sufficient to answer questions you might have without going into excessive detail. Clearly, this is a matter of judgment, and my judgments on these things will likely not match those of all readers. Therefore, if you find more detail in some parts of the book than is of interest to you, I encourage you to skim through those sections until you encounter discussions that do interest you. You can always return later to material skipped upon first reading, should you choose to do so. In this way, I hope the book will be useful to a wide variety of readers with varying interests and concerns.

It is important to recognize that both science and theology are continuously-developing realms of thought. However, due to the constant influx of new scientific data and the inherent conservatism of religions, scientific understandings tend to evolve more rapidly than theological ones. For this reason, some of the science presented in this book, especially with regard to certain specific data, may already be somewhat dated by the time this book appears in print. It is also possible that certain perspectives in some scientific fields will become significantly altered within the next few years or decades. Had I written this book fifteen years ago, for example, there would have been no mention of dark energy and its crucial influence in the history of the universe, because its existence was not yet known. Such is the nature of science, which presses ever forward toward new and better understandings of the world around us. On the other hand, if the past is any indication, most such changes will involve additions to, or minor reworkings of, our present understandings of nature, rather than anything approaching a large-scale and fundamental change in perspective, such as occurred with the discovery of dark energy.

I should also note that currently no absolute consensus exists on certain issues in the various sciences, such as the precise time frame and cause of the inflationary event in the earliest fraction of a second of the universe's history, or the origin and precise pathways of the evolution of life on Earth. For this reason, I have attempted to indicate where current consensus exists, or not, on the topics addressed in this book, what some of the major competing theories are where no consensus exists, and which, if any, seems to be the theory preferred by the majority of scientists working in the field.

Regarding topics involving numbers, such as the dating of prehistoric events or the size or distances of various astronomical entities from us, I have given either the broad range of possibilities or selected the number most often seen in the scientific literature and have typically noted where it is an approximate figure. This lack of consensus in certain aspects of science mirrors the lack of consensus in various areas of theology, such as the nature of God's power and the way it is manifested in God's dealings with the universe. In spite of these uncertainties, I expect that the overwhelming majority of scientific and theological material presented here will remain essentially valid for the foreseeable future.

The structure of the book is as follows. Chapter 1 is a discussion of the notion of worldviews, including a description of what worldviews are, the two fundamental types of worldviews that exist, and why people can and do have such widely different understandings of reality. Chapters 2 and 3 examine the methods employed by science and theology respectively, the general criteria by which worldviews may be judged to be more or less adequate descriptions of reality, the supposed traditional conflict between science and religion, and some issues regarding theological language. Chapters 4 and 5 present the contemporary scientific understanding of the nature of physical reality, including various theories that will be important in understanding the history of the universe presented in the chapters that follow. These include discussions of the nature of space and time, the fundamental constituents of physical reality, and the four fundamental forces of nature (in chapter 4), and descriptions of the theories that explain their nature and interactions, including relativity theory, quantum mechanics, superstring theory, chaos theory, and evolutionary theory (in chapter 5).

Chapter 6 begins the long story of the history of the universe, starting with its origin and subsequent development through the first 200 million years or so of its existence. Chapter 7 presents a discussion of the topically-related Christian understanding of God as Creator of the Universe. Chapter 8 follows the evolution of physical structure in the universe from the rise of the first stars and galaxies up to the birth of our own galaxy, the Milky Way. Chapter 9 recounts the formation of our Sun and Solar System some 4.6 billion years ago. A discussion of the Christian understanding of the nature of God follows in chapter 10, including the nature of God as Infinite Agapic

Spirit and the way God deals with flawed humanity, seeking to help humans find salvation. Chapters 11 and 12 detail the early history of Earth and the origin of life on our planet. Chapter 13 continues the earlier discussion of the nature of God—focusing especially on God as everlasting, omniscient, and omnipotent—and the "problem of evil," which has traditionally been a major objection to theistic worldviews.

Chapter 14 details the evolution of our planet and its life during the Archaean and Proterozoic Eons, from about 4 billion to 541 million years ago.[10] Chapter 15 begins our discussion of the Phanerozoic Eon, from the explosion of new life-forms in the Cambrian Period to the greatest mass extinction of life in the planet's history at the end of the Permian Period, around 252 million years ago. The rise of the "Age of Reptiles" in the Triassic Period and the "Age of Mammals" in the Quaternary Periods are discussed in chapter 16. Chapter 17 examines the way God might be conceived to interact with the universe and its creatures, including God's possible role in the evolution of life on Earth.

Chapter 18 discusses the evolution of humankind from its primate origins through the appearance of the Australopithecines. Chapter 19 recounts the evolution of the members of our own genus, Homo, including our species, Homo sapiens sapiens. It concludes with a discussion of the ultimate future of humankind and the universe. Chapter 20 examines the Christian understanding of humankind and the question of life after death. Finally, Chapter 21 sums up what I believe we can learn from our study of the history and nature of the universe and why the Christian worldview presented in this book provides a rational and satisfying way of explaining why the universe is the way it is and why the history of the universe has unfolded as it has.

I owe a debt of gratitude to many people for their assistance in the creation this book. As always, those of us who attempt to understand any given subject stand on the shoulders of those who have come before us—the famous and not-so-famous whose thought has taught and inspired us and driven us forward to ever richer understandings. I therefore offer my sincere thanks to the many people from whom I have learned, both living and dead, personally and through their writings, and whose insights are reflected in this book. For the sake of flow and the general readers for whom this book is primarily intended, I have seldom referred in the body of the text to these authors. Their names and writings appear in the bibliography.

I also owe special thanks to California Lutheran University, where I have had the privilege of teaching for the past twenty-five years, and

10. See the Geologic Time Scale in Appendix A.

especially to its academic administrators, President Chris Kimball, Provost Leanne Neilsen, and Dean of the College of Arts and Sciences Joan Griffin, whose granting of a sabbatical leave during the Fall of 2009 and other support during the writing of this book has been exceptional. I also want to express my gratitude to a number of people who have read parts of the manuscript and/or have offered valuable criticism and suggestions. Specifically, I want to thank Sharon Andrews, Dave Belknap, John Cobb, Greg Daniels, Terence Hartwick, Hank Keeton, Schubert Ogden, Greg Orfalea, Roger Walters, and my colleagues at California Lutheran University, especially Bill Bersley, Bill Bilodeau, Xiang Chen, Marylie Gerson, Rahuldeep Gill, David Marcey, Brian Rasmussen, Linda Ritterbush, Mike Shaw, Sam Thomas, Victor Thasiah, Nathan Tierney, and Colleen Windham-Hughes. My deepest gratitude goes to my friend and mentor David Lull, who has read the entire manuscript and made many thoughtful suggestions that have helped to make this a better book.

Above all, I must thank my wife Susan, for her critical reading of, feedback on, and proofreading of each chapter, and for her unstinting support during the writing of this book. Without her, I would not have completed this book. For these reasons, as well as my deep love, I dedicate this book to her.

Jarvis Streeter,
California Lutheran University,
Thousand Oaks, CA
July 29, 2012

"I will make sure that your book gets published."

—David Lull

There are not enough words in the history of God's universe to convey the value these particular words held for Jarvis. Remembering them always brought tears to his eyes. They removed a heavy burden, a worry—important for one facing death. They were likely the greatest gift that he received after his diagnosis of terminal pancreatic cancer.

Additionally, they were a gift to me during Jarvis' illness and death and afterwards. I would not have known where to begin such a process, and with all that comes with the death and grieving the loss of a spouse, it is likely something I would have put off indefinitely but would have remained a guilty burden.

I am forever grateful for David's love and regard for Jarvis that he would be willing to take on such a monumental task. It has been perhaps far more than he expected, but he has honored his wonderful gift to Jarvis and to me. My love and thanks go to David and to his wife, Karen, for the friendship and support that that they have shown me through this process. You will both always be very dear to me.

Susan Gresham Streeter
Georgetown, TX
September 30, 2015

Abbreviations

BOOKS OF THE BIBLE

1–2 Chr	1–2 Chronicles
1–2 Cor	1–2 Corinthians
1–2 Kgs	1–2 Kings
1–2 Sam	1–2 Samuel
1–2 Tim	1–2 Timothy
1 Thess	1 Thessalonians
Deut	Deuteronomy
Eph	Ephesians
Exod	Exodus
Gen	Genesis
Isa	Isaiah
Josh	Joshua
Judg	Judges
Lev	Leviticus
Matt	Matthew
Phil	Philippians
Ps(s)	Psalm(s)
Rev	Revelation
Rom	Romans

GENERAL

aBB	after Big Bang
ASV	American Standard Version of the Bible
AU	astronomical unit

Abbreviations

BCE	Before the Common Era (= BC)
bya	billion years ago
ca.	circa, about
CE	Common Era (= AD)
ch.	chapter
cm	centimeter
CMB	cosmic microwave background
cu	cubic
d.	died
EHz	exahertz (1 quintillion cycles per second = 10^{18} Hz)
EM	electromagnetic
esp.	especially
ft	foot
GHz	gigahertz (1 billion cycles per second = 10^{9} Hz)
HB	Hebrew Bible
Hz	*hertz* (1 cycle per second)
in	inch
kg	kilogram(s)
KJV	King James Version of the Bible
km	kilometer(s)
lb(s)	pound(s)
m	meter
MHz	megahertz (1 million cycles per second = 10^{6} Hz)
mi	mile(s)
mm	millimeter
mya	million years ago
µm	micrometer (1 millionth of a meter = 10^{-6} meter)
NASA	National Aeronautics and Space Administration
NIV	New International Version of the Bible
nm	nanometer(s)
NRSV	New Revised Standard Version of the Bible
NT	New Testament
OT	Old Testament
PHz	petahertz (1 quadrillion cycles per second = 10^{15} Hz)
RSV	Revised Standard Version of the Bible
THz	terahertz (1 trillion cycles per second = 10^{12} Hz)
WIMP	weakly interacting massive particle

1

Worldviews

EVERYONE HAS A *WORLDVIEW*. A worldview is the idea you have of the way things are, a conception of what reality is like. For most of us, this worldview is *informal*. We have some basic ideas about the nature of reality but have not attempted to define them more precisely, nor considered carefully whether this collection of beliefs holds together logically, nor evaluated its adequacy in comparison to the knowledge, experiences, and differing worldviews of others. Instead, our worldviews tend to be somewhat amorphous, floating, as it were, in the background of our minds, yet constituting the mental framework within which we live and interpret what we experience throughout our lives. It is the rare person—generally someone more academically- or theoretically-minded—who seeks to make his or her worldview explicit, to express it clearly, and to test it for internal coherence and compatibility with a broad range of human knowledge and experience. The primary aim of this book is the explication of such a *formal* type of worldview, one that is both scientific and Christian.

Our worldviews are built up from our knowledge and beliefs. Conceived simply, *knowledge* is made up of the accumulation of *facts*, which may be defined as what any rational human being must accept as true about the world, assuming his or her senses and mind are functioning properly. *Beliefs*, on the other hand, are based upon our *interpretations* of facts—how we put together the facts that we know in order to come up with more general ideas about the way things are.

Worldviews, then, are systems of belief built up from our interpretations of the facts we have encountered through our experiences in life. The *ideal* way to construct the best possible worldview would logically be to learn all the facts about reality and then interpret them in their interrelations with one another in a fully objective and rational way. This would result in a fully comprehensive and coherent understanding of reality. As such, all knowledgeable and rational people would find it credible and persuasive.

It is obvious, though, that we do *not* all share the same worldview. We hold a considerable variety of profoundly conflicting worldviews, and the reasons for this are not difficult to discern. Indeed, a widespread consensus has developed in recent decades among neuroscientists, psychologists, sociologists, philosophers, and others as to why we hold the views we do, as well as why changing them seems so difficult for us.[1]

In the first place, none of us comes to the process of worldview construction without some understanding of reality already in place. This initial understanding is a product of our earliest experiences as children. Even as infants, we already form some rudimentary sense of what the world is like—more felt than thought—based on our experiences of ourselves and those around us, particularly our parents or other primary caregivers. We quickly learn that we are in control neither of our own bodies nor of our surroundings, and so must rely on others to provide us with what we need to survive. This makes us feel anxious and insecure. Our anxiety will decrease to the degree that those around us provide for our physical and emotional needs, or it will intensify to the degree that they do not. As a result, our earliest *felt* worldview is of reality as generally supportive, relatively benign, or largely hostile to ourselves.

As our mental abilities develop and we progress through childhood, we gradually form a *conceptual* worldview, most of which we learn from others, rather than create by ourselves. This occurs because of what we term the *socialization process*, in which we learn as children how to function in the particular familial and cultural world(s) we inhabit. This socialization includes learning how to view the world and behave in it from those around us, especially our parents or primary caregivers, others in our immediate family, and those with whom we have frequent and intimate contact. This "cultural indoctrination" is both intentional, through specific teaching by those around us, and unintentional, based on our observations of what these others say and how they behave—including when they say one thing and do another.

1. Compare, for example, Newberg and Waldman, *Born to Believe*, esp. chs. 1–5; Polanyi, *Personal Knowledge*; Streeter, *Human Nature*, chs. 1–2, 4–5.

Our first conceptual worldview is thus largely that of those closest to us, whose views we typically believe to be true without it ever occurring to us to question them. We assume that what they believe is the way things are, and that everyone shares this understanding of the world and how to behave in it. This impression is heightened to the degree that others with whom we have contact reinforce that understanding of reality; and, since most people tend to associate with others who are largely like themselves—belonging to the same ethnic, religious, socio-economic, and other such groups—our parents' worldview is typically reinforced throughout our formative years. In this way, each of us emerges from childhood with a very specific worldview, one that differs from others who have been raised by different parents and are not part of the social groups to which we belong.

A second reason our worldviews differ so greatly from one another concerns the body of facts we come to know over the course of our lives. Most of us recognize that every one of us has a unique set of experiences in life and so learns a body of facts different to a greater or lesser degree from those of other people. This means that each of us will have greater knowledge in some areas than in others. Thus, scientists in a particular field, say geology, will know more about geology and about science in general than most nonscientists, and Christian theologians will know more about theology and Christian studies in general than most non-theologians.

The point here is that *nobody* can possibly know in the short span of a human lifetime *everything* about *everything*. Since no two of us know exactly the same body of facts, and since our worldviews are largely determined by the facts we know, it is clear that no two worldviews will be identical. Those of us who would construct a more comprehensive worldview, therefore, will seek to learn from others more knowledgeable in areas in which we are less knowledgeable, in order to have the widest range of facts undergirding our perspectives.

The problem is that there is no such thing as a "naked fact." Every supposed "fact" involves an unconscious *interpretation* of the data we received through our senses. For example, when you look at this page, what you actually see are patterns of black squiggles on a white background. It is only by an act of interpretation that your mind can determine the squiggles to be letters of the alphabet that, when grouped together, form words that we can read, and that have meaning. Wherever interpretation is involved, however, there is room for error. Even the way we interpret what seem to be simple sensory data is not always correct. A stick that is partially submerged in water appears to be bent, when it is not bent at all. The same is true of desert mirages, where the pavement ahead appears to be wet, even though it is not. Also, as most of us know, eyewitnesses to accidents are notorious for giving conflicting descriptions of the "facts," even though each of them based their description on sensory input that presumably accurately took in what transpired. This inherent layer of interpretation, in moving from what our senses perceive to what we interpret those perceptions to mean, only complicates further reaching a consensus on what the facts of reality are. That is one reason why people construct different, and even conflicting, worldviews.

Another problem related to facts is that most of the things we learn about the world are not things we ourselves have experienced but instead come to us from others, including supposed experts or authorities. The problem is that not all "experts" are equally expert in any given field, and that even true experts do not always agree with one another on everything. Many supposed experts unintentionally teach incorrect "facts." Others, who are not really experts, teach us incorrect "facts." Still others intentionally lie to us, in order to manipulate us.

For example, I have heard pastors throughout my life teach that John the Baptist and Jesus were cousins, based on Luke 1:36. Actually, in the original Greek that verse does not say that John and Jesus were cousins (though the King James Version of 1611 did incorrectly translate it that way—just one example where even the Bible as translated into other languages can mislead us). The Greek word used to describe the relationship of Elizabeth, soon to give birth to John the Baptist, and Mary, soon to give birth to Jesus, is *syngenis*, meaning "relative" or "kinswoman," which in the Jewish culture of the time could mean a member of one's immediate or extended family, someone from the larger *clan* of which one's family was a member, someone from the same large-scale *tribe*, or even simply any fellow Jew. Given the breadth of possible meanings, to say that John was Jesus's cousin is a far more specific claim than the meaning of the word *syngenis* can support. The point here is that much of what we "know" has come to us from others, including supposed authorities, and yet these people are not always correct in what they tell us. Unless we have investigated such "facts" ourselves and confirmed their truth (as far as we are able to do so), we cannot be certain that the "facts" we have learned are indeed facts.

Another problem regarding facts arises from the distinction between public and private facts. Some things, say the nature of leaves from a particular type of tree, are accessible to any of us whose senses and minds are functioning properly. It is therefore reasonable to believe that essentially all of us would agree on the basic facts concerning that leaf, including its basic color, shape, smell, texture, and the like. However, some facts, such as the unique personal experiences you have in life, including especially your mental and emotional experiences, are not like these publicly observable facts. Only you experience them. For the most part, they are unique and unrepeatable. You will accept such experiences as facts, though other people, having not had those same experiences, may not. This becomes particularly important for our discussion when considering, for example, religious experiences and their effect on a person's worldview. All these considerations surrounding the notion of "facts" demonstrate that even what we should consider as the proper body of facts to be interpreted in the process of forming a worldview can be quite problematic.

If determining the facts to be taken into account in constructing an understanding of reality is difficult, the process of rationally *interpreting* them in relation to each other is even more problematic. First, because we all begin with some worldview, the one learned as a child from our family and friends, we typically interpret our experiences from the perspective of that worldview, whether or not it is the best one to make sense of those experiences. This is why a person raised in a fundamentalist[2] Christian home will typically have much more difficulty accepting or even considering the possible truth of evolution than a person raised in a non-fundamentalist family. Similarly, a person raised in an atheistic family will typically find it more difficult to believe in the existence of God, or even to consider that possibility, than one reared in a religious home.

In addition, the correct or most logical interpretation of a body of facts is often far from obvious. The same facts can sometimes be reasonably interpreted in more than one way, generally due to the body of known facts being insufficient to decide between rival interpretations. As we will see later, this is a problem often encountered in both science and theology.

Another problem for properly interpreting a body of facts is that some of us have greater mental skills and/or training than others, and so are more able to think rationally about our beliefs. Others are simply not as well equipped for logical thought. I remember an evening where I was engaged in a public debate over evolution and Christian belief, and the discussion turned to the topic of biblical inerrancy, which is, in fact, what is finally at

2. "Fundamentalism" is a term that originated within Christianity in the early twentieth century to designate those Christians who subscribed to the tenets taught in a series of pamphlets, collectively titled *The Fundamentals*. The writers of these pamphlets argued that one could be a real Christian only if one believed in biblical inerrancy (that there are no errors of any kind whatsoever in the Bible) and John Calvin's sixteenth-century doctrine of salvation by penal substitution. According to this doctrine, Christ voluntarily suffered and died on the cross in our place, taking the punishment we deserved for our sins onto himself. In this way, Calvin said, we Christians can be saved—that is, we get to go to heaven when we die. Today "fundamentalism" has a broader scope, referring generally to the belief by members of any religion in the literal inerrancy of their scriptures.

stake in the evolution debate. I argued against biblical inerrancy, to which a man in the audience responded that he could show me twenty passages in the Bible where evidence (from archaeology or other historical records from the time, for example) has confirmed the truth of what was said there, and that therefore the Bible was inerrant. I responded that I could show him a hundred such passages, but that just because a hundred passages were factually accurate did not necessarily mean that all the other passages in the Bible were as well. He argued that it did, failing to understand that he was committing the part-whole fallacy in logic and so was not thinking rationally.

Of course, part of his refusal to think rationally in this case may also have been because he, like most of us, did not want to be wrong. If this is true generally and with regard to minor things, it is even more the case with regard to our understandings of reality, which have been central to how we have understood ourselves and the world, and how we have acted throughout our lives. As a rule, the longer we have held our worldviews, the less likely we are to be open to changing them. Most of us will not change our understanding of reality unless a large body of evidence forces us to change—and even then, some of us will not do so, simply refusing to acknowledge as factual the evidence that indicates the incorrectness of our beliefs. To have to admit to others and to ourselves that we have been wrong makes us feel diminished and vulnerable. It, therefore, lowers our sense of self-worth, our *self-esteem*. Typically, the more weak and threatened we feel in our lives, the more inflexible we will be in our beliefs and practices. Simply put, for many of us the *need* for self-esteem, for emotional well-being, trumps the *desire* to be rational. For some, it is a matter of emotional survival. In the end, most of us tend to believe what we want to believe, rather than what a rational, objective analysis of the facts would lead us to believe.

This underscores the important role psychological and sociological factors play in our ability not only to construct a rational worldview, but also to be open to modifying it in the light of new facts and experiences. These factors are several. We should consider them individually.

One factor is *the nature of our genetic inheritance* and, as just noted, *our earliest childhood experiences* with our parents or primary caregivers, and how they have affected us. If we have been gifted by nature in our physical, emotional, and intellectual makeup, and if our parents or caregivers were responsive and loving in caring for and socializing us, and if we have not suffered any or very many traumas in life, we will tend to be less defensive and so more open to new ideas. Conversely, the less gifted we are biologically, the more trauma we have experienced as children and throughout our lives, and the less loving and attentive our parents or caregivers were to us when we were young, the more defensive and rigid, and therefore less rational and open to new ideas, we will tend to be. Defensiveness regarding our worldviews can even lead the weaker among us to ridicule others for believing differently. That seems like a desperate attempt to try to shore up one's own worldview.

Another factor shaping our rationality and our openness and ability to modify our worldviews will be *the kinds of general experiences we have had throughout our adolescent and adult lives and how we have responded to them.* Some experiences, such as being friends with those who have other worldviews, might prompt us to rethink aspects of our worldviews, assuming we do not feel too threatened by them to do so. However, if we have led sheltered lives or have tended to respond negatively to new experiences and ideas generally, we will tend to maintain our present worldviews, and perhaps become even more firmly entrenched in them, when we encounter other viewpoints.

Still another factor influencing how open and rational we might be in rethinking our worldviews based on our continuing experiences in life is *the nature of the worldview we learned as a child.* If it was one that was positive about other people and groups and inclined us to seek new experiences and appreciate ways of living in and viewing the world other than our own, we will tend to be more open and rational in considering the strengths and weaknesses of our own culture and worldviews. On the other hand, if we have been taught not to trust others or to consider the possible truth of any views that deviate from our own, we will tend to cling to our worldviews, whether it is reasonable to do so or not.

Yet another factor influencing our openness to modifying our worldviews is *the influence of the people with whom we associate* throughout our lives, especially those closest to us and from whom we derive emotional and perhaps even physical and/or financial support. Such people will tend to have a profound influence in shaping what we believe. *Peer pressure,* to cite one example, is a powerful factor in shaping our beliefs, with its power over us proportional to our insecurity. It takes courage and strength to deviate from the group's way of seeing and doing things and stand alone with our own convictions. Whereas earlier in human history, standing alone threatened one's *physical* survival, today—at least in much of the so-called "developed world"—the threat is more often to our *emotional* survival. This emotional component of survival is hardly trivial. We can die, even take our own lives, when the emotional blows become too much to bear. It should come as no surprise, then, that our tendency is to conform to the views and behaviors of the groups into which we were born and socialized, and to hold the views of the groups of which we are a part throughout our lives. Again, the weakest, most insecure of us will be most inclined to such behavior.

Finally, we must also note *the fallibility of memory*, which plays an important role in worldview formation and maintenance. As we have seen, our understanding of the nature of reality is built up from our interpretations of the facts we know based on our experiences, including those things we learn from others. Putting together all these facts is heavily dependent on our *accurate* memory of them—and contemporary neuroscience has demonstrated just how unreliable our memories can be. Study after study has confirmed that we often do *not* remember things accurately, shaping them instead according to our personal desires and needs—and all of this without our ever consciously realizing it. Thus, the fallibility of our memories is another problem in attempting to construct rational worldviews.

All of these factors, and perhaps still others, interact in complex ways unique to each of us, so that as individuals we exist somewhere along a spectrum ranging from more closed and dogmatic to more open and rational with regard to life and the worldviews we hold. From the perspective of creating a rational and comprehensive worldview, it is clear, then, that some of us will be more able, and others less able, to do so. We all start with some worldview born of our childhood experiences and training. The issue is whether we tend to cling irrationally to this worldview no matter what we experience subsequently, or are open to altering that initial view in light of new facts and experiences. If our tendency is toward the latter position, we clearly stand a much greater chance of viewing reality more the way it actually is than as we would like it to be.

It does seem that most of us, in fact, do modify our childhood worldviews to some extent over the course of our lives. Reality has a way of upending some of our most cherished beliefs, forcing us to change at least some of our views, even if we would rather not change them. Change is difficult, though, as we have seen, because it is threatening, particularly for those who are insecure and anxious. Changing our minds confronts us with our own fallibility and the fact that we do not really know—and, indeed, can never know—the truth about existence. This awareness can in turn undercut our confidence in ourselves and the meaning of our lives, which is also anxiety- producing and a threat to our emotional survival. It is, therefore, no wonder that most of us continue to conform to the major features of the worldviews we have been taught, and live within them throughout our lives, rather than seriously entertaining the risky possibility that another person's (or group's) views might be correct. It is easier, and safer, to stay where we are, to continue believing what we already believe, changing only when and if something absolutely forces us to change. This means that a rational worldview is a hard-won achievement for most of us and nearly impossible for others.

TYPES OF WORLDVIEWS: RELIGIOUS AND NON-RELIGIOUS

Perhaps the most fundamental distinction between types of worldviews is whether they are *religious* or *nonreligious* (or *spiritual* or *non-spiritual*).[3] The basic issue here, in my understanding of the essential nature of religion (or spirituality), is whether one believes another aspect or dimension to reality exists in addition to the physical one we experience through our five senses. The nonreligious person, or *materialist*, denies that anything else exists in addition to the physical universe; the religious person believes there is also something else, another realm or dimension of existence, in addition to the physical one, that we may term *spiritual reality*. Important here is the recognition that both religious and nonreligious worldviews, like all worldviews, are *belief* systems, based on interpretations of the facts available to us. As such, we cannot *prove or disprove* either view. We can only argue each view based on logic and the available evidence. Nonreligious (or non-"spiritual") persons believe that the facts they know indicate that no such spiritual reality exists, while religious (or "spiritual") persons believe that the facts they know indicate that it does exist. People who believe that a spiritual reality exists and is *personal* are termed *theists* (ones who believe in God or gods); those who believe in a spiritual reality that is *impersonal* are generally called *non-theists*; those who deny the existence of God or gods are called *atheists*; and those who neither believe nor disbelieve in God or gods or any spiritual reality are termed *agnostics*.

How then might we judge rationally which among these competing worldviews—and, in some cases, within each type of worldview as well—has the greater claim to be the closest to a true description of reality? This is the heart of the problem, because, as we have seen, not all of us will agree either on what the relevant set of facts is or on what the most rational interpretation of those facts is, and this typically for the reasons we have cited above. Thus, regarding the issue of what constitutes a genuine fact, materialists typically claim that the only data that could possibly be considered factual are those that are empirical and scientifically testable, and that anything not available to public scrutiny, such as personal spiritual experiences, is automatically disqualified as potentially factual. This view has been termed *scientism*.

3. Recently, some have wanted to make a distinction between being religious and being spiritual. For them, being religious implies subscribing to the teachings and practices of a specific religion, whereas being spiritual means having one's own ideas about the spiritual world and following one's own spiritual practices.

On the other hand, some conservative Christians, such as fundamentalists, will insist that the only certain facts are those found in the Bible, read literally, and therefore any data from science or any other source is to be considered factual only insofar as it agrees with what the Bible says—as they interpret it. This view has been termed *biblicism.* Interestingly, even though materialists and biblicists stand at opposite ends of the religious belief spectrum, both assert essentially the same thing: that any data are to be excluded from consideration as potentially factual if they do not fit their *pre*conceptions about what can count as a fact. Because it is the interpretation of facts that lies at the heart of worldviews, *pre*judging what one will consider as potentially factual obviously means *pre*judging what might be a possibly true worldview.

Such prejudging does not seem a rational way to construct a worldview. We cannot justly prejudge what the facts upon which to build our worldviews are without having examined them. Rather, to construct the most adequate worldview, we should evaluate all possible data—whether from science, religion, or any other source—on their own merits as to their factuality. As we have already noted, the most adequate worldview will be the one that includes the broadest range of factual data, whether scientific, religious, personal, or otherwise, and interprets those data in the most reasonable way. The broader the range of facts a worldview accounts for—that is, the more *comprehensive* the fact-set incorporated—the more adequate the resulting understanding of reality should be. The fundamental difficulty in the religious versus nonreligious worldview debate, even among open-minded people, will, of course, be agreeing on what the facts actually are.

In addition to the *comprehensiveness* of the fact-set, another general determinant in evaluating any worldview will be, as we have seen, the *adequacy* of its interpretation of that fact-set. Here an essential criterion is a worldview's *internal coherence*—whether its interpretation of the facts is such that each fact fits together with all other facts to produce a worldview in which there are no logical contradictions. A third criterion for evaluating worldviews is *clarity*, or the degree to which the worldview is clear in what it claims and is *understandable* to those who encounter it.

Several disqualifying criteria argue against the plausibility of the worldview. One important such criterion is the use of *ad hoc solutions*[4] to problems within the system. In the next chapter is a scientific example of this in the discussion of the geocentric versus heliocentric cosmologies, where more circles were gratuitously added to the existing circular orbits

4. *Ad hoc* solutions are those that do not arise out of the system—in this case, the worldview—itself. They are like bandages, with different types invented to deal with different problems with the system.

of the planets in the geocentric understanding in order to defend the view in the face of significant evidence against it. In chapter 3 is a theological example of this in the way various ad hoc arguments have been used in an attempt to defend the doctrine of biblical inerrancy in the face of a good deal of biblical evidence against it.

Adherence to these criteria—a comprehensive fact-set with an internally coherent interpretation of those facts and a clear explanation of the worldview without resort to ad hoc additions to prop up the system's weaknesses—does not guarantee that a given worldview will be true. Nevertheless, it does at least provide a foundation from which to begin to evaluate rival worldviews and to expose the inadequacy of any worldview that fails to fulfill them.

As I have said, the worldview presented in this book is both scientific and Christian. It seeks to combine what most scientists take to be the well-founded theories/beliefs of science with what I consider to be the well-founded theories/beliefs of Christianity into a single comprehensive, coherent, and clearly presented worldview. The strength of this approach, in my view, is that it does not prejudge what kinds of truth claims will be evaluated. It attempts to take the perspectives of both science and religion seriously, to evaluate the rationality of the facts and theories they assert, and then to construct an understanding of reality based on what seem to be the best and most relevant facts and theories of both.

The first thing we must do in seeking to accomplish this task is to look at the ways scientists and theologians determine what they believe. This means taking a close look at the methods of each discipline, their strengths and weaknesses, as well as the ways in which scientists, theologians, and philosophers have historically viewed the relationship between the two disciplines. These are the topics to which we turn in the next two chapters.

2

How Science Operates

SCIENCE AND THEOLOGY EMPLOY methods of discovery that are different in some key ways. And yet, they also share some similarities that often go unrecognized, as we will see in the next chapter. According to scientists and philosophers of science, the general form of the *scientific method* consists of several parts. It begins by moving from a question about some feature of nature to a *hypothesis* about it, and then to experimental testing of that hypothesis. If repeated testing supports the hypothesis (or does not falsify it), scientists create a *theory* to explain why this aspect of nature is the way it is and, perhaps, even how it fits with other scientific theories into a comprehensive understanding of nature. Scientists concerned with the same or similar questions test this proposed theory, ideally based on the empirical predictions that the theory makes. Should this subsequent experimentation continue to provide support for the theory, and should the theory seem the most likely explanation of the available facts recognized by the community of scientists working in that field, then that theory becomes widely held. If, over time, continued testing consistently supports (or fails to disprove) the theory, it eventually comes to be accepted as a *law of nature*.

What is crucial to recognize here is that this general method, contrary to the belief common to many nonscientists, does *not* lead to *certain* truth but only *probable* truth—even in the case of the so-called laws of nature. By its very nature, *no scientific theory can ever be proven to be true*. The reason is that proof and certain truth can be found only in the fields of mathematics and logic, since these fields are based on *deduction* from presumed true axioms or premises. If the axioms or premises are true, the new ideas deduced from them *must* be true, and so can be rightly said to be proven.

Neither science nor theology, however, works primarily by means of deduction—though deduction does play an important role in both fields. Rather, both science and theology are fundamentally fact-driven enterprises, in that they proceed by gathering up facts and then drawing what seem to be the most reasonable general conclusions that can be derived from those facts.[1] The greater the amount of supporting evidence, combined with an absence of evidence refuting a scientific or theological theory, the stronger is the *probability* that the theory is true. Thus, few scientists, if any, would claim that what we believe to be true in most areas of science today is absolutely true. Truth is what scientists hope to attain, believing that they draw nearer to it over time by means of the increase of relevant data and the experimental falsification of incorrect theories. Thus, as theoretical physicist S. James Gates Jr. (1950–) put it, "Science . . . is not about truths. It is about forming beliefs that are less false."[2] The same can be said of theology.

A simple hypothetical example can illustrate the process of theory formation in science. Suppose a botanist wanted to know how many petals are on the flowers of a particular species of daisy. She would likely begin by locating such a daisy and counting the number of petals on it. If she counts twenty-four petals on that daisy she may form the tentative hypothesis that all daisies of that species have twenty-four petals. To test her hypothesis, she counts another such daisy's petals and again finds twenty-four. Does that prove that all daisies of that species have twenty-four petals? No, she knows only that the two daisies she has examined have twenty-four petals. However, she may now have slightly more confidence in her hypothesis that all daisies of this species have twenty-four petals.

To continue to test her hypothesis, she counts more daisies. She counts eight more and finds that each of them has twenty-four petals, just like the first two. Has she proven her hypothesis that all daisies of that species have twenty-four petals? No, but her hypothesis seems *more probable* now because she has counted more daisies (i.e., performed more experiments), and all of them have had the predicted number of petals. Therefore, she decides to test the hypothesis further, counts the petals on a thousand such daisies, and discovers that every one of them has twenty-four petals. Has she now proven her hypothesis that all daisies of that species have twenty-four petals?

1. This form of reasoning, moving from individual facts to general truths, is called *induction*, in contrast to *deduction*, in which one moves from general truths to individual facts.

2. "Who is Afraid of Music?" Gates also says this in "The Fine Art of not Knowing" at 2:59–3:04.

Still the answer must be no, and the reason is simple: she has not counted all the daisies of that species that exist. All it would take to demonstrate that her hypothesis was wrong would be to find a *single* daisy that had not been tampered with and that had fewer or more than twenty-four petals.[3]

What she *can* say is that her hypothesis is much more probable than it was before. Yet even if she were to find every such daisy on Earth and count the petals on them all, and in every case find that there were twenty-four petals, she still would not have proven that all daisies of that species have twenty-four petals, for she would not have counted all such daisies there ever were or ever will be. It is still possible, even if not probable, that one of them will naturally have a different number of petals and thereby disprove her hypothesis. Nonetheless, having counted the petals on all the daisies of that species that currently exist on the planet, and finding all of them having twenty-four petals, her hypothesis that all such daisies have twenty-four petals would have attained a very high degree of probability. Therefore, botanists would generally accept it as true. Given this finding, she would then seek to formulate a theory explaining *why* this daisy species has precisely this number of petals, which would likely lead her to studying the specific *genome* (genetic makeup) of the species and looking for the specific gene or set of genes that control petal number.

This is a very simple hypothetical case regarding how science typically operates. It represents the scientific method in its idealized form. In actual practice, however, the process of doing science is almost always more complicated than this.

For one thing, progress in science does not generally take place smoothly, by the gradual accumulation of knowledge, as in our daisy example. That it does is another commonly held misconception about science. Rather, scientific advancement tends to proceed fitfully, as Thomas Kuhn (1922–1996) noted in his classic book, *The Structure of Scientific Revolutions*, with major scientific revolutions (or "paradigm shifts") occurring only occasionally. Within all scientific fields, particular theories reign at any given point in time. Sometimes facts begin to accumulate that do not fit well within the reigning model. Typically, that model is modified somewhat in an attempt to accommodate the new facts. However, sometimes scientists discover additional data that do not fit the modified theory, in which case that model might undergo additional modifications. The reigning theory typically continues to hold up in the face of the discovery of new data that do not conform to it until someone comes up with a new theory that offers

3. What is fascinating about this is that, while no amount of data could ever prove a scientific theory, only a *single* piece of evidence can disprove a scientific theory.

a simpler explanation of the facts. Key here is the idea of *incommensurability*—that is, that the two paradigms have neither common interpretations of many key notions nor common standards by which the two theories could be evaluated. Scientists generally subscribe to the principle of *Ockham's* (or *Occam's*) *razor*, which states that, among competing explanations for something, the simpler explanation is usually true. If continued testing—ideally, of predictions made by the new model—supports the new theory, more scientists will gradually find it superior to the old one until, at some point, this theory becomes the new reigning model, and the paradigm shift is complete.

For a comparison of Ptolemy's Geocentric model of our universe and Copernicus's Heliocentric model, see Niko Lang's image at http://en.wikipedia.org/wiki/File:Geoz_wb_en.svg.

This pattern has been witnessed historically in a number of cases. One example was the shift from the *Ptolemaic* (*geocentric*, or Earth-centered) *cosmology* to the *Copernican* (*heliocentric*, or Sun-centered) *cosmology* in the early Modern Era, about 1600–present (see the QR code above).[4] That the Sun, Moon, and stars moved around the Earth had seemed obvious to people from the time they first looked up into the sky. Ptolemy (90–168) had already noted in the second century CE (or AD),[5] however, that the planets did not move in a consistent, regular path around the Earth but occasionally engaged in "retrograde motion," in which they suddenly began moving backwards across the sky relative to their typical clockwise motion, only to stop at some point and begin moving forward again. In response to this, Ptolemy affirmed the theory that the planets did move in circular orbits around the Earth—the circle being, in his view, the most perfect geometrical form—but that they were tied to smaller orbital circles residing on the

4. You can view the images on the Internet with a smartphone and QR code reader; or you can type the web addresses (in the captions) into your web browser to view the images. Please note, however, due to the dynamic nature of the Internet/World Wide Web, an image might have migrated to another web address—it might even have disappeared from the Web.

5. Scholars have in recent years moved away from the use of BC (before Christ) and AD (*anno Domini*, "in the year of our Lord") to denote the two major eras of history according to traditional Christians usage. They now use BCE (Before the Common Era) and CE (Common Era) to remove the religious connotations of the former scheme, as these seem inappropriate when dealing with those of other faiths or no faith at all. The numbering itself is not affected: 5 BCE is the same as 5 BC, 1216 CE the same as AD 1216.

larger circles (see the QR code above). Thus, when the planets were moving backward, they were really only doing so as part of their movement on the smaller circles, not on the large orbital paths that went around the Earth.

When further, more precise observations in the late medieval period indicated that even this theory did not adequately account for the observed planetary motions, scientists sought to modify the Ptolemaic model further to accommodate the new facts. However, the problems and increasing complexity of the Ptolemaic theory caused some scientists to accept, under the guidance of Ockham's razor, Nicolaus Copernicus's (1473–1543) simpler theory:[6] that the Earth and other planets moved around the Sun in simple, circular orbits. When the observations of other astronomers, such as Galileo Galilei (1564–1642), provided additional support for the Copernican theory, it soon became the new reigning theory. Thus a paradigm shift had occurred in cosmology.

The heliocentric model itself continued to evolve with the appearance of new data—for example, elliptical orbits replaced circular ones—but the model remains the paradigm today and will almost certainly continue to do so. The move from the "catastrophic" to the evolutionary understanding of geological change on Earth[7]—the change from a static to an evolutionary understanding of life on Earth—in the nineteenth century, and, in the twentieth, the move from Isaac Newton's (1642–1726) to Albert Einstein's (1879–1955) understanding of space, time, and gravity are further examples of such paradigm shifts in science.

All of this underscores that scientific theories are inherently historical, that what is believed to be the truth about nature has been at times, and in most fields, later shown to be inaccurate, at least to some degree, and that therefore at least some of the scientific theories we accept today may be shown to be false to a greater or lesser extent in the future. This means that those who proclaim that science presents us with the truth about physical reality are quite simply wrong. Rather, science provides us with the best *current* thinking about the way physical reality is, not a certainly true account of it. This also pertains to all religious theories (i.e., doctrines). Thus, any description of the nature of "reality" is always tentative, historical, and susceptible to later correction.

6. Interestingly, the so-called *Copernican Revolution* in the theoretical understanding of the orbits of the Sun and planets actually owed more to the work of Johannes Kepler (1571–1630) than Copernicus. See Chen, "Revolutionary Change," 283–84.

7. Catastrophism was the view that Earth is essentially static except for the one major geological change/catastrophe in Earth's history, namely, the flood in Noah's time, as described in Genesis 6–9.

The language of science is *mathematics*. In this language, science is most accurately stated.[8] Scientists usually translate that mathematical language into conceptual language in an attempt to understand what the math is telling them, and in order to communicate their findings to nonspecialists. Such translation, however, is not always possible, especially in the more esoteric forms of theoretical physics, such as with certain aspects of quantum mechanics or superstring theory. For a fully developed and accurate understanding of the natural sciences, one must learn this mathematical language. To do so, however, requires an intelligence and mathematical training that most of us simply do not have. Therefore, most of us are unable to understand this language and the science it describes in some significant ways. There is an analogy here with music, in that to be able to read a musical score, one must also learn a new symbol system and its rules of syntax (i.e., a new language).[9] The point is that most of us will never fully understand science in its true, mathematical form, and so our understanding can only be conceptual, with all the translation problems entailed in conveying meaning from one language into another.

In addition, mathematical descriptions of nature will sometimes lead scientists to new discoveries about nature—rather than the more typical other way around—as it did for one of the early quantum physicists, Paul Dirac (1902–1984). Dirac was trying to describe the electron mathematically in light of Einstein's relativity theory and quantum mechanics. What he ended up with were two mathematical equations that could be used to describe the electron, one for which the electron had the expected negative charge, but the other for which the electron had a positive charge. Everyone knew electrons had negative charges, and no one had ever encountered one with a positive charge, yet the equations were saying that it should exist. In addition, the equations indicated that, if a positively charged and a negatively charged electron collided, they would *annihilate*—literally destroy each

8. A number of scientists, especially physicists, such as Albert Einstein, Eugene Wigner (1902–1995), and John Polkinghorne (1930–), have wondered why mathematics, a theoretical system whose foundation lies in purely abstract human thought alone, should describe the actual nature of physical reality. Why should natural laws be inherently mathematical? For a number of scientists, this mathematical character of natural laws seems quite unexpected, even inexplicable.

9. It is interesting that numerous studies have found that those who have learned to read musical notation and play musical instruments are typically also adept at mathematics. This should not be surprising, since music involves mathematics—for example, the way the frets on a guitar divide the lengths of strings into parts is mathematically determined, so that pressing the string against any fret produces a particular note in the standard Western musical scale. Some physicists have noted that their mathematical description of the universe is analogous to a musical score, both of which can be said to be beautiful.

other, so that nothing would remain of them except the amount of energy produced by their destruction.[10] Therefore, Dirac, *based on mathematics alone*, had discovered the anti-electron, or *positron*, and, thus, the existence of *antimatter*, as well as the fact that, when matter and antimatter collide, they annihilate, resulting in the production of energy. The existence of antimatter was confirmed experimentally shortly thereafter, as was Dirac's prediction of the energy produced when matter and antimatter annihilate. In this way, mathematics led scientists to new discoveries about physical reality.

Even more surprising than this is that the *elegance* or *beauty* of the mathematical equations is typically, if not always, an indicator of which equations accurately describe the nature of reality. Dirac reportedly altered his focus as a theoretical physicist from doing physics to searching for beautiful mathematical equations, and when he found them, he looked at nature to see what they described.

Mathematical beauty is something the mathematician and physicist recognize, but it is harder for the layperson to grasp. That equations can be mathematically beautiful involves their having certain qualities, such as simplicity, symmetry,[11] depth of meaning, and perhaps unexpectedness. One example of such a mathematically beautiful equation is Einstein's $E=mc^2$. It is simple, in that it employs only several terms—energy (E), mass (m), and the speed of light (c)—and it is an incredibly concise equation. It is profound and unexpected, in that it says that energy and mass are essentially the same thing in different forms, such that a little bit of mass is equivalent to a lot of energy and vice versa. It is also symmetrical, in that it indicates the conversion can go in both directions, mass to energy or energy to mass. The number of examples of this relationship between beauty and truth in the equations that describe the nature of the universe is significant.

Many physicists today continue to be driven by their belief that mathematically beautiful equations are the ones that describe nature, as we will see (in chapter 5) regarding scientists researching superstring theory. As string theorist Brian Greene has noted, "[I]t is certainly the case that some decisions made by theoretical physicists are founded upon an aesthetic

10. What happens in this mutual annihilation, according to Einstein's $E=mc^2$, is that the mass of the electron and positron are converted into energy, hence the presence of new energy where the two particles had formerly been.

11. When people think of symmetry, they typically think of *bilateral symmetry*, which is the kind of symmetry in which one half of a thing is essentially a mirror image of its other side—such as in humans looked at from the front or rear. However, many other types of symmetries exist, such as *rotational symmetry*, in which a thing (such as a ball) looks the same no matter how you rotate it. Physics uses a host of symmetries to describe the world. See, e.g., Zee, *Fearful Symmetry*.

sense—a sense of which theories have an elegance and beauty of structure on par with the world we experience . . . So far, this approach has provided a powerful and insightful guide."[12] That mathematical beauty should be tied to scientific truth is even stranger than the fact that nature is inherently mathematical.

Another complexity in the actual practice of science is that the correct interpretation of relevant scientific data is not always clear-cut. Different scientists may interpret the same facts in quite different ways, and at times they do, which produces two or more competing theories. One example of this is the ongoing analysis of a roughly four-billion-year-old Martian meteorite found in 1984 in the Allan Hills region of Antarctica. The team that first analyzed the rock interpreted certain features they found, especially the presence of unusual organic molecules and mineral deposits, as indicators that Martian microbes once lived in that rock. Were this true—that there had once been life on Mars—it would be perhaps the greatest discovery in the history of science, and one that would significantly alter our understanding of the universe and our place in it, as well as the possibility of finding life elsewhere in the cosmos.

Subsequent analyses by other researchers, however, have resulted in interpretations of those mineral and organic deposits as products of certain *chemical processes*, not of *living entities*. Today the issue is still not settled, though it seems that more scientists are skeptical of the claim that the meteorite indicates the former presence of life than those who accept it. The point is that even well attested facts—in this case, mineral and organic compounds found in the rock, whose presence no one questions—can result in two quite different interpretations of what these facts mean. Similar disagreement can also be found in how the mathematics of a theory should be translated into conceptual form, as we will see in our discussion of quantum mechanics (in chapter 5).

Another complexity of science as actually practiced is that scientists cannot always directly observe the entities or processes they investigate. For one thing, some objects may be too small to observe and their movement so rapid that their existence and behavior can only be inferred from other evidence. For example, subatomic particles are impossible to see, even with our most sophisticated microscopic instruments. Therefore, the existence of such particles can only be inferred, such as by the bubble tracks they make in the cloud chambers of particle accelerators. This difficulty of interpretation in determining the existence and nature of things not directly observed

12. Greene, *Elegant Universe*, 167.

is fundamental, not only to physics, but also to a number of other scientific fields.

Another, related complexity is known as the "background problem." This is the problem that the presence of certain data in an experiment that are irrelevant to the thing being studied complicate the interpretation of the experiment. The presence of tracks in cloud chambers left by stray particles (such as cosmic rays) that are *not* a result of the particle collision is one such example. The judgment of the experimenter determines what constitutes data relevant to the experiment and what is irrelevant background data. Not all experimenters' judgments might agree. Even if they do agree, their judgments might not be correct. Thus, the "background problem" is yet another complexity in the practice of science.

The idealized scientific method also fails to account for the fact that a fundamental distinction must be made between scientific theories that are directly testable and repeatable in the present and those that deal with historical events and processes, which are neither directly testable nor repeatable. For example, if a scientist wants to test some aspect of the standard theory of particle physics, she can do so in the present by undertaking experiments in a particle accelerator, which will tend to support, modify, or refute the reigning theory. Other scientists can replicate such experiments in different places, both now and in the future.

However, when studying features of the geological or evolutionary past, such as when and how layers of material were deposited in some ancient rock formation, or which ancient primate fossils might have been directly ancestral to human beings, the scientist obviously cannot observe or replicate in the present the laying down of those layers of material or the evolution of one species into another that occurred millions of years ago. Thus, in the evolutionary example, while fossil evidence can determine the degree of *morphological similarity* (similarity of form) between two ancient primates—assuming that scientists possess sufficiently complete skeletons of both—and perhaps that one came before the other, it cannot tell us if the one in fact evolved into the other. They might well have been similar but unrelated species. This does not mean that evolution (or geology) is in some way a deficient form of science, but rather that different sciences must be studied in ways appropriate to their nature, and that in some cases, as when dealing with unique historical events, different means of inquiry must be used, and these may not always yield definitive results.

Another deviation from the simple, idealized conception of the scientific method is the fact that theory sometimes *precedes* observation when scientists seek to understand certain phenomena, as is the case with contemporary superstring theory. Here, theoretical solutions to several problems within Big Bang cosmology and the Standard Model of particle physics have been offered based solely upon formulations that solve the problems in mathematically elegant ways. Yet, owing largely to the unimaginably small size of the strings postulated as the most fundamental constituents of the universe, there is to date neither experimental evidence nor even any conceivable experiments proffered by string theorists that might provide direct evidence for the truth or falsity of this theory.[13] Here, hopefully, indirect experimental evidence will someday help determine whether the theory is probably true or not.

Related to this is an even more fundamental issue for science, which is that scientists necessarily rely on certain foundational assumptions about nature, which they cannot demonstrate to be true, yet upon which many of their fundamental theories are based. One example is the belief that the same natural laws have operated from the beginning of—and everywhere in—the universe in the same way as they do today in our solar neighborhood. While such assumptions are accepted because they seem to be reasonable, and some astronomical evidence suggests they are true, it remains the case that these assumptions still cannot, and perhaps never can, be validated. That such assumptions cannot be validated means that there is no way of *knowing* that they are true; they can only be *believed* to be true.

Relying on indemonstrable foundational assumptions about nature might not seem to be much of a problem, but an examination of the history of both mathematics and science shows that some assumptions believed true at one time have later been shown to be false. For example, before the invention of non-Euclidean geometry it was assumed that the shortest distance between two points is a straight line, and that the sum of the angles in a triangle always equals 180°. These definitions are true in Euclidean geometry, because that geometry deals with flat surfaces, such as a sheet of paper on a flat tabletop.

However, when scientists began to consider curved surfaces, such as the surface of the Earth, and what geometry would look like on large scales there, they quickly learned that the shortest distance between two points is a curved line, and that the sum of the angles of a triangle will always be greater than 180°. In fact, in the case of a right triangle, whose base stretches along

13. While this is true regarding direct evidence, some experiments may have an indirect bearing on the truth of this theory. Superstring theory is tied to the theory of *supersymmetry*, and, as at least some superstring theorists have noted, should supersymmetry be proven false, it would suggest that superstring theory, at least in some formulations, is wrong. Experiments at the Large Hadron Collider in Geneva may be able to confirm or disconfirm the existence of some of the lighter proposed *superpartner* particles (see chapter 5), which would be experimental evidence relevant to superstring theory.

Earth's equator and its top is at one of the poles, the sum of its angles is 270° (three 90° angles). An example from science of an assumption later shown to be wrong was Isaac Newton's assumptions that space was unchanging and time was unrelated to space, both of which Einstein later showed were false. Thus we cannot claim *certainty* regarding the truth of the assumptions scientists hold at any given time. The truth of foundational assumptions remains a matter of *belief*.

Perhaps the most fundamental assumption of all science is *methodological naturalism* or *methodological materialism*: namely, the assumption that nature is a closed system, such that things in nature and their behavior can be explained fully and only by other things in nature. Many scientists think that this assumption is necessary to the conduct of science, in that, if natural things were to be explained by things outside nature, such as God (as traditionally conceived), then that might well put an end to science. Scientists might not investigate anything very deeply, since God would be the easy answer to every question about the nature and behavior of physical things.

What is essential to understand here is that this *assumption* of *methodological* materialism does not assert anything whatsoever about *whether* something "beyond" or in addition to the physical universe, such as a spiritual dimension of reality, exists or does not exist. To assert that nothing exists but the physical universe is to hold the *belief* called *philosophical materialism*. Although certain atheistic scientists sometimes state or imply otherwise, science itself does *not* assert anything one way or the other about the *existence* of spiritual reality, let alone prove or disprove its existence. Science, by virtue of its method, can *only* study the *physical* universe and the physical things in it and, therefore, has nothing to say about spiritual realities, such as God.

Thus, *methodological* materialism seems in the eyes of many scientists and philosophers of science to be a necessary assumption of science, but holding the view of *philosophical* materialism/atheism is not. The latter is a metaphysical *preconception* some scientists bring to the *interpretation* of science. It is not a consequence of science itself. The existence of a large percentage of scientists who are religious, shown consistently in objective surveys of scientists for at least a century,[14] bears witness to this fact—as

14. According to a "survey of scientists who are members of the American Association for the Advancement of Science, conducted by the Pew Research Center for the People & the Press in May and June 2009, . . . just over half of scientists (51%) believe in some form of deity or higher power; specifically, 33% of scientists say they believe in God, while 18% believe in a universal spirit or higher power . . . Finally, the poll of scientists finds that four-in-ten scientists (41%) say they do not believe in God or a

well as that religion is *not* in the process of dying out among scientists, as some inimical to it have suggested.

Perhaps the aspect of science as actually practiced that would be most surprising to the nonscientist is that science is not always a wholly *objective* or *rational* enterprise. This really should not come as a surprise, given that science, like every other human enterprise, is done by human beings, and we cannot simply put aside all of our personal foibles, our wants and needs, and other personal characteristics, when we engage in scientific research.

On the positive side, *imagination* and *intuition* often supplement conscious, rational thought. These have been so important in the history of science that Einstein famously said that imagination was more important for a scientist than knowledge. An example of this is the way Stephen Hawking (1942–) intuited how radiation might escape from a black hole via quantum tunneling—an idea that most, if not all, astrophysicists now accept as true.

On the other hand, the objectivity that the scientist would employ in her work is at times negatively affected by factors individual and social, including psychological, political, and economic concerns, as well as the metaphysical commitments already noted. For example, scientists tend to be skeptical of new theories, preferring rather to accept current, seemingly well-established ones. There is good reason for this, since most novel ideas in science prove to be wrong and, in some cases, spectacularly so, as with the claims for cold fusion some years ago.[15] Although such conservatism is often a sound scientific practice, it can also retard scientific progress by preventing some novel ideas from receiving the attention they deserve—at least for a time. Murray Gell-Mann (1929–) recently made this point with regard to some scientists' skeptical reception of his radical new *quark theory* in the early 1960s.[16] Standing on the wrong side of the theoretical divide in science can result in the reduction of one's stature in the field and thus a limitation of future prospects for professional advancement and research funding. Given these realities, it seems prudent for the scientist to go along with the predominant view of the scientific community, even if it is not necessarily the best for scientific progress.

Interestingly, though, there is also a certain countervailing tendency to this theoretical conservatism among scientists, because one is most likely

---

higher power." Masci, "Scientists and Belief."

15. Several physicists claimed that they had been able to cause nuclear fusion to occur without pumping in the enormous amounts of energy (heat) understood to drive such fusion reactions in the cores of stars. Other scientists were unable to reproduce the claimed result following the same procedure as the original physicists, and so most physicists have ceased to be interested in the idea of cold fusion.

16. For the theory of quarks, see the detailed explanation in chapter 4.

to gain recognition from one's peers—and all that goes with it psychologically, professionally, and economically—when one *disproves* a currently held theory. This is why scientists like Galileo, Darwin, and Einstein are famous both within and without the scientific community. They risked going against the standard scientific thought of their time to propose new theoretical models neither obvious nor immediately accepted at the time by other scientists in their fields, yet their theories subsequently came to be accepted as fundamental truths about nature. Thus, there is a significant incentive to be a maverick in science, despite the personal and professional risk. Nevertheless, the fact that going against prevailing theories more often than not turns out badly for the scientist helps explain why the majority tendency is to side with current theories.

In addition, philosophical and/or religious views can influence what theories a scientist chooses to believe and/or how he interprets their meaning. For example, it is widely known that cosmologist Fred Hoyle (1915–2001) preferred his steady-state theory of cosmic origins to the Big Bang theory partly because he was an atheist who believed Christians could use the Big Bang theory in support of their doctrine of God's creation of the universe "out of nothing."[17] Indeed, despite the accumulation of ever more evidence supporting the Big Bang model and the absence of evidence in support of his steady-state theory, Hoyle evidently continued to subscribe to his theory to the end of his life.

On the other side of the religious divide, Michael Behe (1952–), a conservative Christian biochemist and Intelligent Design[18] spokesperson, has interpreted certain chemical processes in our bodies to be "*irreducibly complex*"—so complex as to be beyond any reasonable possibility of having evolved naturally. This, he claims, is proof that God must have established them—meaning, of course, that God must exist (though, like other Intelligent Design proponents, he prefers not to speak of God as the one responsible, only the Designer, in the hope that Intelligent Design might be taught in public schools as science). While scientists have shown how such biochemical processes have evolved, Behe continues to believe in Intelligent Design. Every scientist and theologian, like every other person, has certain core beliefs about the nature of reality, and these will at times determine which theory she believes and/or the meaning she thinks that theory should be given.

17. In Hoyle's understanding, matter, in the form of hydrogen, continued to be created in the everlasting, expanding universe. Most people, including Hoyle, interpret Genesis 1, in the Bible, as indicating that the universe had a beginning, before which nothing existed, and that God caused it to begin.

18. For an examination of the Intelligent Design movement, see chapter 7.

With personal and professional stakes such as these, it should also come as no surprise that arguments over competing scientific theories not only exist and are vigorous but at times even turn vicious, with some scientists openly demeaning, not only the views of other scientists, but also the scientists themselves in ways that would probably shock the general public. This, unfortunately, is also seen occasionally in theology. Why should this surprise us? Again, scientists and theologians are only human and, like us all, share similar core virtues and vices, hopes and fears, strengths and weaknesses All of these may vary in degree and manifest in different ways from person to person, and yet they will typically influence to some extent the way they think about their fields of study. What makes the scientific method so powerful in pointing toward empirical truth, however, is the fact that it is a *communal* enterprise. In the end, the work of many other scientists from around the world, all of whom are susceptible to their own weaknesses, usually negate the experimental and theoretical flaws, and personal characteristics, of individual scientists. For this reason, science can claim an objectivity for its findings that few other academic disciplines can.

One more issue is possibly the most fundamental question regarding science and its description of the world: namely, whether science provides us with knowledge of reality or only reality as it appears to us. This is a philosophical debate that goes back to the ancient Greek philosophers and has remained a topic of dissention among certain scientists and philosophers of science to this day. The core question is whether the things we perceive by means of our senses and the tools that amplify them (microscopes, telescopes, and the like) provide us with an accurate description of reality, of the way things actually are. The position that affirms this claim is called *realism*. It is been rightly pointed out, on the basis of neuroscience as well as philosophy,[19] that the only way to verify the realist position would be for us to stand outside of ourselves and compare what our senses tell us about reality with reality itself. This is obviously impossible. We can only know what our senses, as interpreted by our brains, tell us about the world.

Thus, any view about whether what our senses tell us is an accurate description of reality can only be a matter of belief. Most practicing scientists are realists and have good reasons for believing that science really does tell us about reality. I share this belief. However, the point is that scientists do not *know* this. It is not a fact, but rather a *belief*. Since, as we saw above, scientific theories also are not facts, but beliefs, even though with evidence and rational arguments to support them, the typical view of nonscientists

19. See, e.g., Newberg, *Principles*, 214–18. An example of a philosopher who critically discusses "realism" is Alan H. Goldman, William R. Kenan Jr. Professor of Humanities and Professor of Philosophy at the College of William & Mary.

that science tells us the actual truth about reality is, again, simply false. As we have noted, science tells us what scientists in any era *believe* about physical reality based on the (always incomplete) data they have at the time and the assumptions that undergird those beliefs. The best we generally can do is to formulate our beliefs about reality based on the best current evidence and the most objective and rational interpretation of that evidence. Nonetheless, it is important that we realize that *all* our beliefs—scientific, theological, and otherwise—are always provisional and potentially incorrect, not certain truths.

3

How Theology Operates

THE METHODS OF SCIENCE AND THEOLOGY COMPARED

PRETTY MUCH EVERYTHING THAT I have just written about science and scientists also applies to theology and theologians. Theologians have the same strengths and weaknesses, rationality and irrationality, wants and needs, and hopes and fears as scientists by virtue of their being human, too. Science tries to overcome the more detrimental effects resulting from personal idiosyncrasies by continued testing to attain a broad consensus on theories among those within the field. So also theology, facing these same problems, appeals to professional theologians worldwide for the testing of the theories (i.e., theological doctrines) proposed by individuals or church bodies. Such consensus-building generally takes time within the scientific community. It also takes time—and usually a longer time, given typical religious conservatism—to build a consensus among theologians for any new perspective.

In science, one must at times infer the existence and nature of something that cannot be directly observed, such as dark matter and dark energy. The same need for inference is sometimes seen in theology. In particular, in addressing the questions of the existence and nature of God—by definition a spiritual, not a material, entity—theologians are obviously dealing with that which cannot be directly observed using our five senses or any mechanical extensions thereof. Thus, they must revert to other means, such as the inference used by scientists, to speak, for example, about the existence of God.

One approach to addressing this question is by the method theologians call *natural theology*, in which they seek to infer the existence and even, in some cases, certain characteristics of God from observed features of the natural world, under the belief that, as with art, the creation will tell us something about the one who created it. The main way theology addresses such things, however, is through *personal experiences*—our own and those of people from the past, the latter usually including certain foundational experiences by the originators and early adherents of a religion. In the case of Christianity, the Bible includes accounts of such experiences and interpretations of what they mean.

Another similarity between science and theology has to do with their needs to conform their methods of investigation to the different kinds of things they study. Physics and chemistry, for example, employ methods of study different from those of the historical sciences of evolutionary biology and geology. Similarly, theology employs different methods appropriate to the study of different aspects within its purview, such as the interpretation of the Bible, the reconstruction of historical events in church history, Christian doctrinal teaching, and Christian ethics. In addition, just as the historical sciences study unique, unrepeatable events from the past, so too theology studies these kinds of events, such as claims to having received revelations from God. Also, just as in science, the best interpretation of the facts that undergird a theological belief is not always clear. In addition, imagination and intuition play an important role in both science and theology in order to come up with new perspectives and ideas that can move the discipline forward. Thus, some of the same general kinds of complexities that characterize science also characterize theology.

Furthermore, as in science, so too in Christian theology, paradigm shifts sometimes occur. When the overwhelming majority of Christian thinkers come to believe certain doctrines, they become "orthodox teachings," "doctrines," or even "dogmas" of the church.[1] Yet, just as in science, virtually none of these doctrines or dogmas can ever be said to be

1. The distinction between orthodox teachings, doctrines, and dogmas has generally not been well defined by the church. *Doctrines* are typically understood to be *orthodox* (i.e., correct) *teachings* of the church and, as such, are to be believed by all Christians. However, the doctrines taught by the different Christian traditions (the Protestant, Catholic, and Eastern Orthodox) and the various denominations within the Protestant tradition (such as Lutherans, Baptists, Presbyterians, Anglicans, Methodists, and the like) are not always the same. The term "dogma" is usually applied to any doctrine that the church as a whole has defined as a fundamental belief that all Christians must hold.

absolute truths, since sometime in the present or future new evidence or a reinterpretation of existing evidence might arise that would either modify or invalidate them. A classic case of this is the *doctrine of original sin* as promulgated by the highly influential early Christian thinker, Augustine (356–430 CE). Though unknown to most Christians, this doctrine has been gradually dying within the Western church (it has never been accepted in Eastern Orthodox Christianity) for at least the last century and a half. While there are many reasons for this death (which we will discuss in chapter 9), the final nail in the coffin for the original sin doctrine was Charles Darwin's (1809–1892) *theory of evolution by means of natural selection*, which, at least for some in the church, put an end to the idea of humans having suddenly appeared out of nowhere and having been perfectly good before their first sin, after which they "fell" and thereafter could do nothing but sin. Today few mainline theologians, myself included, accept this traditional Western doctrine—at least in its Augustinian form.

In this matter, the church, ironically, was following the teachings of Augustine himself, who insisted that when well-supported science indicates that a Christian teaching is wrong, that doctrine needs to be changed or eliminated from Christian teaching. Augustine's reasoning was that, since theology is based on our human interpretations of the Bible, which is a complex book whose texts can be and have been interpreted in many different ways, we cannot *know* that our interpretation of any given passage is the right one. Thus, well-demonstrated science should be accepted as true when science and theology conflict. If Christians persist in believing things shown by science to be wrong, Augustine claimed, intelligent people would think that Christianity is absurd. Such is still the case today. The belief, on the part of many thoughtful people, that Christianity is nonsense is partly, perhaps even largely, attributable to the claims made by some, especially conservative Christians, that contradict our best science.

An example of a theological paradigm shift unrelated to scientific progress occurred with the rise of modern historical study in the late eighteenth and early nineteenth centuries, with its new concern for factual precision. This led to a new comparative study of the biblical texts to determine whether they all agreed in their narratives of events. If they did, that would tend to support the historicity of those accounts; if not, the factuality of those accounts would be thrown into question. This "historical-analytical" or "historical-critical" study of the Gospels, for example, led to the realization that the four Gospels contain a number of contradictions regarding the life and teachings of *Jesus* (ca. 5/4 BCE—ca. 30 CE), such as when Jesus was crucified. In the Gospels of Matthew, Mark, and Luke, Jesus was crucified on the Jewish holy day of Passover, whereas in the Gospel of John,[2] Jesus was crucified on the Day of Preparation, one day before the Passover. This explains the absence of a Passover meal during which Jesus instituted the Lord's Supper[3] in John's account of Jesus's final meal with his disciples. Similarly, the Gospels do not agree when Jesus's crucifixion occurred. According to Mark 15:25, it occurred at "the third hour" (9:00 AM), while John 19:14 says that it happened at "the sixth hour" (noon).

This awareness of historical contradictions in the Bible has led the mainline Christian denominations to alter their traditional view of the complete historical accuracy of everything in the Bible to one in which the various historical accounts have to be tested against one another and against historical information from nonbiblical sources to determine their accuracy. This is a paradigm shift that most conservative evangelical and all fundamentalist churches have yet to make as they continue to believe in the complete and literal inerrancy of the Bible.

Finally, the role of belief is important in both science and theology. As we have seen, certain beliefs are foundational in science, including the beliefs that the fundamental assumptions of science are true, that the theories put forth by science are true, and that what is described in those theories is the truth about actual reality (as far as it can be known at any given time) and not just its appearance. Belief in theology is similarly seen in belief in Christianity's foundational assumptions (above all, that spiritual reality exists and is fundamentally personal—i.e., that God exists—and that Jesus revealed this God decisively for humankind), that its theories (doctrines) are true, and that what they describe is reality as it actually is.

In all these ways, then, science and theology share some significant similarities. Still, important differences do remain, the chief of which is that science deals with the more accessible and tangible physical reality, while theology deals fundamentally with the spiritual realm, the very existence of which is uncertain. In the view of many people today, this is what gives science the sense of possessing greater trustworthiness than religion.

2. The four *Gospels* (lit., "good news") in the New Testament are named for persons considered by second-century church officials to have been their authors. The earliest Gospel manuscripts had no titles or author names affixed to them. All four are believed to have been written ca. 68–100 CE. In this book, the traditional names of the Gospels (Matthew, Mark, Luke, and John) refer to the writings, not to their supposed authors.

3. The Lord's Supper, also known as the Eucharist ("thanksgiving") or Holy Communion, is a ritual commonly practiced in most Christian churches. For some church bodies, it is simply a symbolic remembrance of Jesus's sacrificial death; for others, it is a means by which God's grace (divine favor; for example, forgiveness) is given to those who participate in the ritual.

Although science may well have the advantage in this regard, religion can be said to have its own advantage in terms of the breadth and existential importance of the questions with which it deals. Science deals well with the world as presented to us by our five senses and the way we understand that world intellectually, but that is not the only way we experience the world. We also experience it emotionally, aesthetically, morally and, for some of us, spiritually, too. That is, we experience the world as it impacts us in emotional ways, such as making us happy or sad, or certain features of it as being beautiful or ugly, or certain events as being morally right or wrong, or as spiritually meaningful or not. Scientists have not typically done a good job in addressing these other ways of experiencing and knowing, at times even disavowing their possible validity as sources of knowledge about reality, because they refer to things that our five senses cannot know.

When scientists have spoken, *as scientists*, of things such as values, their answers have typically been based on aspects of evolutionary science—that is only to be expected, since naturalistic answers are the only ones science can provide. Such answers, however, have seemed overly simplistic and reductionistic, inadequate to the fullness of these experiences to those who have had them. To such people these are factual experiences, and writing them off as meaningless or unimportant reflects a shortsighted view of reality. Thus, philosopher Charles Hartshorne (1897–2000) has asserted, "If anything is unscientific, it is the denial of aspects of existence because they seem inconvenient for our methods."[4] Theology offers the possibility of providing answers to such questions because it takes a wider view of reality and knowledge; and even while theology cannot provide us with certainty, it can at its best offer reasonable and intellectually satisfying answers to such questions.

Somewhat related to this is the fact that science, dealing with the nature of the physical universe, cannot speak to the larger and existentially more important human questions such as, Who am I really? Am I anything more than just another animal? Is there a deeper (i.e., not merely biological) meaning of my life? What is the meaning of my experiences of values such as truth, goodness, and beauty? Science, by virtue of its nature, can say nothing about meaning. That has led some scientists and others to claim that life and the universe are meaningless. This is the same kind of unwarranted jump some make, as we have seen, in moving from methodological materialism to philosophical materialism. In neither case, however, does the first claim entail the second. It is also the case that science cannot answer some of the questions that science itself raises, such as why natural laws

4. Hartshorne, *Creative Synthesis*, 4.

exist at all, why they are fundamentally mathematical, and why they are "beautiful." Here again, as we will see, theology can offer reasonable answers to these questions.

THE METHOD OF THEOLOGY

The formal method according to which Christian theology proceeds is rather like the scientific method, at least in its general goal of learning the truth about reality, as well as the other commonalities we have noted. However, because theology focuses on the spiritual realm, its general method must be different in significant ways from the scientific method. The chief difference is the existence of *authority* as a relevant factor in determining religious belief. Christian theology, like the theologies of other theistic religions, understands the divine to be *personal*—that is, having some of the same kinds of characteristics, such as thought, will, and love, that human persons do, though not in exactly the same way (as we will see in the discussion of theological language at the end of this chapter). On the other hand, God is *not* understood to be *a person* in the sense of what we normally mean when we use that term, which implies being confined in space and time and having a physical nature and appearance. God is a purely spiritual being who does not possess a physical nature and is infinite in space and time.

Because God is personal, theology relies primarily on the self-communication of God to understand what God is like, just as we rely on the self-communication of other people in order to understand what they are like.[5] This divine self-communication is traditionally termed *revelation*, or *God's self-revelation*, and is the foundational authority for most Christian thought and practice. Such revelation is understood to occur between God and individuals, in certain historical events—especially in the person of Jesus—and in nature. Members of theistic religions believe that the foundational revelations of their religion are contained in its *scriptures*, which for Christianity is the Christian Bible.

The concept of God's self-revelation is clearly not without its problems. Two in particular have been discussed at great length in the modern era. The first is how to *know* whether a purported revelation from God is

5. You will find my prose in discussing God to be rather stilted at times, in that I use the terms God and Godself consistently, while refusing to say He or Himself. This is because, as spirit, God is neither male nor female. Humans exist in these two forms for the purpose of sexual reproduction; God, as spirit, does not reproduce. To speak of God as "He" is only to reinforce the traditional—incorrect—belief that God is male. This traditional usage only serves to support the patriarchy that has done so much to undermine the value of women in the past and, sadly, in the present as well.

really a revelation from God and not simply a delusion or wishful thinking on the part of the one who claims to have had the revelation. The second problem concerns the more fundamental question regarding the *possibility* of revelation—that is, how such revelation can take place between a spiritual entity like God and material entities like us. Given what we have seen above regarding differing views of scientists on certain theories, you should not be surprised that there are significant differences of opinion within the Christian theological community on both of these key questions.

Regarding the question of how we can know that a purported revelation has really come from God, one relevant fact is that not only is making such a judgment not obvious today, but also making this judgment was no easier in biblical times (though many Christians think it was). Even in the Bible, we can occasionally see people struggling with this issue. For example, 1 Kings 22 recounts the story of King Ahab of Israel (d. ca. 850 BCE), who is considering going to war against the king of Aram and asks his court prophets—people who claimed to speak for God—whether he would be victorious or not. They tell him that God has revealed to them that he will be victorious. But then Ahab calls on another prophet not related to the court, Micaiah ben Imlah (ninth century BCE), who prophesies disaster should the king go to war.

The general lesson to be learned here is that a number of people claimed to have revelations from God in ancient Israel, but they did not always agree on what God's message was. Thus, Ahab had to decide for himself whom to believe and, unfortunately for him, he chose to believe the court prophets, went out to battle, was defeated, and died. Thus, the not-uncommon idea among Christians, that determining true from false revelation was clear in biblical times, is shown to be wrong by this example, as well as others from both Testaments. Usually it was only after the fact that one prophet or group of prophets was understood truly to speak for God, and this only because his, her, or their prophecies about the future were the ones that had come true.[6]

Some biblical writers cited wondrous acts done by people who claimed to speak for God as signs of the validity of their message. However, even in biblical times such "signs and wonders"[7] were no guarantee that the person

6. Not all prophets foretold the future; rather, the essential character of the prophet was that he or she relayed to Israel what God had to say. Only at times was there talk of the future, and even then most of the time it was delivered in the form of "if you continue doing this or that, this is what will happen to you."

7. The term people use today for what the Bible calls "signs and wonders" is *miracle*. Various Greek terms in the Bible (*thaumasia*, *teras*, and *dynamis*) have been translated as "miracle." In the biblical worldview, which had no notion of nature as a mechanical,

performing them spoke for God (see, for example, Mark 3:19–30; John 9; and Rev 13:11–18 in the NT, where such wondrous signs are believed to be performed also by evil persons in service of the devil). Today claims for anything miraculous are even more contentious and so can provide little support for one claiming to have received a revelation from God.

In determining what might count as a true revelation from God, there are at least several *disqualifying factors* we might cite, some related to the *person* purporting to have a revelation and others regarding the *content* of the supposed revelation. Regarding the personal factors, the very sensible general rule is that persons who are mentally ill, habitual liars, heavy substance abusers, con men, or other such persons known to be less than trustworthy sources of information are, as a rule, not to be believed in their claims to have had revelations from God.[8] On the other hand, those whom we know to be rational and generally honest people, and particularly those whom we know most intimately in this regard, would typically be thought to be more trustworthy in their claims to have had a divine revelation. Thus, if your parent, say an engineer who had little interest in religious matters and was known to be highly rational and committed to honesty, were to claim to have had a revelation from God, you would be more likely to believe him, or at least believe that he might have had a revelation. Were he to suffer ridicule and perhaps other indignities, such as losing his job, on account of this claim and still maintain that he had received a divine revelation, you would have an even stronger reason to think his claim might be correct and to investigate it further.

This leads to the second major disqualifying factor in determining the likelihood of a person having indeed had a revelation from God: the *content* of the purported revelation. For example, many, perhaps most, Christians would not likely consider claims thoroughly inconsistent with important biblical and historic church teaching to be true revelations from God. Thus, should someone claim a revelation in which God called for the hatred and killing of certain individuals or groups of people, Christians, especially those who rightly understand that God loves all people and wants us all to do the same, would certainly reject that purported revelation. On the other hand,

---

closed system, these terms had a different meaning from the one "miracle" has in a Newtonian worldview. We will discuss the subject of miracles in chapter 17.

8. Interestingly, the same kind of analysis was practiced, according to the NT Gospels, by those who rejected the claims of some to be speaking for God. Thus, in the view of some, Jesus was demon-possessed (Matt 12:24; Mark 3:22; Luke 11:15) while in the estimate of some others he was mentally ill (Mark 3:21). The fact that he had been executed by the Roman government as a criminal was another factor in his rejection by some, especially gentiles (non-Jews)—a charge which the author of Luke, himself a gentile, sought mightily to refute in his Gospel (see, e.g., 23:13–25, 37).

a purported revelation encouraging love, compassion, peace, and harmony among people would be more likely to be considered possibly genuine.

Ultimately, this problem of determining who has truly received a divine revelation cannot be solved definitively. I think that, in practice, people tend to accept a purported revelation as real when they are struck in a profound, existential way that the content of that supposed revelation is true, when from the depths of their being they respond to what has been said with a profoundly felt affirmation. This kind of experience is in my view what we commonly refer to as *intuition*—a sense that something is true (or false) even though we cannot clearly produce the exact reasons why we think so. Having an intuition that something is a true revelation is thus only one example of the important role that intuition plays in our lives generally and, as we have seen, in the professional lives of scientists as well.

As we now know, *conscious* thinking is only one type of thought process that goes on in our minds, and *conscious reasoning* only one kind of rational thought. From Freud onward, psychology, and increasingly neuroscience, has indicated that, rather like an iceberg, the vast majority of which lies beneath the water's surface, so our conscious thought is a small part of our mental processes, most of which take place beneath the surface of consciousness, in what we generally refer to as the *subconscious* or *unconscious*. There, it is thought, a great deal of information from all aspects of our experience is gathered, stored, and used by our subconscious minds in ways somewhat like, and in other ways perhaps somewhat unlike, our conscious thought. Unlike conscious and linear reasoning, as in logic, science, theology, and most other disciplines of thought, this subconscious reasoning may operate at least to some extent in a nonlinear fashion, producing a sense or intuition of what *seems* to be true or false, or right or wrong.

John Henry Newman (1801–1890) suggested this kind of subconscious reasoning when he wrote of the "*illative sense*," in which our minds consider a rich calculus of input in ways that occur apart from our conscious thought and which we cannot fully comprehend. As William James (1842–1910) noted, in the ordinary course of life, when considering what we will believe, "we find ourselves believing, we hardly know how or why."[9] This subconscious, probably informal reasoning, this intuition, is what guides us in life more often than conscious deliberation, and especially so with regard to our most important existential decisions, such as whether to marry someone. We do not make lists of pros and cons and decide the issue depending on which list is longer. We have a sense, an overall impression, whether we should or should not marry the person, and it is this intuition

9. James, *Will to Believe*, 9.

that decides the matter. The same is true with other major decisions in our lives—what career we should choose, whether we should become friends with this or that person, whether there is a God, whether we think what someone is saying is right or wrong, and so on. As a primary guide for our most important decision-making in our lives, intuition is an exceedingly important tool.

Intuition has sometimes been called our "sixth sense," implying that intuition operates in a way analogous to the way our five senses operate. It challenges the view of some who, as we have noted, think that the *only* way of knowing about reality is by means of the five senses. Those senses are our connections to the external, physical world, and do tell us about it, but they seem to tell us precious little about our internal, mental and emotional worlds. It may be that our intuition is the sense that tells us about our subconscious mental worlds, our feelings, and the way these provide us with information about reality analogously to the way our five senses do.

The danger of seeing reality as that which is discerned only through our five senses is that not only does it insist that there are no other ways of knowing but also that these senses and the mind's interpretation of them describe the true and only essence of who we are. Feelings and intuitions must finally be viewed as insignificant, not integral to who we are—primitive holdovers, perhaps, of our animal past. Such a view is reminiscent of the seventeenth-century age of Enlightenment's view of human nature, in which reason was seemingly the only human aspect that mattered. The inadequacy of this view was the fundamental teaching of the Romanticist era that succeeded it in the early nineteenth century. Romanticism sought to articulate a more fully rounded portrait of humanity that took into account the feeling and intuitional aspects of our nature, believing they are fundamental parts of us and every bit as valuable as our rationality.[10] This seems to reflect our commonsense understanding of ourselves: that we are not simply disembodied brains. We are richer beings than that.

The significance of this for the Christian understanding of human nature and experience as I conceive it is profound. This understanding values the five senses and what they tell us about external, physical reality. It also values what intuition can tell us about our subconscious experiences, including spiritual reality. I believe it is at this subconscious level that God is "felt" and known, and that these subconscious feelings and experiences are

10. This, somewhat ironically, was also one of the main themes of the science fiction television and movie series *Star Trek*, in which Spock the Vulcan argued for the value of pure reason without emotions while Captain Kirk and the other humans argued for the importance of emotions—the truth of which Spock eventually came to accept.

the basis not only for the belief of many of us that God exists but also for our understanding that God is a being of perfect love.

It is our intuitions, then, that are the channels by which that subconscious experience and knowledge of God can become conscious, so that our already-existing subconscious relationship with God might become one into which we choose to enter consciously and intentionally. In so doing, Christianity teaches, we will find what is most important to us, even though we seldom realize it—living in loving relations with as many others as possible—and in so doing attain the personal fulfillment that is our deepest desire.

Yet for all the importance of our intuitions, we also know that they can be wrong. One's intuition of truth or rightness does not guarantee that something is true or right. As we have seen, everything we experience, think, and feel is the product of interpretation, and our interpretations can be and sometimes are wrong. Intuitions can be highly irrational at times, less influenced by facts than by hopes, attitudes, and erroneous beliefs. This is why conscious reasoning must serve as a check on our intuitions, including our religious or antireligious ones, testing with all our rational powers the truth of what our intuition has told us. This is what is done in science, and it is also what is or should be done in theology.

One thing is clear, though: the experience of a deep and powerful intuition does profoundly influence a person. Should many other people share similar intuitions with the same basic meaning in response to the same or similar experiences, we might well believe that they have perhaps sensed something true, possibly even that they have received a revelation from God. I suspect this is what happened to at least some of the people who experienced Jesus and other great religious teachers throughout history. What these people taught and how they lived resonated deeply within those first witnesses as good and true, just as it still does today for those who follow them and the religions they inspired. The same can be said of the way some people respond to the teachings of certain contemporary religious teachers as well. The intuition that there is something more to reality than just the physical, based on our subconscious experience of the divine, is, I believe, the ultimate reason for the origin of religion in the earliest humans and its persistence throughout history.[11]

11. Some philosophers and social scientists have proposed other sources for the rise of religion. One of the most frequently proposed is fear engendered by selfish priests who wanted power and privilege and so invented religion to get it. Ludwig Feuerbach (1804–1872) proposed that it was people's estrangement from their good side that caused them to take all their good qualities and project them onto a fictitious God "out there." Karl Marx (1818–1883) thought religion was invented, or at least used, by the

The second problematic question—how revelation is possible, given that we are material creatures and God is a spiritual Being—has a long and complex history, which is too involved to recount here. Suffice it to say that the solution to this problem has typically hinged on the claim that humans have, in addition to their material bodies, a spiritual aspect, usually referred to as their soul, spirit, or mind, and that it is this *spiritual* aspect of the person that permits her to receive a revelation from the *Spirit* that is God by means of some sort of mind-to-mind connection.

Despite these two problematic aspects of the doctrine of revelation, it is a fundamental belief for all the theistic religions, Christianity included, that such revelation from God has indeed taken place in human history and that it constitutes our most fundamental source of knowledge about the existence and nature of God. But what is the content of this self-revelation of God, and where can it be found? The traditional Christian answer is that it can be found in two places: nature and the Bible. God's revelation in nature is termed *natural* or *general revelation*, meaning that it is available to all people of all times and places, either through observing the natural world outside ourselves, or looking within ourselves and seeing some trace of the divine in human nature and experience. The kind of theology that emerges from this consideration of general revelation is termed, as we have noted, *natural theology*. In this kind of theology, the primary emphasis has been on arguing for the existence of God and only secondarily, and much more tentatively, attempting to describe one or more of the aspects of God's nature, such as that God is one, intelligent, good, enormously powerful, and/or loving.

Significant attention has been paid to natural theology within Christianity since the Middle Ages, as exemplified by the writings Thomas Aquinas (1225–1274) and the "*five ways*" by which he believed features of the natural world reveal the existence of God. Such arguments have become increasingly important in Christian *apologetics* (defenses of the faith) in the modern

---

dominant capitalist class as a way to enslave the masses and make them work under inhuman conditions in the belief that by suffering without complaint they might earn heaven after death. Sigmund Freud (1856–1939) argued God was invented to address our fears about life once we learned our biological father could not protect us. For sociologist Emile Durkheim (1858–1917), religion was invented to create social cohesion among the members of the group in which one lived. Friedrich Nietzsche (1844–1900) believed that religion was the creation of the impotent in order to prevent superior people from asserting themselves. While each of these theories may have some truth regarding the reasons for belief of some people, that any of them accounts for the rise of religion *per se* seems to me far too facile an explanation to be true. The intuition that there was "something more" than mere physical things seems to me the more likely explanation.

era with the rise of religious skepticism and its attendant agnosticism and atheism. There are essentially two types of natural theology arguments for God's existence based on features of the natural world—the *Cosmological Argument* and the *Teleological Argument* (also known as the *Argument from Design*), plus one argument based on the human experience of morality, the *Moral Argument*. We will discuss these arguments in later chapters as they become relevant to reflecting on our scientific understanding of the history of the universe.

The degree to which such arguments are convincing varies considerably. Some people are convinced by some or all of these arguments and others are convinced by none of them. It is probably true that, for the most part, whether you find them convincing or not says as much about you and your prior philosophical and theological beliefs as it does about the arguments themselves. That is to say, the more the conclusion of the argument coheres with your already-existing beliefs, the more you will tend to find the argument convincing. On the other hand, some atheists and agnostics, such as perhaps the twentieth century's most consistent and influential proponent of atheism, philosopher Anthony Flew (1923–2010), have found one or more of the arguments for the existence of God (in Flew's case, the cosmological argument) convincing and so have become theists. Some theists, on the other hand, have found the arguments of atheists and nonreligious people convincing—especially the argument from the existence of evil in a world supposedly created by a good and loving God—and so have given up their religious beliefs, becoming either atheists or agnostics (e.g., Richard Rubenstein [1924–], a Jewish theologian whose post-Holocaust theology entailed the denial of an omnipotent God and discarded classical Jewish theology, but who did not consider himself an atheist).

While some Christian theologians have employed natural theology in constructing or defending their worldviews, virtually all Christian theologians have, again, insisted that we can know very little about God apart from *special revelation*—that self-revelation of God given to particular individuals or groups in particular places and times throughout history. Thus, the primary source for Christian theology must be the special revelations that were received from God during the formative period of the Jewish and Christian religions, written down, and collected into the Christian Bible. Essentially all Christians agree that, in some way, the Bible contains revelation from God; the chief debate is to what degree and in what form. Two basic positions exist, which constitute the major distinction between the two general types of Christians today—*evangelicals* (conservative to moderate) and the so-called *mainline* (moderate to liberal) Christians.[12]

THE QUESTION OF BIBLICAL AUTHORITY

Most conservative evangelicals believe that the Bible is the inerrant Word of God—that everything the Bible says, whether in the realm of theology, morality, history, or science, is what God has literally revealed to Jews and Christians living in biblical times and is, therefore, unquestionably true. On the other side are the moderates and liberals, who typically believe that the Bible contains some divine revelation but also some of the common thinking of the people who lived when the various biblical books were written. Like any question rationally considered, the evidence for each view must be examined before a judgment can be made. Deciding this question is essential, because how one answers it will determine to a great degree one's basic theological position.

The primary textual support for the inerrancy claim comes from 2 Tim 3:16, which some twentieth-century conservative evangelical scholars have translated from the Greek into English as, "All scripture is *God-breathed* [the universal English translation of this Greek term, *theopneustos*, before the rise of fundamentalism in the twentieth century was "*inspired by God*"] and is useful for teaching, reproof," etc. However, the Greek can also be translated, "Each/every writing inspired by God is also useful for teaching, reproof," etc., because the verb "to be" does not appear in the text and so must be inserted by the reader at the place where it makes sense, as was the custom in Koine Greek. The problem is that the verb can be inserted in either of two places in this Greek text, resulting in the two different translations just given. Thus, either translation is permitted by the Greek text, with one saying that all Scripture is inspired by God and the other suggesting only that some religious writings (and not necessarily scriptural ones) are inspired by God, while others are not. Better recent English translations of the Bible, such as the New Revised Standard Version (NRSV, 1989), have

12. These two broad groups contain a diversity of opinions within them, including on the question of biblical inerrancy. Some more moderate evangelicals seem to have altered their views on biblical inerrancy of late, whereas the most conservative among them, the fundamentalists, have not. Also, some Christians who consider themselves conservatives are members of mainline traditions or denominations. Finally, a minority of Christians questions or denies that the Bible contains any revelation as such but believe that it witnesses to the religious experiences of people from the past; they find the Bible inspiring in their own religious lives, although not divinely inspired in any of its parts.

footnotes stating that there are two valid but different readings possible for this text. Not surprisingly, the recent translation by conservative evangelical Christian scholars, the New International Version (NIV) of 1984, does not. Could it be that in this case the theology of the translators determined how they translated this verse? And if so, what does this do to our confidence that the Bibles we read have been properly translated?

It is important to remember that the NT as an officially recognized collection of documents did not exist at the time 2 Timothy was written, so the only Scripture to which this verse refers is the HB/OT. Whether the author of the letter would have considered all the books that would eventually constitute the NT, including his own letters, as divinely inspired Scripture cannot be known. Thus, to assume that the *one* translation which *might* be construed to support the biblical inerrancy doctrine is *the* correct one, and to make the theological leap that being God-inspired or "God-breathed" implies the inerrancy of Scripture, and to claim further that "Scripture" includes books the author never included (i.e., the NT) is to assume a good deal more than the biblical text itself can fairly reasonably be said to support. Since this single verse is the cornerstone of the claim for biblical inerrancy, the doctrine cannot be said to be very well justified biblically.

In addition to 2 Tim 3:16, other scriptural references have been cited as evidence for the doctrine of biblical inerrancy. For example, the ancient Hebrew prophets generally proclaimed, "Thus says the Lord . . ." before speaking the words that they understood to have been given to them from God; Jesus claimed to reveal what God wanted him to say to the Jews of his time and place; Paul (ca. 5/15–64 CE) at times claimed to have received his teaching directly from the risen Christ; and John (late first century CE), the author of the book of Revelation, claimed that divine revelation was the basis for the visions he reported in that book. Second Peter 1:20–21 is also typically cited in support of the claim of biblical inerrancy, although these verses speak only of prophecy in the HB/OT and so say nothing about the whole of Scripture being divinely inspired. Thus, even if one grants that in all these places we find true divine revelation—and this claim is problematic for a variety of reasons—this still does not mean that *everything else* in the Bible is also divine revelation. To assert this, as we noted above, is to commit the part-whole logical fallacy. Thus 2 Tim 3:16 remains the central text in support of the biblical inerrancy argument.

On the other hand, there are also other biblical texts that claim explicitly that what is being said is *not* from God or the risen Christ. In two places in Paul's first letter to the Corinthian Christians (1 Cor 7:12 and 25), he specifically says that he does *not* have "a word from the Lord" concerning the topic addressed but is giving his own opinion. To claim that what Paul says in these verses is inerrant, therefore, is to claim what Paul himself explicitly denied. Based on these texts alone, then, the doctrine of biblical inerrancy is refuted—and refuted by the Bible itself.

The literal inerrancy of the Bible also cannot claim significant support in the history of Christian thought. Indeed, because of problems with taking the biblical text literally, early Christian interpreters argued that many texts could not be taken in the literal sense at all. Rather, the *true* meaning of scripture lay in its *metaphorical* or *allegorical* sense.[13] This led to a variety of interpretations of biblical texts, which moved Augustine to claim, as we have seen, that we have no way of knowing which, if any, of the interpretations offered of a given text is the correct one.

Martin Luther (1483–1546), a biblical scholar whose writings launched the Protestant Reformation, also rejected the idea of biblical inerrancy. He believed that there were two types of material in both Testaments: *human writings*, which present the kind of God humans think God ought to be, and *divine revelation*, which presents the kind of God that Jesus and, after him, most NT authors, taught—a God who is loving and forgiving of all people and who offers salvation as a free gift to all.

Even John Calvin (1509–1564), the Reformation scholar who championed the inerrancy of scripture more than any theologian until at least the late nineteenth century (and whose teaching remains foundational for the contemporary inerrantist movement), occasionally claimed in his biblical commentaries that some laws or practices found in the Bible were only reflections of those commonly held at the time when the book was written and so were irrelevant for people in his time. Thus, the biblical inerrancy position not only cannot claim significant support from the history of Christian thought, but even its greatest proponent did not believe that everything in the Bible was inerrant revelation from God.

The problems for the claim of biblical inerrancy, however, do not end here. For, even were one to accept the claim that the Bible is inerrant, the question would still remain, *which* Bible is the inerrant one? Hundreds of translations of the Bible are extant today in nearly every language spoken on Earth. Even with regard to English translations alone, there is no agreement on what the proper *translation* of the HB/OT is from the original Hebrew

13. A *metaphor* is a form of nonliteral speech in which one thing is compared to another, such as, "Your eyes are deep pools of water." An *allegory* is another such nonliteral form in which each element in the story stands for some specific other thing, such as a story in which a father represents God and one son represents a sinner while another son represents a righteous person, a castle represents Heaven, and so on. Some early Christian theologians divided these general types of interpretations into further categories, which we need not address here.

language in which it was written, nor on what is the proper translation of the NT from the original Greek language.

Part of the reason is that, as most people who speak or have studied more than one language know, there are no exact equivalents of words in different languages. Words can have several meanings, such as the Hebrew term *ruach*, which can mean "wind," "breath," or "spirit," and different English translations do not always use the same one of these three words to render it. For example, in the second phrase of Gen 1:2, some English translations read ". . . the Spirit of God . . . ," while others read ". . . a wind from God . . ." In addition, words carry connotations, and these too are not the same for a given word and its equivalent in another language. So, given all this, which translation of the Bible (in whatever language one chooses) is the inerrant one? And how could one tell? In fact, *every* translation of the Bible into *every* language is simply the best educated guess, made by the scholar or scholars who translated it, as to which words best capture the meanings of the original texts.

This problem is even more acute when one looks at the Hebrew and Greek manuscripts from which the translations are made. Here two problems are evident. First, there is a number of words, especially in the HB/OT, whose meaning we simply do not know. The reason is that some words appear in the Bible only once or a few times; and since the Bible did not come with an ancient Hebrew dictionary, scholars cannot determine exactly, if at all, what such words mean. Words that appear repeatedly can be checked against one another and their meanings determined by what makes sense given all the readings. In the case of words that appear only once or twice, however, this method cannot be used. In this case, translators can only look to *cognate languages* (languages similar to the one being translated) from the same time and general region of the ancient world and attempt to determine a Hebrew word's meaning, based on its similarity to known words in the cognate language. In some cases, even this cannot be done either, and so the translator simply makes an educated guess at what the word might mean, not knowing whether this is what the Hebrew word means. The problem for us is that translators do not always tell readers when this is the case (NRSV almost always does tell the reader), and so most people think they are reading what the text actually says, when what they are reading is only a scholar's best guess as to what the original Hebrew or Greek word meant. Again, not all scholars agree on the meanings of such infrequently-used words.

The second problem concerning the Hebrew and Greek manuscripts of the books of the Bible is that we do not have a single original copy (or *autograph*) of *any* biblical book, nor, apparently, anything even close in time to an autograph. What we have are copies of copies of copies of copies (and so on) of these books—produced over the span of centuries—and these ancient copies of the Hebrew and Greek biblical books do not completely agree with each other on the wording of these books. In fact, we have some five thousand such ancient manuscripts or fragments of manuscripts of the NT alone, and there are significant differences in many of them regarding what the text of any given passage says. An entire scholarly field, *textual criticism*, attempts to sort out the problem of the relationships between these different ancient fragments and manuscripts and to determine especially which are the best ones. Once this is done, these scholars make educated guesses as to which seems to them to be the most likely earliest text of any given passage or book.

Here again, biblical scholars are not always in agreement about what they think the earliest and best texts are. If you were to look at any scholarly editions of the Hebrew OT or Greek NT, you would see at the bottom of the page a section set apart from the biblical text, comprising a complex set of symbols, letters, and numbers. These indicate alternate readings of portions of the text found in various ancient manuscripts and which manuscripts support which readings. The reading in the text is thus that particular editor's best guess as to what text might be closest to the original. The point here is that we do not *know* what the original texts of the various biblical books are; we have only, once again, scholars' best—and differing—guesses as to what texts are closest to the original writing. Thus, the Bibles that we read in our various modern-language versions are differing best-guess translations from differing best-guess reconstructions of the original biblical books. To claim inerrancy for any one of these best-guess versions of the Bible is thus frankly absurd.

But this is not all. To claim that we can learn God's absolute truth from the Bible means, in practice, that not only must we know which ancient biblical *manuscripts* are the inerrant ones, and which *translation* of these manuscripts is the inerrant one, but we must also know which of the many *interpretations* of the Bible is the inerrant one—meaning we must also know who the one inerrant *interpreter* of the Bible is. Determining what that inerrant interpretation or who that inerrant interpreter of the Bible is—especially given the differences of interpretation of Scripture that even biblical inerrantists hold—as well which ancient manuscript and modern translation is the inerrant one, seems to be a problem without any possibility of solution.

In the end, though, apart from all these general considerations, the case for biblical inerrancy must finally rest on the contents of the Bible itself. If the entire book has been authored by God, then logically there can be *no errors or contradictions within it whatsoever.* As in science, where a single contrary fact is enough to demonstrate a theory to be false, so here a single example of an error in the Bible would be enough to prove the wrongness of the idea of biblical inerrancy. This is precisely where the inerrantist case falls apart for any person who carefully examines the Bible and views the evidence it provides objectively, for we can find not just one error in it but a multitude of them.

As we noted above, the Gospels do not agree on either the day or the time of Jesus's crucifixion, and we can cite a plethora of other examples. Genesis 1 says that God created the world by *speaking* things into existence and created vegetation first, then animals, and lastly humans, male and female, at the same time. Genesis 2, on the other hand, says that God created by *molding* all these creatures from the dirt and created a (male?) human being first, then plants and animals, and finally created a woman from the human being/man.[14] Thus the two stories disagree on both how and in what order God created. Joshua 13:2–4 says Gaza, Ashkelon, and Ekron were *not* taken in Joshua's conquest of the land of Canaan; Judg 1:18 says they were. Likewise, Judges says Jerusalem (1:21), Megiddo (1:27), and Gezer (1:29) were *not* taken under Joshua, whereas Josh 12:10, 12 and 21 say they were. First Samuel 17:38–49 says David killed Goliath, whereas 2 Sam 21:19 says Elhanan killed him. Second Samuel 24:1 says God incited David to take a census of Israel, whereas 1 Chr 19:18 says Satan did. Second Samuel 10:18 says David slew *700* Aramaean chariot teams and 40,000 *horse* soldiers, whereas 1 Chr 19:18 says he slew *7,000* chariot teams and 40,000 *foot* soldiers.

In the NT, Matt 1:16 says the father of Joseph, Jesus's father, was named Jacob; Luke 3:23 says his name was Heli. In fact, the entire genealogies in Matthew and Luke are very different from each other, not only in the names of Jesus's ancestors, but also in how many generations of them there were from Jesus back to David. Another example is that Matt 28:2 says Jesus's tomb was closed when the women arrived on Easter morning, whereas Mark, Luke, and John all say it was already open (see, e.g., Mark 16:4). Matthew 27:3–10 and Acts 1:18–19 give very different accounts of what Judas

14. Some biblical scholars have pointed out that the traditional reading of Genesis 2, that a male human was created first, then the female, who was created from one of his ribs, may not be the correct interpretation. This is because, until the creation of the female, the Hebrew noun used for the created person was *adam*, which means *human being*, not *ish*, which means a *man*. Only when it comes time to speak of the male and female as separate persons do the terms *ish* and *ishah* appear. This suggests that what God originally created was a sexless human being, whom God later divided into a male and a female, so that the human being would not be alone. Perhaps this was the way the author of Genesis 2 said what the author of Genesis 1 said in a different way—that humans were originally created in the image of God (who, as spirit, is sexless).

Iscariot (d. ca. 30 CE) did after his betrayal of Jesus (a case we will examine in greater detail below).

This selection of contradictions could be multiplied many times over. Many additional contradictions come from the four Gospels, which differ significantly from each other in the *chronologies* (order of events) they give for Jesus's life as well as in content, the most obvious of which concerns the fundamental focus of Jesus's teaching. According to the Synoptic Gospels (Matthew, Mark, and Luke), at the core of Jesus's preaching was his claim that the reign of God was coming and was already beginning through his ministry, whereas according to John's gospel Jesus's main message was about himself—that he was the Son of God who reveals God to Israel. Jesus also speaks very differently in John from the way he does in the Synoptic Gospels. Indeed, if you did not know better, you would believe that John's author and the Synoptic authors wrote about two quite different people.

Some contradictions also appear in a comparison of how Paul recounts his life in his letters with how author of Luke and Acts recounts Paul's life. For example, Paul *swears* in Galatians 1–2 that he went to Jerusalem only *once* prior to going there years later to attend the so-called *Apostolic Council* meeting (ca. 48 CE), whereas "Luke" says Paul had visited Jerusalem *twice* during this period (Acts 9:26–30 and 12:25). Also, Paul says he had seen only Cephas (Peter) and James (Jesus's brother), who were two leaders of the early Christian community (Gal 1:18–19), whereas Luke says he preached openly in the city and was widely known because of it (Acts 9:28–29).

In addition to such biblical contradictions, there are also errors of a factual nature, such as scientific errors. For example, Leviticus 11 says rabbits chew the cud (they do not), bats are birds (they are not), and all winged insects (excepting grasshoppers, crickets, and locusts) have four legs and feet (they do not). Mark 4:3 has Jesus saying the mustard seed "is the smallest of all seeds on earth," yet biologists have found seeds that are considerably smaller than the mustard seed. There are also historical errors in the Bible, such as the story of Joshua's making the walls of mighty Jericho fall down and taking the city by means of battle, although archaeology has demonstrated clearly that Jericho was little more than an uninhabited ruin at the time.

Yet another type of problem that makes taking the Bible as the inerrant word of God unreasonable is that certain of its moral laws are such that very few, if any, Christians today would accept them as actually coming from God. For example, Exod 21:2–11 lists the proper conduct for Hebrews who own other Hebrews as slaves, implying that slavery—even selling one's own daughter into slavery (21:7)—was authorized by God. Deut 20:14 adds that it is also permissible to enslave a defeated people's women and children.

Tattoos are forbidden (Lev 19:28), as is eating shellfish (Deut 14:9–10) or meat with blood in it (Deut 12:15–16)—which is what you get in the grocery store—and wearing clothes of blended wool and linen (Deut 22:11). If a man has sex with a virgin, he must pay her father a bride price (Exod 22:16), and any woman who grabs a man's testicles when he is fighting with her husband is to have her hand cut off (Deut 25:11–12). Stoning to death is required for young women if they are not virgins when they wed (Deut 22:13–21), stubborn and rebellious sons (Deut 21:18), children who curse their father or mother (Exod 21:17), anyone whose ox gores someone to death (Exod 21:28–32), any man who engages in sexual acts with another man (Lev 20:13), anyone who has sex with an animal (Exod 22:19), is married and has sex outside of marriage or is unmarried but has sex with a married person (i.e., commits *adultery*, Lev 20:10), as well as anyone who speaks blasphemy against God (Lev 24:13–16). Sex with prostitutes is biblically acceptable, as the case of Jacob and Tamar indicates (Gen 38:15–16). Also, according to Gen 19:8 and Judg 19:24–30, allowing one's daughter to be sexually abused in order to protect the same from happening to strangers visiting one's house is morally commendable. Having many wives and concubines, as in the case of the patriarchs in Genesis, is also permissible.

It is significant that even biblical inerrantists hardly ever assert that we should adopt the ethical stances noted above, which indicates to any thinking person that, despite claims to the contrary, they too, like Calvin, do not really believe that such commandments are from God. In fact, the most typical inerrantist response I have encountered is to claim that, since Jesus fulfilled the HB/OT laws, they are no longer valid (this is based on their dubious interpretation of Matt 5:17–20, where Jesus says, "Do not think that I have come to abolish the law or the prophets; I have come not to abolish but to fulfill. For truly I tell you, until heaven and earth pass away, not one letter, not one stroke of a letter, will pass away until all is accomplished."). But this claim, too, is problematic, in that inerrantists certainly believe that some HB/OT laws, especially the Ten Commandments and those against same-sex sexual acts, remain valid. Why these HB/OT laws are still considered to be binding while others are not—when *all* are equally commanded by God, according to the inerrantist position—is a problem that, to my knowledge, no inerrantist has ever reasonably defended.

The essential message of all these considerations is that, even if one wanted to interpret 2 Tim 3:16 as asserting biblical inerrancy (despite the manuscript, translation, and interpretive problems with that claim), it is the biblical text itself that disproves this claim and does so decisively, with its internal contradictions, scientific errors, and ethical precepts that no Christian (and few people of any type today) would accept as moral, let alone would attribute to the God of perfect love to whom Christians claim the Bible bears witness.

These facts do not mean that inerrantists, confronted with this evidence, would renounce their belief in biblical inerrancy. Rather, they typically adopt one or several strategies for denying that such evidence counts against biblical inerrancy. One strategy is to distinguish between the biblical text as we have it today and the autographs of the various biblical books. The inerrantist argument here is that our Bibles today have contradictions in them but the original autographs did not. Therefore, they claim, the Bible in its original form was inerrant. The problem with this defense, of course, is that, since we do not *have* any of the original manuscripts, we have no way of knowing whether they contained errors of any kind. Thus, the assertion of the inerrancy of the autographs is a belief without any evidence to support it. However, even if the autographs were inerrant, the Bibles that we have today clearly are not, and this for the reasons we have noted above, and so practically one gains nothing by asserting that the autographs were inerrant.

A second typical biblical inerrantist strategy is to deny that the textual contradictions or errors are in fact contradictions or errors at all. A typical response is to say that the texts each claim things that should be combined to give us a more complete truth. For example, as we noted above, Matt 27:3–10 provides an account the death of Judas Iscariot, who betrayed Jesus for thirty pieces of silver. It says he repented of his betrayal of Jesus, brought the money back to the chief priests at the Jerusalem Temple, and went out and hanged himself. Then, it says, the chief priests conferred with each other and decided to use the blood money to purchase the potter's field as a place to bury foreigners—hence, it is subsequently called the Field of Blood. However, Acts 1:16–19 tells quite a different story of Judas' death. There no mention is made of Judas repenting or giving the silver back; rather, it says he bought (or acquired) a field with the money. Then, "falling headlong [or "swelling up"—the Greek is imprecise here], he burst open in the middle and all his bowels gushed out" (1:18). Because Judas died there in this way, this story says, the field came to be known as the Field of Blood.

Inerrantists have insisted that both stories are true. What happened, they say, is that Judas repented and gave the money back to the priests, who then bought the field with it and gave it to Judas. Judas then hanged himself in that field, but the rope broke and he fell down and cut himself open on a rock, or some such thing, and his bowels gushed out and he died. In this way, they claim, both stories can be seen to be true, and so the Bible's inerrancy is maintained.

The problems inherent in this defense are typical for other such attempts to conflate conflicting stories. First, this explanation necessarily ignores certain textual claims. The Matthew text clearly has Judas hang himself before the priests confer and purchase the field, and the reason cited for calling the field the Field of Blood is different in the two stories. Second, the scenario proposed—that Judas, though so remorseful he decides to kill himself, nonetheless waited for however long it took the priests to find a field for sale, buy it, and give it to him before hanging himself, and that the rope broke as he hanged himself and he happened to hit something beneath that sliced his midsection open—must seem far-fetched to anyone looking at the two stories objectively.

Third, and in my mind most damaging, is that this kind of defense creates two serious *theological* problems, neither of which inerrantist Christians seem to realize exists, nor would they want to maintain what necessarily follows if they did. First, if their explanation of what actually happened by combining the different accounts is correct, then how can we trust what *any* single passage of the Bible says? Anyone reading the Matthew text would surely believe Judas felt remorse for betraying Jesus and therefore killed himself; but anyone reading the Acts account would believe Judas had no remorse, kept the money, bought a field with it, and then had an accident. If the inerrantists' scenario is true, then people would be wrong in their commonsense interpretations of these two Bible passages. Everyone would have to know the Bible cover to cover to see if what any given passage said was the truth. Moreover, if you cannot trust that all the passages you read are true accounts of what happened, then how can you trust that what you think the Bible says in any other passage is actually true? The implications of this would be devastating for Christian interpretation of the Bible and for faith.

The other theological problem that arises if the inerrantist interpretation of the Bible is correct has to do with the nature of God. If, for example, in the Judas stories the inerrantist scenario describes what really happened, then why did not God, as author of the Bible, just say so in both places, rather than seeming to lead people astray with two such different accounts? Either God has trouble simply stating what actually happened or it seems God is trying to trick us. A typical inerrantist response is that God includes such seemingly contradictory accounts to test us, to see if we will believe in God's Bible as inerrant despite the fact that it does not seem to be.

This only creates another theological problem, one that goes to the very heart of the Christian understanding of God. Is God a self-giving, loving Father as Jesus claimed and as Christianity has historically taught: One who above all desires a loving relationship with everyone and, therefore, desires the salvation of us all? Or is God rather the Great Test-Giver, testing us to see if we will believe the Bible is inerrant despite all the evidence to the contrary and thereby make the grade and earn heaven? Would even a halfway decent *human* father who loves his children constantly test them to measure their worthiness to receive good gifts from him? If this inerrantist perspective of a testing God is true, then Jesus was quite wrong in teaching about an infinitely self-giving God who loves all of us and works for our salvation, whether we deserve it or not. Such a testing God would not, in my opinion, be worthy of admiration, let alone worship.

From all of the above, I think the reasonable person must conclude that the evidence is overwhelmingly strong *against* the claim of biblical inerrancy—in fact, about as strong a case as could be imagined. Indeed, and on this basis, I consider the claim that the Bible is not inerrant to be a fact. The textual evidence for it is weak, and the textual evidence against it is quite powerful. Further, the theological problems created by the doctrine would tend strongly to negate what Jesus and the Christian tradition have claimed about the nature of God as perfect love. Biblical inerrancy as a theory, we must say, with its unconvincing, *ad hoc* explanations for its problems, fails the test of Ockham's razor. The simpler, more reasonable position, and the one that underlies the understanding and interpretation of the Bible in this book, is the one held by Luther and all contemporary biblical scholars who are not conservative evangelicals: namely, that the Bible does contain some divine revelation, but that it also contains human ideas, some of them errant, which are cultural artifacts from the ancient world that need have no bearing on Christian belief or practice today.

This view, however, creates its own problem: How does one distinguish what is human thought, and therefore potentially untrue, from what is divine revelation, and therefore certainly true? Here we can apply several criteria.

One criterion employs what we have already noted above: that the Bible contains a variety of contradictions, scientific and historical errors, and immoral commands. Wherever one encounters such contradictions in the Bible, one must analyze them closely to see which, if either, of the two contradictory texts appears more probably true according to reason and any knowledge archaeology or other ancient literary sources bring to bear. Thus, in the case of *historical* assertions, we employ the tools of historical analysis that historians apply in determining historical veracity (including archaeological evidence); or, in the case of assertions about the nature and operation of the world, we bring to bear the methods and knowledge of science. In neither case can we arrive at certainty, since historical and scientific assertions can only ever attain, as already noted, the status of being probably true based on our current knowledge. Where we find historical or scientific errors or immoral commandments, we simply reject them and conclude that they are not divine revelation. While this cannot decide every question we have about which parts of the Bible contain divine revelation, it can at least provide a solid start in that direction.

A second and most important criterion that you can bring to bear on the question of what texts might contain revelation from God concerns recognizing the *type of literature* you are reading, given the fact that the biblical authors did not write history or theology in the ways we do today. Both history and theology are typically expressed in the Bible by means of *stories* that were not intended to be taken as literally true. Therefore, you need to take these stories as they were intended—to teach a lesson about God, the world, ourselves, and/or how we should relate to God (these points will be fleshed out below). That is, you should view the biblical authors as intending to convey truths of *theology and ethics*; therefore, you should look for their teachings on these subjects, rather than concerning yourself with the truthfulness of the history or science recounted in the texts.

Another criterion for judging whether a text contains revelation is what it teaches about God. Here again, I find Luther's distinction to be pivotal. If the text defines God in terms of typical human expectations of what God ought to be, such as some self-centered potentate from the ancient world, then it is not to be viewed as true revelation; but if it is one that teaches the kind of unexpected God of absolute mercy and love as taught by Jesus and the Christian church, past and present, then the text should be accepted as at least potentially one that contains divine revelation.

Yet another criterion for judging whether a biblical passage might contain a revelation from God is *biblical coherence*. Biblical coherence means that any given passage agrees with the fundamental teachings of the Bible—that is, those that appear in a wide variety of books written in different times and places. It is the consistent biblical themes, rather than single verses or passages here or there, that should be considered "biblical teaching" on any topic. For example, any particular biblical commandment must be considered in relation to the foundational teaching of Jesus that the entirety of the legal teachings of the Torah can be reduced to the two commands to love God and love others—even one's enemies—as one loves oneself. Any particular law that violates that fundamental principle must be judged to be a human idea, not a divine commandment.

Added to this, however, must be the recognition that the Bible contains writings produced over many centuries and that therefore biblical thought undergoes change as it moves from earlier to later writings. The earliest *traditions* in the HB/OT go back perhaps as far as 2000 BCE, though the oldest *written texts* probably do not reach back much before 1000 BCE at the earliest. The latest NT writings probably date from 125 CE or so. Thus, the biblical books were written over a period of about a thousand years or more and contain ideas that span perhaps as much as two thousand years.

We would naturally expect some ideas to change over so long a period of time, and this is precisely what we see in the Bible. For example, concerning theological teaching, the first of the Ten Commandments has God proclaim to Israel, "you shall have no other gods *besides/before* me" (Exod 20:3; emphasis added). What is interesting to note is that the commandment does *not* claim that there *are* no other gods besides Yahweh, Israel's God, only that they are not to worship them. This is not surprising given the period in which the text was written, where each people believed they had their own god who protected them and to whom they sacrificed. It was only some seven hundred years later, with the prophet scholars refer to as Deutero-Isaiah[15] (sixth century BCE), that we find the first biblical writing to assert that in fact there *are* no other gods, only Yahweh (Isa 45:5). This means that the "*radical monotheism*" that we associate with Judaism was not part of early Hebrew thinking but arose only later in Israel's history.[16]

We see the same kind of development in Israel's ethical thought, for example, when we compare Exod 21:2–6 with the later Deut 15:12–18. The Exodus commandment says that when a Hebrew male takes another Hebrew male as a slave because of the debt owed to him, the slave shall only serve six years and be freed in the seventh. The Deuteronomy text says this rule applies to female as well as male slaves, and includes the injunction that the male shall not be sent out empty-handed, but that his owner should give him some of his wealth when he frees him. This addition likely arose when, over time, it was recognized that a man who was enslaved for debt, if sent out with nothing, would likely go into debt again almost immediately in order to pay for his food and other necessities. This might well end up

15. Most biblical scholars recognize that the book of Isaiah is a compilation of the writings of either two or three different authors, the first being the eighth-century BCE prophet Isaiah and the other(s) being follower(s) of Isaiah's teachings who lived some two centuries later yet attributed their writings to him and presented them alongside their master's prophecy. Such practice was common in the ancient Near East and Western worlds, as seen, for example, in some letters attributed to Paul in the NT by his followers and those attributed to *Pythagoras* (ca. 570–ca. 495 BCE) by those who followed him in the ancient Greek "Pythagorean School" of philosophers.

16. Editor's note: If 1 Cor 8:4–6 means that *only pagans*—not Paul, other Jews, and the Corinthians—*believe* that idols are gods, his view would be more like Isa 45:5. On the other hand, the phrase "*as in fact there are* many gods" (emphasis added) shows that Paul acknowledged the *existence* of other gods, but insisted that Jews and Christ-believers were to have only one God, the Creator, so that his view would be more like Exod 20:3. That would indicate that "development" need not be strictly linear.

with him becoming a debtor-slave once again, generating a continuing cycle of indebtedness and slavery; and so the Exodus law was modified to help prevent this from occurring.[17]

The point here is that what constitutes "biblical teaching" on any subject must be understood both as what is taught across the whole biblical canon and also within the larger framework of the progressive development of biblical thought and practice throughout the period during which the books were written.[18] It seems reasonable that we should consider the later teachings, including those of the NT,[19]and those that did not change over time[20] to be those that would form the likeliest candidates for expressing divine revelation.

17. This practice of sending prisoners out with some money once they have served their time continues to this day, at least in the United States.

18. Editor's note: I raised questions about this section of Jarvis's manuscript, and Jarvis acknowledged that he had to address them. However, it was at the end of his life when he did not have the capacity to revise what he had written. Here are my questions: Do the later writings always reflect "progressive development"? If you mean we should look for evidence of a trend toward "progress," what are the criteria of "progressive development"? Do the later writings always trump the earlier ones, as you say in the next sentence?

19. Editor's note: I do not think that Jarvis wanted to imply that later developments are always progress. For example, in the Pauline letters, many scholars over the centuries have argued that the disputed letters (Colossians, Ephesians, 2 Thessalonians, and the Pastorals) conform to conventions of the dominant culture more than the undisputed letters (Romans, 1 and 2 Corinthians, Galatians, Philippians, 1 Thessalonians, and Philemon). If these scholars are correct, would that be progress or decline? One could raise similar questions about the history of traditions in the OT, and about the history of Jesus traditions. For example, I think that Jarvis would have agreed with Schubert M. Ogden (his doctoral dissertation advisor), who has argued for the priority of the earliest "apostolic witness to Jesus" over later developments. Jarvis did not believe that the disputed Pauline letters would be better candidates for "divine revelation" than the undisputed letters would be, or that 1–3 John would be better candidates than the Gospel of John, or that 2 Peter, Jude, and Revelation would be better candidates than the rest of the NT.

20. Editor's note: I pointed out to Jarvis that Richard B. Hays uses this argument to support the abiding "truth" of the condemnation of same-sex sexual acts, which does not change from the time of Leviticus to Rom 1:18–32, 1 Cor 6:9–11, and 1 Tim 1:9–10. See Hays, *Moral Vision*, 379–406. In addition, the Bible, from Genesis to the NT letters, consistently affirms, or at any rate fails to condemn, the buying and selling of slaves. The later disputed Pauline letters restrict the role of women, in contrast to the earlier undisputed letters (except for 1 Cor 14:33–36, which many scholars believe is a later interpolation from the period of the disputed letters). Jarvis was aghast that his hermeneutical argument could lend support for any of these views. Instead of rewriting what he had written, I propose the following as something more consistent with his overall perspective. Something like the following "kerygmatic" norms provide criteria for evaluating biblical materials that meet Jarvis's two criteria (consistency across the biblical writings, later "developments"): Jesus's witness to God as Perfect Love, Paul's

OTHER AUTHORITIES FOR CHRISTIAN THEOLOGY

In all of our reflections on the idea of divine revelation, we have been using the traditionally accepted fundamental authorities for Christian theology as defined by the Wesleyan Quadrilateral: Scripture, tradition, reason, and experience.[21] According to this widely held perspective, *Scripture* is the foundational Christian authority, but the Christian *tradition,* generally understood as the history of Christian reflection on the meaning of Scripture by the church over the past two millennia, is also to be taken into account. Such reflection has yielded different results at times, expressing the variety of perspectives within the church and its many particular church bodies, and so has not always spoken with the same voice. Indeed, there are probably at least as many different theologies and theological emphases as there are church bodies. What all Christians have shared through the centuries is the conviction that there is a loving God, and that Jesus revealed God decisively as such and in some way facilitated human salvation. Beyond this, there has been widespread disagreement on how this core belief is to be properly expressed, as well as how Christians are to live according to God's will.

As I have already suggested, orthodox beliefs in Christianity are like orthodox beliefs in science: they represent the prevailing perspectives at a given point in history. Where the church has failed, in my opinion, is when it has claimed some of these beliefs to be *dogmas*—things certainly true for all time that all Christians must believe. Thinking that certain doctrines determined at some point in the history of the church by certain individuals or groups are infallibly true is as irrational as claiming the Bible to be inerrant. Theology is the *human* attempt to discern what the truth of Christianity is, and, being fundamentally the product of human interpretation of the Bible combined with whatever else we know about reality, it is therefore inevitably fallible. Christian theology needs to be open in the way that science is open: holding current views as long as they are reasonable, changing them when they are no longer so, and never claiming dogmatically to have found the absolute and certain Truth about God and God's dealings with the world. If it did that, I think people within as well as outside the church would have more respect for Christianity.

---

witness that Christ died for all without exception, and that the Spirit brings freedom, along with the witness of Israel's prophets from Jeremiah to Jesus to God's "preferential option for the poor."

21. Editor's note: Albert C. Outler, the leading American Wesley scholar of the last century, coined "the Wesleyan Quadrilateral." Jarvis provides Outler's order here; below, however, he discusses experience before reason.

Another traditional authority for Christian thought and practice is *experience*, broadly construed to include both individual and general human experience.[22] General human experience includes all we know from history, science, and the other academic disciplines, all of which are to be taken, following Augustine's lead, as valid authorities for Christian theology. Throughout the history of Christianity, new knowledge, such as discoveries in astronomy and biology, has provided new perspectives from which to understand and interpret biblical teachings, and these have altered aspects of Christian thought. Christian theology must be consistent and coherent with everything else we have collectively discovered about reality.

In addition to such general knowledge provided by the sciences, history, and other disciplines, the unique experiences of individuals have also resulted in new questions and sometimes new perspectives in theology. Mystical experiences—understood as unmediated experiences of the divine, often enabled by certain spiritual practices—have at times been influential in the history of Christian thought. Personal spiritual struggles as well, such as those of Martin Luther, have led to new insights, which in turn have given rise to theological perspectives that have had a lasting impact on the church. At bottom, then, it is such novel experiences, both individual and general, that keep theology open, growing, and deepening its reflection on the nature of God and the divine-human relationship.

Finally, *reason* has also consistently been a theological authority from the very beginnings of Christian reflection. Reason is the tool by which we *analyze* and determine what are the relevant facts from Scripture, tradition, and experience and then *synthesize* them to form a Christian theology or worldview. People will rarely, if ever, accept anything as true unless there are rational reasons, in their judgment, for doing so, and this is also true of Christians. While such reasons may at times be ultimately irrational—and typically for the kinds of reasons noted in the first chapter regarding human beliefs generally—they must nonetheless *seem* rational to the individual or group in order to be accepted. The question for theology, then, is not whether people *believe* something to be rational, but whether in fact that belief *is* rational, and it is logic and facts that must decide the issue, as they did above with the question of biblical inerrancy.

22. Editor's note: In "The Wesleyan Quadrilateral in Wesley," Outler describes his understanding of the relationship among the four authorities this way: Scripture "is illuminated by the collective Christian wisdom of other ages and cultures between the Apostolic Age and our own (i.e., tradition)"; "the disciplines of critical reason . . . rescue . . . the Gospel from obscurantism"; and "the requirement of 'experience'" is that "Biblical revelation must be received in the heart by faith" (11), that is "the assurance of grace and its prospective triumphs in this life" (17).

Thus, the method of Christian theology involves integrating these four authorities—Scripture, tradition, experience, and reason—into an understanding of reality in which each is given its place and informs the others. In this way reason, experience, and tradition question Scripture; reason, experience, and Scripture question tradition; reason, Scripture, and tradition question experience; and experience, Scripture, and tradition question reason. The product is *a* Christian understanding of reality because, as should be obvious by now, there is no single, unanimously agreed upon Christian theology, nor has there ever been one. Even in the earliest church, we find differing understandings of what it means to be a Christian. For example, Paul's letter to the Galatians, written around the year 51 CE—a mere twenty years or so after Jesus's death—already indicates a contentious split within the church over whether gentile converts were required to be circumcised and obey Jewish law. That Jesus and all of the earliest Christians were Jewish, and these followers of Jesus constituted one of several distinct groups within Judaism, were core reasons for the debate.[23]

Subsequent centuries have provided many more instances of such disagreement, and this is why there are so many different church bodies today. Given this diversity, it is incumbent upon me to say again that this book delineates one (my own) understanding of Christian theology—though numerous other theologians today also hold something very much like it. This theology provides, I believe, the most rational and persuasive Christian understanding of reality that can be fashioned at this point in history given the knowledge available to us today.

In concluding this section on theological method, we should summarize what we have said earlier in the chapter about the method of science and its relationship to the method of theology. While they do have several similarities, they are fundamentally different enterprises due to the focus of their studies, physical reality in the case of science and spiritual reality in the case of theology. Because of this, we noted, authority plays an important role in theology in a way it does not in science. However, we need to examine this point a bit more closely, for it is not the case that authority is totally lacking in science. For one, there is the scientific method, and anyone doing science takes that as authoritative.

23. Editor's note: The term "Christian" appears in the NT only in Acts 11:16 and 26:28. The Gospels speak of "disciples" and "followers" of Jesus. The undisputed Pauline letters speak of "saints," "those who are in Christ," and those who believe/have faith in Christ and participate in Christ's faithfulness. Christianity does not emerge as a religion distinct from Judaism until the second century CE at the earliest, with the creation of the Apostles' Creed, and possibly not until the fourth century at the latest, with the beginning of the ecumenical councils.

More importantly, we have seen that, in both science and theology, reason and experience are fundamental "authorities." What distinguishes religion is that it allows for more types of experience to be authoritative—especially, spiritual and other value-experiences of the individual that are not experienced by other people. Science must explain such experiences as having natural causes, most of which are explained (away) by psychological factors. Theology, however, is open to considering other causes, particularly the possibility of spiritual causes.

The other main difference is that theology considers Scripture and tradition to be authorities as well. Why are they considered authoritative? As we have seen, all Scripture and tradition are based on experience, specifically, the experiences of certain people from the past that were written down and so constitute the heart of both Scripture and tradition. Both scientists and religious people give authority to their own experiences; what makes religions different is that they give authority to other people's experiences in a way that science does not. Why do religious people do this? It is because those experiences of people from the past resonate with their own spiritual experiences in the present. For example, that love is the central feature in finding fulfillment in life is something claimed by ancient religious thinkers of all the major religious traditions; that it fits with the experiences of those in the present is why their teachings are given credence. That so many people do give credence to such teachings is the reason the great religious traditions exist and have continued to exist for millennia.

THE HISTORICAL RELATIONSHIP BETWEEN SCIENCE AND CHRISTIAN THEOLOGY

There is a widely held misconception that science and Christian theology have traditionally been enemies. This "conflict thesis" appeared in two books from the later 1800s and continues to be promoted today, not only by those inimical to religion generally and Christianity in particular but by many conservative Christians as well. Historians of science have decisively refuted this conflict thesis since it was first promulgated, but this refutation has done little to dissuade those who, in support of their own agendas, continue to promote it.

The first book to argue that science and at least some forms of Christianity were inherently inimical, John William Draper's (1811–1882) *A History of the Conflict between Religion and Science* (1874), was more an anti-Catholic polemic than a work of serious science- or history-writing. As historians of science have frequently noted, Draper often misquoted the people he cited and fabricated certain "facts," such as that people at the time of Columbus still believed the world was flat. In fact, most did not, though many educated people today continue to hold this mistaken belief.

Despite its manifold flaws, Draper's book became a best seller. The reason probably has more to do with the appeal to certain Protestants of the book's central claim: that the Catholic Church had throughout the modern era sought to stifle science. Protestantism, Draper claimed, had not historically engaged in this warfare against science and was in fact a "twin sister of science." A number of reputable scholars immediately criticized the book upon its publication. They pointed out quite clearly the falsehood of the book's main thesis and the supposed "facts" cited in support of it.

Following closely upon the heels of Draper's book, Andrew Dickson White (1832–1918) published *The Warfare of Science with Theology* (1876), eventually to be followed by his massive, two-volume *A History of the Warfare of Science with Theology in Christendom* (1896). Though a historian by training, White committed a number of serious errors in history writing, even if less egregiously than Draper. Historians of science pointed out at the time that White, too, failed to check his sources carefully, sometimes misquoted them, took quotations out of context, and evidently even invented some "sources." He continued to assert some of Draper's factual errors, including the one regarding Europeans viewing the world as flat in the time of Columbus. He also enlarged Draper's thesis to claim that the church generally had sought throughout the modern era to stifle science and to dictate to scientists what views they could hold. Historians of the time pointed out not only that White's reasoning was at times illogical, but also that he committed the scholarly sin of ridiculing those who held views different from his own. This latter point remains an appropriate caution in the current science-and-religion debates.

The central claim advanced by Draper and White of a persistent conflict between science and religion was easy for historians of science to disprove, and a number of them did and continue to do so. First, they pointed out, throughout the early modern era, all of the great scientists were committed religious, mostly Christian, believers, including Copernicus, Galileo, Newton, and their coworkers. Indeed, this had also been the case for scientists from the early centuries of the Common Era. Augustine taught that there were two "Books" which God had authored to instruct humankind—the Book of Scripture and the Book of Nature—both of which were "written" by God and so supported a single, coherent view of reality. When contradictions between the two arose, either the science or the interpretation of the Bible was in error. This view of the two Books of Scripture and Nature continued to be held and referenced by most, if not all, early modern scientists.

That the Bible might contain errors was not problematic for Augustine, who believed that divine revelation in Scripture had to be "*accommodated*"—i.e., "dumbed down"—by God because of our limited human mental abilities. Humans could never finally understand fully the nature and work of God. Also, as we have noted, Augustine believed that not everything in the Bible was meant to be taken literally, because God had often intended to convey deeper metaphorical and allegorical meanings in these texts (which, we saw above, was the common view from before Augustine to at least the Reformation more than a thousand years later). For this reason Augustine—unlike modern biblical literalists—did not believe that Genesis 1 was to be taken literally, and so he rejected the idea that the world had been created in seven days (as modern "young-Earth creationists" believe) or even in seven eons (as modern "old-Earth creationists" believe). Science for Augustine was an aid to Christian theology, not its enemy; so even prior to the advent of modern science, Christian theologians saw the two as not just compatible but as allies in the search for truth.

Secondly, even today the compatibility of science and theology continues to be asserted by many scientists, especially by scientists who are themselves religious. As they and many philosophers of science have pointed out repeatedly, only two cases in the past five hundred years might possibly be construed—as they were by Draper and White—to be evidence for a conflict between science and religion/Christianity. One concerned Galileo and his defense of Copernicus's heliocentric theory of the universe, and the other concerned Darwin and his theory of the evolution of life on Earth by means of natural selection. We need not go into all the facts of these two cases, as a single fact will suffice to repudiate Draper, White, and those who follow them. In both cases, the theories were new and were by no means initially accepted by all scientists at the time, and so at least some of the church's leaders were simply following the scientific consensus until that consensus changed. Also, in Galileo's case, he was a professed Christian. It is also true that, in the case of Darwin, there were not only some scientists who agreed and others who disagreed with his theory of evolution, but also some theologians who accepted and defended Darwin's theory, such as Henry Ward Beecher (1813–1887) and Lyman Abbott (1835–1922), as well as those who rejected it. Thus, even in these two instances, no reasonable case can be made for the existence of a fundamental conflict between science and Christian thought.

The other possible relationships between science and religion, besides conflict, have been termed the independence, dialogue, and integration views. The *independence view* argues that science and religion are two distinct areas of study and thought: science studies and reflects upon physical reality, and religion studies and reflects upon spiritual reality. Evolutionary biologist Stephen Jay Gould (1941–2002) referred to science and religion as two "non-overlapping magisteria,"[24] arguing that, because each has its own distinct area of study and neither area properly includes the other, the two must exist independently, having nothing to do with each other.

The *dialogue view* sees interaction between science and religion as valid when both are dealing with the same topic, such as the origin of the universe. The claim here is that each may have a perspective that could be useful to the other. Regarding the origins of the universe, for example, science can elucidate *how* it occurred and religion can suggest *why* it occurred. This view is perhaps primarily useful for scientists who are also religious persons.

The *integration view* stands at the opposite end of the spectrum from the conflict thesis regarding the relationship of science and religion. This position seeks to synthesize the findings of science and religion to form a single coherent worldview. The logic is that, since there can ultimately be only one true description of reality, whatever is true in science and whatever is true in religion cannot be in conflict with each other and so can be integrated to provide the best understanding of reality we can have at any given time. It is this integrative perspective that undergirds the worldview presented in this book.

THE PROBLEM OF THEOLOGICAL LANGUAGE

One final problem needs to be addressed when considering Christian theology, and that is the problem of language, which has several aspects. The first concerns the language of the Bible. When people read the Bible today, they typically and understandably do so with twenty-first-century American presuppositions. Most, unless they have learned differently, tend to read the biblical accounts as factual and as properly taken at face value. However, there are several problems here. First is the fact, as we have already noted, that not all biblical accounts were intended to be taken literally, just as not all literature today is intended to be taken literally. For one thing, there is a lot of poetry in the Bible. The Psalms, which are poems that were essentially the lyric sheets of ancient Hebrew songs, contain *ideas* the ancient Hebrews believed to be true, but, as poetry, the authors did not intend the songs to be taken literally. Poetry *by definition* is nonliteral language, and no rational

24. Gould, *Leonardo's Mountain,* 274. See also Gould, "Nonoverlapping Magisteria"; and *Rocks of Ages.* Editor's note: Gould also wrote, "If there is any consistent enemy of science, it is not religion, but irrationalism" (*Ever since Darwin*, 141).

person takes poetry literally today; thus, no one should interpret biblical poetry literally, either. It is interesting to note that when the prophets speak a word from God, they do so in poetic form. The creation story in Genesis 1 is also in Hebrew poetic form. These facts should, therefore, be reflected in how we interpret these writings. Conservative evangelicals, however, typically interpret both the prophets and Genesis 1 literally.

Another problem in dealing with biblical language has to do with how the ancient Hebrews expressed their beliefs about God and God's relationship with humankind. Today people describe God using terms such as eternal, omnipotent, infinite, and the like—in other words, by using abstract concepts. One indispensable key to understanding the Bible is that this type of thinking and speech did not exist for most of the biblical authors. It appears only sparingly in some NT writings. This type of thought and expression was invented in the West by the pre-Socratic Greek philosophers around the sixth century BCE. Only slowly, over many centuries, did it become integrated with the language used by the common people. Until that happened, people throughout the ancient Near East and Europe had to describe their beliefs in another way. They did so by telling stories that illustrated the nature of the gods or God, the world, and the ways the gods/God interacted with humankind. The Greek word for such stories was *mythos*, from which we get the English term "myth."

"Myth" has come to refer in modern English parlance to an ignorant falsehood. Ancient people, however, considered myths to be true to the extent that they correctly illustrated the nature of the divine, humans, their relationship to the divine, and other questions about the nature of life in the world—all personal, *existential* (having to do with one's existence) concerns. The myths were, in other words, theology in story form, as distinct from today's theology in abstract conceptual form. Thus, *literally* they were false, because they were not meant to be taken literally (even if some people did take them in that way), but they could be—and many Jews and Christians would say they are—*theologically* true.

This is highly relevant to our purposes in this book in that biblical scholars today (apart from inerrantists) recognize that, for example, the creation accounts in Genesis 1 and 2 are myths in the proper sense of that term—theologies in story form. Through their stories, they teach both that God is the creator of the world, beneficent, and very powerful, and that humans are special creatures in God's creation, to whom God has given the task of caring for the Earth and its other creatures. However, as we noted above, the stories do contain *literal* contradictions. A key difference is in their descriptions of the *way* God made the creatures. In Genesis 1, God simply speaks and they come into existence, whereas in Genesis 2 God molds each of them from the soil, one at a time, and then breathes life into them. This particular difference in the stories illustrates two different aspects of God. In Genesis 1, God is an enormously powerful yet somewhat remote God, whereas in Genesis 2 God is close to and intimately involved with God's creatures. In the abstract language of modern theology, we would say that the first story portrays a *transcendent* God, while the second portrays an *immanent* God. These stories show that ancient Hebrews understood God to be both, transcendent *and* immanent.[25]

The lesson in all this is that the Hebrews, like all the other peoples around them, stated what they believed about God, the world, themselves, and the relationship between God and themselves by telling stories that illustrated these things, i.e., they did theology by creating myths. The fact that some interpreters (especially inerrantists) interpret them literally is no fault of the biblical texts but rather of those who interpret them without this knowledge and so lack a proper understanding of the type of literature they are reading.

Yet another crucial fact concerns another type of biblical literature that we have already addressed—the historical writings. Today we assume that historical writing has as its primary objective the accurate presentation of historical facts, with attendant theories that attempt to explain those facts. However, this notion of history is a relatively recent phenomenon, created roughly around the end of the eighteenth century in Europe, and especially by German scholars such as Leopold von Ranke (1795–1886). This does not mean that history as a discipline did not exist prior to this time, rather, *how* history was conceived was different. "History" to these earlier people—and this means all the way back to and including ancient Israel—was essentially about telling interesting stories with *some* basis in fact but told mainly to teach people moral, theological, or other important lessons about life. While it is true that some ancient "historians," such as Herodotus (ca. 485–422 BCE), seem to have been more concerned than others to be factually correct, for most writers historical accuracy was much less important than getting their point across. In this way, ancient "histories" were more like today's historical fiction than modern history, in that while they do contain some historical facts, they also contain invented events and dialogue in order to get their didactic points across.

This kind of history writing is therefore what we typically see in the Bible. The primary intent of the biblical history writers was to teach lessons,

25. It is interesting to note that, in both creation stories, God relates to what already exists—in Genesis 1 to the dark and watery chaos, and in Genesis 2 to the desert and the creatures God creates, especially humankind, with whom God has a special relationship.

primarily theological and ethical ones. Comparative study reveals, for example, that the Gospel authors occasionally changed, or would pick and choose, the stories that were passed down to them concerning the life and teachings of Jesus in order to make the points about him they wanted to get across to their audience. For example, in Mark, the main source used by the later authors of Matthew and Luke, the Roman centurion standing at the foot of the cross says, when Jesus dies, "Truly, this man was God's Son" (15:39). Matthew follows Mark and has the centurion say precisely the same words (27:54). In Luke, however, the centurion's words have been changed to "Certainly this man was innocent" (23:47; a note in the NRSV offers "righteous" as an alternative to "innocent"), reflecting that Gospel's concern to show to its law-abiding Gentile audience that Jesus was innocent of the crimes against the state that had been charged against him and therefore was no criminal. In this way Luke sought to remove an obstacle to the readers becoming followers of Jesus.

Luke also replaced Jesus's cry of forsakenness on the cross, "My God, my God, why have you forsaken me"[26] (Mark 15:34) with "Father, into your hands I commend my spirit" (Luke 23:46) to create a picture of Jesus faithfully trusting in God to the end, even while undergoing a torturous death (note the possible translation of "righteous" as an alternative to "innocent" in the NRSV translation of Luke 23:47). Changing Mark's words did not matter to him, nor would it have mattered to other "historians" of the time, because historical accuracy was not their main concern. Similarly, John eliminates Jesus's baptism by John the Baptist,[27] which would be inconsistent with the portrait of Jesus as an almost superhumanly faithful person, who needed no baptism for the forgiveness of sins. These few examples should be sufficient to demonstrate that biblical history writing, like virtually all history writing in the ancient Near Eastern and Western worlds, was more concerned to teach lessons than to preserve history as it actually happened. This understanding can help us greatly in interpreting what the original messages of the historical narratives in the Bible were intended to be.

Recognizing all these factors also enables us to understand *why* the literal contradictions we find in the biblical text exist. The writers of the biblical texts were certainly aware of at least some of them, but neither they nor their readers (or hearers—few people in the ancient world could read)

26. This is a quotation of the first verse of Psalm 22. This psalm is a type of psalm called a *lament*, in which the speaker laments the sorrowful state of Israel and, typically, calls on God for aid.

27. Editor's note: John 1:19–34, esp. v. 32, echoes stories of Jesus's baptism (Mark 1:9–11; Matt 3:16–17; Luke 3:21–22), but it is possible that the author of John did not know these stories.

cared about them, because they were irrelevant to the theological and ethical points the authors were trying to make. Sometimes these contradictory texts were even placed right next to each other, such as Genesis 1 and 2, or intertwined with each other, as in the case of the story of Noah's ark in Genesis 6–9.[28] Their conception of historical writing simply did not insist on factual accuracy. Thus, again, as with the presence of myth in biblical writing, the problem is not the Bible's, but rather those modern readers who do not understand the nature of the biblical writings and therefore misconstrue them.

In addition to these problems with interpreting biblical language properly, there is another kind of problem inherent in all theological language when it attempts to speak of God. The Christian tradition has universally recognized that human language is inadequate to speak of God, fundamentally because human language was created in order to speak of the things we encounter in the world—finite things. God, on the other hand, is by definition *not* a finite being and therefore cannot be accurately described using terms that speak of finite, worldly things. Our limited language reflects the finitude of our minds, which are inherently incapable of understanding the God who is qualitatively (not just quantitatively) so different from us as to be essentially inconceivable. It is like an amoeba attempting to understand a human being; it is simply impossible given the rudimentary nature of the amoeba's mental capabilities compared to our own. This is the traditional Christian doctrine of *divine ineffability*, though something like it is also found in most strands of the major theistic traditions of the world.

This recognition has had two consequences for Christian theological language. The first is the use of what has been termed *apophatic* or "*negative*" theological language. Those who employ this type of language have typically claimed that, given the inadequacy of finite human thought and language regarding God, the only way we could completely avoid misrepresenting God would be to say nothing at all. However, since that does not get us very far in our quest to have at least some sense of what God is like, proponents of negative theology have said that the next most appropriate procedure is to say what God is *not*, rather than to make positive affirmations about what God *is*. In this way, certain speech about God is ruled out: God is *not* finite (is *infinite*), *not* a bodily being (is *incorporeal* or *immaterial*), and so on.

28. In the case of the Noah story, this intertwining in every part of the story, except the dimensions of the ark, is why it reads so oddly, with constant repetition (with slight variations according to the style of each author), and why, for example, there are two different numbers given for how many animals were taken into the ark: one pair (male and female) of all animals (e.g., 6:19–20) versus seven pairs of all clean (i.e., edible) animals and one pair of unclean (inedible) animals (e.g., 7:2–3).

Another way to make essentially the same point is by using "*omni-*" language, which can be understood to employ a pair of negatives in making assertions about God. Thus, to say God is *omniscient* (all-knowing) means that there is nothing God does not know, or to say that God is *omnipresent* (present everywhere) means that there is no place where God is not present, and so on.

Apophatic theology has been most strongly advocated in the Christian tradition by mystics, who claim to have had a direct experience of God and, as a result, have recognized the futility of attempting to describe God in human language. Thus, Thomas Aquinas, one of the greatest theologians in the history of the church, after he had such a mystical experience of God, is said to have ceased writing theology altogether, leaving several works unfinished because he said that, compared to the reality of God, all his attempts to describe God were like straw—that is, trash.

Yet the Christian tradition as a whole has felt compelled to make *some* positive assertions about God as well, though in so doing it has generally claimed that whatever it says about God must be understood *analogically*. An *analogy* assumes both similarity and dissimilarity between the things being compared. For example, when it is asserted that God loves us, it means that God's Love for us is analogous to the love we humans have for one another, being like our love in some ways—for example, in its essential concern for the well-being of the beloved other. On the other hand, God's Love is unlike human love in that it is perfect, untainted as most human love is by self-centeredness, and infinite in scope, applying to every creature in all times and places. Thus, even in positive assertions about God there is always a negative element as well: God's attributes do not contain any of the negative aspects found in comparable human attributes. God's attributes must be stripped of all the finitude and weakness associated with comparable human attributes and then multiplied into infinity. It is with these cautions, then, that we will proceed to speak of God in subsequent chapters.

4

Space, Time, and the Fundamental Particles and Forces of Nature

INTRODUCTION

THROUGHOUT HISTORY, HUMANS HAVE wondered about the nature of the universe in which we live, and over the millennia many thinkers have suggested theories that attempted to describe and explain it. Our current scientific understanding of the nature of physical reality is complex and at times difficult to comprehend. Nonetheless, having at least a basic understanding of the nature of the universe is necessary if we are to make sense of its history.

Much debate has concerned *the size of the universe*. Today we understand it to be unimaginably vast in scope, though exactly how large it is we really do not know. In fact, there is some debate among cosmologists as to whether the universe might even be infinite, with most seeming to think we never will know. The reason is that the universe is, and has always been, expanding rapidly outward in all directions, and that expansion has resulted in a universe so large that we can only see a part of it—what we call the *visible universe*—even with our most sophisticated telescopes. Because light's speed is limited to 299,792 kilometers per second (km/sec), or 186,282 miles per second (mi/sec)—which is the maximum velocity anything within the universe can have—we can see only those stars and galaxies whose light has reached us since the universe first began. Stars and galaxies so far away that their light has never and will never reach us must remain forever unknown to us. Current estimates of the diameter of the visible universe are about 93 billion *light-years* (the distance light can travel in one year).[1]

This also means that we can know nothing about what may exist outside our universe, if indeed anything does. As we have noted, a number of cosmologists have recently suggested that our universe may not be the only one, that others too might exist, or have existed in the past, or will exist in the future. If this *multiverse* does exist, the true scope of physical reality would be immensely larger than anyone could even begin to comprehend.

In addition, we cannot really know anything about *the shape of our universe* with certainty. To speak about the shape of something requires knowing its boundaries, and we cannot see the universe's boundary (if it has one). Nonetheless, cosmologists tend to think of it as an enormous sphere that has radiated out in all directions since the Big Bang. Some have calculated that the radius of the entire universe (not just the visible part) must be at least 150 billion light-years, while others have suggested a diameter of as much as 10 sextillion[2] times larger than the visible universe. On the other hand, if the universe should turn out to be unbounded, the universe could actually be *smaller* than the visible universe, because things that appear farther away could be duplicate images of nearer galaxies whose light has traveled around the universe. Most cosmologists, however, seem to think the former view is correct and that the visible universe is only a small part of the entire universe.

Interestingly, we can still see light from the early universe in the form of the cosmic microwave background (CMB) radiation. That light was emitted some 13.37 billion years ago, once the fog that filled the earliest universe became transparent, letting photons (particles of light) stream out freely through the universe. Before that, light could not travel in a straight line through the dense fog of ionized particles (particles carrying a positive or negative charge due to the loss or addition of an electron in their outer orbital shells) that had existed since the Big Bang. Since the data indicate that this transition from fog to clarity within the universe occurred some 380,000 years after the Big Bang (aBB), the universe must be approximately 13.75 billion years old.

We can see back in time to nearly the beginning of the universe because the various types of telescopes we use are in reality rather like time machines. While they cannot transport things back in time, they can show us what the universe looked like in the past. The more powerful the telescope,

1. Ninety-three billion light-years is the distance of the most remote things in the universe (which are the earliest things created in the universe during the Big Bang), given the expansion of the universe.

2. 10,000,000,000,000,000,000,000 (10^{23}).

the farther back in time it can see. The reason for this is the finite speed of light. We can see things in the sky because we see the light emitted by them, in the case of stars, galaxies, and the like, or reflected off them, as is the case of planets, moons, and smaller celestial bodies. Either way, that light has traveled for a certain amount of time to reach us, and so we are seeing these objects as they were—a short time ago, in the case of nearby objects, and a longer time ago for more distant objects. Thus, even in the case of the relatively nearby Sun, which is only some 149.6 million km (93,000,000 mi) away, the light that it emits has taken eight minutes and nineteen seconds to reach us. When you look at the Sun, then, what you are seeing is the Sun as it was more than eight minutes ago, not as it is in the present.[3]

The nearest star to us, *Proxima Centauri*, is 4.2 light-years away, meaning it has taken 4.2 *years* for its light to reach us.[4] Therefore when we gaze out to the farthest reaches of our most sophisticated telescopes' range—currently that of the Hubble telescope—we are literally seeing more than thirteen billion years into the past. It is this curious fact of nature that has helped us understand the history of the evolution of our universe.

As regards the *contents* of the universe, current data reveal that the matter that we can see (typically called ordinary matter) is only a small portion of what actually constitutes our universe. The remainder is made up of mysterious dark matter and dark energy, which, along with the nature of ordinary matter, will be discussed below. Ordinary matter in the universe is basically uniformly distributed, or *homogeneous*, when observed on the largest scales; it is also *isotropic*, which means that the universe looks essentially the same in all directions and from whatever place within the universe one observes it. On smaller observational scales, however, this is not the case. Maps of the distribution of matter on these scales reveal a filament-like web of galaxies, galaxy clusters, and super-clusters, some even in sheet-like formations, with large empty voids in between, rather like the structure you see if you cut open a solid, low-density rubber ball. An example of a

3. It is odd to realize that the Sun could already have exploded and you would not know it had occurred for another eight-plus minutes.

4. Actually, calculating this number is somewhat more complex and becomes more important the farther away an object is, due to the continual stretching of the space-time fabric throughout the universe's history. The number just given assumes a static universe, which works reasonably well for closer objects because the space-time fabric has expanded insignificantly during the time it took for their light to reach us. But the farther away some astronomical object is from us, the more space has expanded during the time it has traveled, and therefore we must determine the amount of that expansion to know how far into the past we are in fact seeing. Physicists have devised formulas that take this expansion over time into account to give more accurate estimates of the distance and time the light has traveled to reach us.

sheet-like galactic formation is the "Great Wall," which is more than half a billion light-years long, 200 million light-years wide, and a more modest 15 million light-years thick. The more recently discovered Sloan Great Wall is even larger, the largest known structure in the universe. According to one common estimate, the visible universe contains about 100 billion galaxies, each containing some 100 billion and perhaps even hundreds of billions of stars. The *visible* universe is thought to contain something like 10^{23} stars, along with the planetary systems that surround some, perhaps most of them. How many stars and planets there are in the *entire* universe, no one can possibly estimate.

For an illustration of the large-scale structure of the universe from Earth's perspective, including galaxy clusters, superclusters, and filaments, see Thomas Jarrett's image at http://commons.wikimedia.org/wiki/File:2MASS_LSS_chart-NEW_Nasa.jpg.

Also, since the universe is indeed expanding, this means that in the past the universe was smaller. And if one runs time backward, with the universe becoming continually smaller the farther back one goes, this means that at some point in the past the universe was infinitesimally small, with all its material jammed into a single point, which we term a *singularity*. The explosive expansion of the universe from this single point was the origin of the idea of a Big Bang at the beginning of our universe.

THE NATURE OF SPACE AND TIME

From ancient times to the early twentieth century it was believed that *space* was empty, unchanging, and everlasting. It was seen as the stage on which the actors of the universe, the stars, planets, and Earth, played their roles, but it was assumed to be unaffected by them. *Time* was understood to be an entity distinct from space and the things that inhabited it, continually flowing from the past to the present to the future for as long as the universe has existed. This was the understanding that undergirded Newton's theories of mechanics, or the motions of things and their interactions with the fundamental forces of the universe. Thus, in space an inch was always an inch and in time a second always a second, no matter what was occurring within the universe. This understanding is called "classical mechanics."

At the beginning of the twentieth century this view underwent a major revision due to the work of Albert Einstein and mathematician Hermann

Minkowski (1864–1909). Minkowski used Einstein's Theory of Special Relativity of 1905 to show that space and time should not be thought to exist independently of each other but, rather, that they form a single, interrelated, four-dimensional entity called "space-time." Most people were familiar with the three dimensions of space—length, width, and height—but Minkowski argued that time is also a dimension, the fourth dimension, which is inseparably conjoined with the three dimensions of space. This makes sense if one considers that to locate something in space, one must not only locate *where* it is in three dimensions, but also, since everything in the universe is in motion, *when* it will be at that location. Thus, we must not think of space and time as different and unrelated entities, but rather as different dimensions of a single "fabric" of which the universe is made.

The strange part, according to special relativity, is that not everyone necessarily experiences space and time in the same way. Rather, one's perception of space and time will vary under conditions humans do not normally experience. At great velocities—for example, for a person in a spacecraft traveling at a significant percentage of the speed of light—space and time will be experienced by that space traveler as precisely the same as they would if the spaceship were traveling at a much lower velocity or standing still. But for a person standing outside the spacecraft, the craft would appear to shrink in length (called the "Lorentz contraction") the closer it came to the speed of light, while its mass would increase. Thus, if it were possible for the spaceship to travel at the speed of light (which is not possible, since only something without mass, such as a photon, can travel at that speed), to an observer the ship's length would become zero and so would vanish (into two dimensions) while its mass would become infinite—clearly not fitting in with our common-sense ideas about the behavior of physical things. Also, time would pass more slowly for the person on the spaceship and faster for the person on the ground (this is called "time dilation"). Thus, if one twin went on a space journey and traveled at a high speed relative to her twin who remained on Earth, the one on Earth would be significantly older[5] than her twin once she returned.

This same phenomenon applies to all observers moving relative to each other, Einstein claimed, though in our usually rather-slow-moving world, the change is so small we do not normally notice it. However, these ideas have been tested frequently and rigorously on faster-moving objects, such as a spacecraft circling the Earth at the relatively modest rate of some

5. How much older she would be depends on the speed of the spacecraft and the duration of the space journey. Einstein's equations can be used to calculate how much older she would be, given these factors.

27,400 km/hour (17,000 mi/hr)[6], and have in each case matched the predicted values. The spacecraft indeed appears from Earth to have shortened slightly, and identical watches on the spacecraft and on the ground, despite functioning perfectly and identically, register different times (the ground-based watch being ahead of the spaceship-based watch) once the ship has landed. Thus, Einstein's special relativity theory is correct. Space and time are not experienced to be the same for all people but are experienced differently if they are moving at different rates with respect to each other. They are therefore relative (hence the term "relativity") to those who experience them—and both are equally correct within their spatio-temporal frames of reference. (The only exception to this rule is light, which always and for all observers, no matter their frame of reference, travels at the speed of light.) The important lesson we learn here is that space and time are clearly *not* absolute and invariable, and therefore scientists before Einstein, such as Newton, were quite wrong in their assumptions about them.

With Einstein's subsequent theory of general relativity (1916), which relativizes gravity also, the picture of space and time became even stranger. Einstein claimed that the space-time fabric bends and warps as things pass through it, and that gravitational attraction was due to this warping of space-time, not an inherent tendency in things with mass to attract each other, as Newton and all others since him had thought.

This is a hard notion to picture. One possible way we might begin to imagine this four-dimensional space-time and the new idea of gravity associated with it is by first thinking of space-time as an ordinary bed-sheet that has been stretched tightly on all sides. If you were to roll a heavy object across it—say, a billiard ball—you would see the sheet sink for a certain distance around wherever the billiard ball went, and the depression (force of gravity) would be greater the closer to the ball it was and would gradually taper off the farther one went out from the ball—just as Newton's equations said gravity does. This deformation of the sheet by an object with mass is like the deformation of space-time when a massive object passes through it.

For an illustration of Earth warping the space-time fabric, shown in two-dimensions, see http://commons.wikimedia.org/wiki/File:Spacetime_curvature.png.

6. The speed required for any object to stay in orbit around Earth depends on its orbital distance from the planet. At this velocity the object would orbit at a distance of 200 km/124 mi above Earth's surface. Note that I have capitalized the term "Earth." That is because it is the name of a planet and so should be capitalized like every other named planet (e.g., Mars).

To image the reality of space-time more nearly you need to think of its fabric not as an essentially two-dimensional sheet but rather as an immense, three-dimensional piece of fabric made of stretchable fibers.[7] If you were to move an object, say a small ball, through the fabric, you would see the fibers it encounters in all three dimensions bend, forming a trough around the entirety of the ball as it passed by. As the ball moves, the fibers in front of it spread apart while those surrounding the ball have formed the kind of trough seen in the two-dimensional model, only now in all three dimensions. Once the ball has passed, the fabric returns to its original shape.

This is the way gravity and space-time interact, according to Einstein, and numerous experiments since he proposed this theory have lent support to his claim. As an object with mass moves through space-time, its fabric is warped by the object, and the greater the mass of the object, the greater the warping of space and time around it. If another object is moving in a straight line and encounters the warped space near a massive object, its path will be bent according to the way space is warped by the massive object. This altering of the path of things near massive objects is true even for massless photons. In Newton's theory, a photon's path would not be affected by a massive body it encountered as it shot through space, since in that theory attraction is determined by the masses of both objects and photons have none. In Einstein's theory, however, it is space itself that warps, and so photons, though massless, have their paths through space bent when they encounter massive objects.

Einstein's prediction has been empirically verified by astronomical observations for almost a hundred years now. It is seen clearly in the phenomenon called "gravitational lensing." In gravitational lensing, an object such as a star, which when viewed from Earth lies behind a massive object such as a galaxy, will appear to us to be multiple stars, or an arc of light, because the light from the star is bent by the galaxy's warping of space-time around it. Fittingly, one such object has been named "Einstein's Cross." In light of general relativity, then, Newton's idea of gravity as an attractive force between objects having mass is wrong; gravity is the warping of space-time. For Einstein, the simplicity and power of the theory convinced him that it must be true; it was too beautiful to be wrong.

For gravitational lenses or cosmic mirages due to the curvature (warping) of space, see http://hubblesite.org/newscenter/archive/releases/2000/07/image/c.

7. Currently a number of theorists are working to determine the nature of space-time's fabric, and some are suggesting that space-time is not fundamental but that it might be composed of some kind of ultra-microscopic particles.

For an image of a gravitational lens called "Einstein's Cross," see http://hubblesite.org/newscenter/archive/releases/1990/20/image/a.

But space is not the only part of space-time that is warped by objects within it; time is, too. It runs more slowly when, say, a spaceship is traveling near a massive object than it does farther away from it. This phenomenon is called "gravitational time dilation." The fundamental lesson here is that objects within space-time both affect and are affected by space-time, and that neither space nor time is absolute. We live in a universe that is relational through and through.

Another fascinating thing about space-time is that it is not static, as was traditionally thought by most astronomers. Rather, it is expanding in every direction, as it has been ever since the Big Bang, but for about the past four billion years it has been expanding at an ever-accelerating rate (due to the effect of dark energy, as will be discussed below). This means that the space-time fabric itself is being stretched in all directions, carrying with it all the matter in the universe. If we continue with our analogy of space-time as a fabric, this means the fibers are all stretched outward as the universe expands, causing the spaces between them and all they contain to become larger. The result is that galaxies are all moving away from each other and the space between them is increasing. Gravity, however, counters this drawing apart of galaxies when they are sufficiently close together, which accounts for the formation of large-scale structures in the universe, such as galactic clusters and super-clusters, filaments, and sheets. The rate of the universe's expansion is given by the "Hubble constant" or "Hubble parameter," which is currently calculated to be about 72 km/sec (43 mi/sec) per million parsecs (one parsec being 3.26 light-years, or about 30 trillion kilometers/18.6 trillion miles). Besides being thoroughly relational, the universe is also an extremely dynamic place.

THE FUNDAMENTAL CONSTITUENTS OF THE UNIVERSE

The fundamental constituents of the universe are ordinary matter and energy, and perhaps also what many physicists have termed dark matter and dark energy. Ordinary matter and energy, according to Einstein's $E=mc^2$, are simply two forms of what is termed *matter-energy*. You might think of matter as condensed energy and energy as unconfined mass (which is what happens in an atomic bomb—the conversion of mass to energy). We will look at each of these ordinary and theorized forms of matter and energy in turn, beginning with that about which we know the most: ordinary matter and energy.

Until a little more than one hundred years ago, scientists thought that atoms were the most fundamental building blocks of all matter in the universe and that they were indivisible. Atoms were believed to be of varying types, as laid out in the Periodic Table of Elements, first created by Dmitri Mendelev (1834–1907) in 1869. By the early 1930s physicists had learned that the atom was typically composed of a nucleus, made up of protons and neutrons, surrounded by electrons that traveled around the nucleus like planets around the Sun. Since then, scientists have realized that even protons and neutrons are not fundamental particles but are themselves composed of yet smaller entities. In addition, a host of other fundamental particles have been discovered, and together they constitute what is called the Standard Model of Particle Physics. Here our discussion becomes rather technical.

For "The Standard Model of Particle Physics," see http://commons.wikimedia.org/wiki/File:Standard_Model_of_Elementary_Particles.svg.

The Standard Model (see the QR code above) provides a schema for describing the twelve fundamental matter particles, called *fermions,* and the four force-carriers that govern the interactions between them, termed *gauge bosons*. The fundamental fermions—as distinguished from the particles made from combinations of such particles, which are also called fermions—are divided into two groups: the relatively heavy *quarks*, comprising the *up*, *down*, *charm*, *strange*, *top,* and *bottom* quarks (these are sometimes called the six *flavors* of quarks), and the relatively light *leptons*, including *electrons*, *muons*, *tau particles*, and the three species of *neutrinos* (*electron neutrino*, *muon neutrino,* and *tau neutrino*), together called the six *flavors* of leptons. Each fermion also has a corresponding *antimatter* particle, sharing the same mass as its matter counterpart but with an opposite electrical charge. Thus, each quark has a corresponding *anti-quark*, the electron has a corresponding *anti-electron*, normally called a *positron*, and so on. All fermions have spins of ½, -½, 1½, -1½, and so on and obey the Pauli Exclusion Principle, which states that no two particles can occupy the same *quantum state* at the same time.[8] This means, among other things,

8. A *quantum state* is the state of a *quantum system* as described by a set of variable *quantum numbers* that specify the properties of a given particle. Thus, in the case of an electron, its quantum state is described by four quantum numbers, n, l, m_l and m_s, that

that no two fermions can occupy the same space at the same time: When they collide, they bounce off one another.

Together these twelve elementary particles are classified into three *generations*, each with two quarks and two leptons (see the chart below). Within each of the three generations, one of the leptons has an electrical charge of -1 (the electron, muon, and tau particle) and the other lepton has no charge (the three neutrinos). Quarks also have electrical charges, though these are not whole number charges like the electron, muon, and tau particle. Instead quarks have fractional charges of either +⅔, in the case of the *up-type quarks* (those on the top tier of the quarks of all three generations—i.e., the *up*, *charm*, and *top* quarks), or -⅓ in the case of the *down-type quarks* (the *down*, *strange*, and *bottom* quarks).[9] These charge numbers indicate the kinds of interactions the various fermions can have with each other, since when they combine the total particle must have a charge of +1, 0, or -1. Each successive generation of particles has greater mass than the previous generation (except, perhaps, the nearly massless neutrinos, which are very difficult to detect and study, so that there is significant uncertainty about their nature, including their true masses). Finally, it is important to know that the quarks and leptons in generations two and three decay almost instantly into generation one quarks and leptons, with the exception of the neutrinos, and this is what happened in the earliest universe as energy levels dropped as the universe expanded and cooled. Thus, all the ordinary matter that the universe contains today, and has contained since the early universe, is made up of only six particles—up and down quarks, electrons, and the three species of neutrinos. Together these six particles combine in various ways to form all the particles known to science (more than three hundred). The Standard Model is an elegantly simple system.

Quarks combine to form various kinds of composite particles called *hadrons*. Hadrons consist of two *families*: *baryons*, composed of three quarks; and *mesons*, composed of a quark and an anti-quark. Thus, a proton is composed of two up quarks and one down quark (their combined charge is ⅔ +⅔ -⅓ = +1, the charge of a proton), and a neutron contains an up quark and two down quarks (their combined charge is ⅔ -⅓ -⅓ = 0, meaning it has no electrical charge). Mesons occur only in high-energy environments, such as the very earliest universe just after the Big Bang or those produced

indicate, respectively, its energy level, its orbital angular momentum, the projection of that angular momentum along a particular axis, and its spin.

9. The top and bottom quarks were sometimes in the past termed the *truth* and *beauty* quarks. Calling the top quark "truth" was never very popular, but referring to the bottom quark as "beauty" was. Even now, particle accelerator complexes that produce bottom quarks are sometimes referred to as "beauty factories."

in *particle accelerators*—huge devices that smash subatomic particles such as protons or neutrons together at nearly the speed of light in order to break them into their constituent parts. Mesons exist for only a small fraction of a second before decaying—in the case of charged mesons, into electrons and neutrinos or, in the case of uncharged mesons, into photons. Mesons come in a variety of types, depending on which quark and which anti-quark they are made of.

Quarks also have two other distinctive properties, *spin* and *color charge*. Spin refers to the *angular momentum* (or *rotational momentum*) of a particle around its axis. Spin on the quantum scale is somewhat like the spinning of a top except that the particles that display it are understood to be points. Fermions all have spins of half-integers, such as ½, 1½, 2½, etc. Color charge is somewhat analogous to electrical charge, except that where electrical charge concerns only a single aspect, whose values are either + or -, color charge has three aspects, which are symbolized by the three primary colors, red, green, and blue—hence the use of "color" for this type of charge. Because the six quarks come in three-color charges each, there are a total of eighteen different quarks (or thirty-six when one also includes the anti-quarks). Color charge concerns only the way quarks and *gluons* (one of the four fundamental forces in nature and the one that holds quarks together) interact as specified by the theory of *Quantum Chromodynamics*. One result of this theory is that only particles whose net color charge is zero can exist freely in nature. One way to achieve no net color charge is for a quark and its anti-quark to combine, in which case there is a meson. The other is for three quarks (or three anti-quarks), one of each color charge, to combine, in which case their color charges cancel out. These are the baryons, such as protons and neutrons.

The three quarks contained within protons orbit around their center at a mind-boggling 10^{24} times per second, while the protons themselves rotate some 10^{20} times per second and the electrons that orbit them do so around 10^{15} times per second. Clearly, atoms are in constant, feverish motion no matter what material object they inhabit.

Leptons, as we have noted, either carry no electrical charge (the three neutrino types) or a charge of -1 (the electron, muon, and tau), with the latter three having antimatter partners with charges of +1. Leptons have mass and spin, like quarks, but none of them carries color charge. Thus, there are two types of electrons, one with spin up and the other with spin down, and so also two types of positron. Second- and third-generation charged particles—muons and tau particles (or *tauons*)—decay quickly into electrons, just as their quark counterparts decay into first-generation quarks (up and down quarks), so it is electrons that are the fundamental negatively charged particles in the universe. Electrons thus orbit nuclei whose protons and neutrons are made up solely of up and down quarks. It is electrons, traveling in distinct orbital shells (see below), that determine all chemical interactions. Second- and third-generation neutrinos, however, do not decay, and, since they interact only rarely with other ordinary matter particles, they are difficult to detect as they stream freely throughout the universe, including passing through most material objects. Billions pass through our bodies each second. All leptons, like quarks, have antimatter partners, though there is debate regarding whether neutrinos are their own antimatter equivalents or not. Finally, it is the understanding of contemporary physics that quarks are the only particles influenced by all four fundamental forces. Charged leptons are affected by all forces except the strong nuclear force, and neutrinos are affected only by the weak nuclear force and gravity.

THE FUNDAMENTAL FORCES THAT GOVERN THE INTERACTIONS OF ORDINARY MATTER PARTICLES

According to the Standard Model, it is *gauge bosons* that are the force-carrying particles that create the four fundamental forces in nature. As indicated in Figure 5 above, they consist of the *photon*, which is responsible for the *electromagnetic force*; the *gluon*, which is responsible for the *strong nuclear force*; and the *Z and W bosons*, which are responsible for the *weak nuclear force*. Another boson, the *graviton*, is hypothesized to be responsible for the fourth fundamental force in nature, the *gravitational force*, though its existence has yet to be confirmed.[10]

Unlike fermions, with their half-integer spins, bosons have integer spins (0, 1, 2, etc.). Thus, unlike fermions, they do not obey the Pauli Exclusion Principle: many of them can occupy the same quantum state at the same time. This means that many of them can occupy the same space at the same time, and that they do not bounce off one another when they collide but instead pass through one another (provided they have no charge). All force carriers have energy—i.e., exist as packets (or quanta) of energy—and it is this energy that is exchanged between the particles, functioning, as it

10. According to the Standard Model, a sixth boson, the *Higgs boson*, also exists (see the QR code on p. 74 above)—a particle whose existence was only recently confirmed. These bosons are not force-carriers, but when many of them are present together, a field is created, the *Higgs field*. According to current theory, this field is present throughout the universe, interacting with the various matter particles and thereby giving them their specific masses. Other types of non-force-carrying bosons also exist, including, for example, mesons.

were, as messengers that tell the particles how they must interact (either in attraction or repulsion).

The strongest of the four forces is the (appropriately named) *strong nuclear force* (or simply the *strong force*), and so the strongest force-carrying particle is the one that mediates that force, the *gluon*. Gluons interact only with quarks, the protons, and neutrons, which are made of quarks, and other gluons. Thus, gluons (of which there are eight types, but we need not go into the distinctions between them) are what bind together quarks into protons and neutrons and bind protons and neutrons into atomic nuclei; without them, nothing more complicated than free-floating quarks would exist. The strong force is created by the constant exchange of huge numbers of gluons. Like all force-carrying bosons, gluons have no mass but have a great deal of energy. For example, the mass of the quarks that make up a proton accounts for only about 2 percent of the total mass of the proton; the rest is from the mass/energy of the gluons.[11] In reality, then, in terms of their energy density, protons are made mostly of gluons, not quarks. Also, gluons cannot exist independently, except in the enormously high-energy conditions of the very earliest universe or in certain very powerful particle accelerators.

Though the force they carry is enormous—some one hundred times greater than the electromagnetic force, the next strongest force, which it must be to counter the repulsive force of multiple positively-charged quarks and protons close together in an atomic nucleus—it operates only at extremely short distances. Thus, unlike all other gauge bosons, the gluon's force gets stronger with distance, not weaker, acting rather like a rubber band as it is stretched. The maximum range of gluons is roughly 10^{-15} meters, which is the maximum size of a stable atomic nucleus. Beyond this, the electromagnetic force takes over, causing the protons to repel one another. Interestingly, the strong force is stronger at lower energies and weaker at higher energies, which is the exact opposite of the electromagnetic and weak forces. Like all leptons, neutrinos and electrons do not feel the strong force. Gluons can be split into quark/anti-quark pairs, and these can in turn annihilate, producing energy in the form of a gluon. Also, quarks can spontaneously emit a gluon and reabsorb it.

11. According to Einstein's famous equation, $E=mc^2$, mass (or more precisely, the *rest mass*—the mass of the particle at rest) and energy (E) are really the same thing existing in different forms, rather like steam and ice. Mass being turned into energy is what happens at the center of stars and in atomic bombs. This is also the reason why, in the Standard Model diagram, the units of mass are given as the energy of the particle (measured in electron volts [eV], millions of electron volts [MeV], or billions of electron volts [GeV]) divided by the speed of light in a vacuum, squared (because light can be slowed down when it passes through various media, such as water).

The *weak nuclear force* (or simply the *weak force*) is responsible for the radioactive decay of unstable nuclei into stable ones. The weak force is mediated (or carried) by the three massive (about one hundred times the mass of the proton!) gauge bosons, the W+, W- and Z. This makes the weak force unique in that it is mediated by two different fundamental particles. Because of their great mass, they can act only over an extremely short distance, roughly a thousand times less than that of the strong force, or some 10^{-18} meters. Unlike the strong force, the weak force gets stronger as the distance between the particles it governs decreases. Quarks, leptons, and neutrinos all feel the weak force. In fact, the weak force is the only means by which neutrinos interact with other neutrinos.

Because the weak force employs massive particles whose force is relatively weak, it acts slowly in the process of *radioactive decay*, also termed *alpha decay*. In the case of uranium, for instance, complete decay takes about a billion years. In this process, a helium nucleus (also called an *alpha particle*) is ejected from a radioactive element's nucleus, thereby transforming the uranium[12] into a new element, *thorium*, with two fewer protons (90) and two fewer neutrons (144). The half-life of uranium-238—the time it takes for half of the uranium atoms to decay into thorium—is approximately 4.5 billion years.

Beta decay is another radioactive process in which a so-called *beta particle*, either an electron or a positron, is emitted from an atom by the action of the weak force. There are two main types of beta decay. In *beta minus decay*, the nucleus of an element is converted to that of the next higher element on the Periodic Table by the decay of a neutron into a proton, which entails emitting an electron (e-) and an electron neutrino. In *beta plus decay*, the nucleus of an atom is converted into that of the next lower element on the Periodic Table by converting a proton into a neutron and emitting a positron (e+) and an electron antineutrino. All these types of decay may not seem particularly important, dealing as they do with the decay of radioactive particles, but the weak force helps to keep the Sun burning, and therefore neither we nor any other form of life as we know it would be here without it.

The *electromagnetic force* governs only the interactions of charged particles, such as quarks, protons, and electrons, and is mediated by the

12. In nature, uranium exists almost exclusively as uranium-238, which contains 92 protons and 146 neutrons, although traces of U-235 and U-234 (the number of protons being the same in each but with three and four less neutrons, respectively). All of these isotopes (different forms of the same element, distinguished by their containing different numbers of neutrons) are unstable and so decay by the action of the weak force.

exchange of *photons*—massless, uncharged particles of light[13]—between the charged particles. Electromagnetism is a single force yet it appears in two inseparable aspects, *electrical fields* and *magnetic fields*. When an electrical field undergoes a change it creates a magnetic field, and when a magnetic field changes it generates an electrical field. It is the combination of the two that is responsible for the electromagnetic force. The electromagnetic force is very strong at short distances—though not as strong as the strong or weak forces—but rapidly loses strength as the distance between the charged particles grows. Without the electromagnetic force, all molecules and atoms would dissolve and, again, all that would exist in the universe would be subatomic particles.

Unlike all other forces, which operate in only a single direction, the electromagnetic force operates in two, creating either attraction or repulsion between charged particles. Thus, a negatively charged electron is repelled by another electron but attracted to a positively charged positron or proton, while a positively charged proton or positron is repulsed by other protons or positrons. We also see this manifested in the attraction or repulsion of the two ends of bar magnets, depending on whether their poles have the same or opposite charges. Interestingly, at very high energies photons and the weak force W and Z bosons behave in essentially the same way and so unite into what physicists term the electroweak force. This unity last occurred early in the first fraction of a second of the universe's history.

For an illustration of electrical and magnetic fields moving through space at 90° angles from each other, see http://ls7pm3.gsfc.nasa.gov/whatsRM/em.html.

The electromagnetic force is responsible for the *electromagnetic radiation* that permeates the universe. Light, a form of electromagnetic radiation, was understood to behave like a wave but later came to be understood also to behave like a particle, the photon. Which behavior it manifests depends on the kinds of investigations of it being conducted. As a wave, electromagnetic radiation travels through space in oscillations whose wavelength is the distance between one peak (or trough) and the next one and whose frequency is the measure of how many peaks (or troughs) pass by a stationary point in a given time (see the QR code above). Wavelength is inversely proportional to frequency, meaning that the longer the wavelength, the less

13. Photons, along with neutrinos, are the most ubiquitous particles in the universe. There are about 400 million photons per cubic meter in the universe, which means that there are about a billion photons for every proton or electron that exists. A 100-watt light bulb emits some 10^{20} photons per second!

is its frequency. The frequency of a wave is measured in Hertz (Hz), where one Hz is equal to one oscillation per second. A magnetic wave exactly matches the electrical wave but is shifted by ninety degrees relative to it in the third dimension, so that the two waves are perpendicular to each other as they travel in sync through space. The *velocity/speed* of electromagnetic radiation in the vacuum of space is always the speed of light, though as it passes through other media, such as water, it is slowed down. The *energy* of the radiation is proportional to its frequency: the higher its frequency, the stronger the radiation. Note also that, because of the wave-particle duality of physical things, we can speak of waves of radiation possessing a given energy or, equally correctly, of photons possessing that same energy. Thus, we can say, electromagnetic radiation consists of waves of energy composed of photons.

There is a range of wavelengths and frequencies of electromagnetic (EM) radiation, from the *long waves*, which can be thousands of meters long, to gamma rays, which are less than an *angstrom*[14] in length. There are no clear breaks between the different named types of EM radiation, so the ranges listed here must be understood to be only approximate.

The lowest-frequency, and therefore least energetic, type of EM radiation is called *long waves* (see diagram below), whose wavelengths range from kilometers (miles) to as short as approximately 10 m (32.8 ft). Frequencies for long waves stretch from fractions of one Hz up to some 50 MHz (i.e., 1 *megahertz* = 1 million cycles per second). When these waves pass through air, water, or other media that can be compressed, they produce sound that can be heard by various animals, depending on the range of their hearing. Humans can hear sounds between about 20 Hz (deep bass) and 20,000 Hz (20 kHz, high treble), whose wavelengths range from about 17 m (55.8 ft) to about 17 millimeters (mm, or ⅔ in).

The next, more energetic category of EM radiation is *radio waves*, which have the frequency range of radio and television signals (see chart below for the designations of the various frequency ranges. Note that AM radio broadcasts are in the lower, long-wave spectrum). The least energetic radio waves have frequencies of about 50 megahertz (MHz, or 106 Hz) and wavelengths of about 10 m, while the most energetic have frequencies of about 1,000 MHz (1 *gigahertz*, GHz, or 109 Hz) and wavelengths of about 0.5 m/50 cm (19.7 in).

Microwave radiation constitutes the next (higher-frequency) type of EM radiation. Microwave frequencies range from about 1 GHz to about

14. One angstrom is equal to about 0.1 nm (nanometer, 10^{-9} m), or 10^{-10} m, or about 3.94×10^{-9} in.

300 GHz, with wavelengths of about 50 cm (19.7 in) down to 1 millimeter (.04 in). Lower-energy microwaves are used in Wi-Fi devices, while higher-energy microwaves are used in microwave ovens to cook food.

Still more energetic is *infrared radiation*, with frequencies ranging between some 300 GHz and 380 THz (*terahertz*, or trillions of cycles [10^{12}] per second) and wavelengths between about 1 mm and 0.75 µm (*micrometers*, or millionths [10^{-6}] of a meter, which equals 3.94×10^{-5} in—about the size of a pin's point). A significant part of the Sun's radiation comes to us as infrared radiation, which we feel as heat. Infrared radiation is also emitted by a wide variety of other warm things we encounter, including light bulbs and the human body, whose infrared radiation is mostly in the 10-µm wavelength range.

For an illustration of the electromagnetic spectrum, see http://commons.wikimedia.org/wiki/File:Electromagnetic-Spectrum.png.

Beyond infrared radiation is the *visible light* part of the spectrum, whose frequencies lie between about 380 and 740 THz and wavelengths range from about 0.74 µm (or 740 nm [*nanometers*] or billionths of a meter) to 0.39 µm, or roughly the size of a single-celled organism. It is this part of the electromagnetic spectrum that our eyes respond to, including the whole spectrum of colors from the longer-wavelength red, at about 740 nm, to the shorter-wavelength violet, at about 390 nm. The Sun emits a great deal of energy in the visible light spectrum, and it is this light, reflecting off objects, that enables us to see them.

Ultraviolet radiation operates in the 740 to 30,000 THz (or 30 PHz [*petahertz*], 10^{15} Hz, or quadrillions of cycles per second) frequency range, with wavelengths of between 390 and 10 nm, or about the size of an average molecule. While UV radiation is invisible to humans, some birds and insects can see in the lower range of its spectrum. The Sun emits significant amounts of UV radiation, which, because of its relatively high energies—3 to 124 eV—is harmful to most living things. Were it not for the ozone (O_3) layer in our atmosphere, which uses the UV energy radiation to break down O_3 into normal oxygen molecules (O_2) and so prevents the UV radiation from reaching Earth's surface, the planet's land surface would be a barren desert.

X-ray radiation has wavelengths between roughly 10 and .01 nm, about the size of an atom, and frequencies between approximately 30 and 30,000 PHz (or 30 EHz [i.e., *exahertz*, 10^{18} Hz], or 30 quintillion cycles per second). Energies of X-ray radiation range between 124 eV and 124 keV, with those on the less energetic end of the spectrum often called *soft X-rays* and those on the more energetic end called *hard X-rays*. Soft X-rays cannot penetrate matter very well, while hard X-rays penetrate certain solids and liquids and are the type used in making X-ray and CT images of the interior of the human body and in crystallography studies. Because of their high energies, extended exposure to X-rays can lead to cancer and other bodily damage. Only a minority of our X-ray exposure comes from medical X-rays and CT scans; the vast majority comes from natural sources.

Gamma radiation is the highest-energy EM radiation, with wavelengths under about .01 nm (about the size of an atomic nucleus), frequencies over 30 EHz, and energies above 124 keV. Gamma rays are created on Earth by the radioactive decay of certain unstable elements, by interactions in the atmosphere caused by *cosmic rays* (the "ray" term is rather confusing, as rays are mostly made up of protons, with a small minority being alpha particles [helium nuclei] and most of the rest nuclei of heavier elements), and, less often, by lightning strikes. A significant amount of gamma radiation comes from a variety of processes that occur in space, while some is generated artificially on Earth in nuclear reactors and particle accelerators. Gamma rays easily penetrate the bodies of living things and so can cause various types of damage, including DNA damage and cancer.

The fourth and final force of nature, the *gravitational force*, is that which causes everything in the universe to draw together. As we have seen, gravitational attraction is caused by the way in which massive objects warp the space-time around them, thereby causing everything to fall in toward them—even massless things such as photons. Thus, gravity is the only one of the four forces that acts on *all* particles. In absolute strength it is the weakest of the four forces; if the gravitational force were assigned a value of one, the weak force's strength would be 10^{25}, the electromagnetic force's strength would be 10^{36}, and the strong force's strength would be 10^{38}. Nonetheless, because its effects are felt throughout the universe, gravity is the main determinant of the large-scale structure of the universe.

According to current theory, gravity is thought to be mediated by particles called *gravitons*. They should be massless particles of spin 2. The problem is that the energies needed to detect gravitons are so ridiculously high that for all practical purposes it would be impossible to detect them. On the other hand, since gravitons, like photons, are believed to function both as particles and waves, experiments are currently being conducted to try to find evidence for *gravitational waves* which, if found, would confirm the graviton's existence. Another problem concerning gravitons is the fact (which we will explore below) that, in their current formulations, general relativity, which operates on large scales in the universe, and quantum mechanics, which deals with the behavior of particles on the atomic and subatomic scales, conflict in seemingly irreconcilable ways. A *Quantum Theory of Gravity* that would overcome this conflict is currently being sought by physicists, and several possible solutions have been offered, including *Superstring Theory* (see chapter 5), one of which might finally prove to be the key to the *Theory of Everything* which is the "Holy Grail" of physics.

Thus, there are four fundamental forces, with differing strengths and properties operating under their own unique conditions. We have already noted that at certain high energies the electromagnetic and weak forces unite into a single force, the electroweak force. We have also noted how the strong force gets weaker at higher energies while the weak and electromagnetic forces get stronger. At some very high energy, attained only in the very earliest fraction of a second in the history of the universe, all three of these forces were of equal strength, meaning they all functioned in the same way and therefore were unified as a single force. Physicists expect that at some still higher energy and still earlier moment in the history of the universe, gravity was united with these other three forces into a single force. The nature of this single force, however, remains unknown and is the subject of continuing research.

DARK MATTER AND DARK ENERGY

As we have seen, until fairly recently scientists thought that everything in the universe consisted of protons, neutrons, electrons, and, subsequently, quarks, leptons, and bosons—what we now term *ordinary* or *baryonic matter* (or the forms of energy these represent). But today the picture is radically different—so different, in fact, that many physicists believe this ordinary matter to constitute only about 4.6% of the mass-energy content of the universe, with most of the other 95.4% being composed of other and mysterious types of matter and energy, which they call *dark matter* and *dark energy* (the rest consists of photons, .005%, and neutrinos, .0034%). Dark matter is thought to comprise about 23% of the mass-energy of the universe, while dark energy makes up the remaining 73.4%. The term "dark" is used to indicate that physicists do not know what these new kinds of matter and energy are.

Dark matter was the first to be theorized to exist, and this stemmed from a problem noted in the rotation of galaxy clusters. In theory the galaxies within a galaxy cluster should rotate faster around the gravitational center of the cluster the closer they are to it. Thus, galaxies on the outer edges of the galaxy clusters should revolve more slowly around the cluster than the inner material. This is precisely the pattern we see in our own solar system, where, for example, the innermost planet, Mercury, travels much more quickly around the Sun than do the outer planets, such as Uranus and Neptune.

But in the 1930s, Fritz Zwicky (1898–1974) noted this was not the case, but that the outlying galaxies in galaxy clusters revolved in a way not consistent with the expected result. Then, in the 1970s, it was noticed by Vera Rubin (1928–), who pioneered the study and measurement of the rotational motions of galaxies, that stars, gas, and dust on the outer edges of galaxies were all moving faster than they should be given their distance from the galactic core. The only reasonable explanation seemed to be that there was much more mass surrounding these galaxies than could be accounted for by this ordinary, visible matter—stars, gas and, dust. The same phenomenon was noted when examining the orbital motions of galaxy clusters.

These findings were supplemented by studies of gravitational lensing. Measurements of the amount of gravitational lensing by stars and galaxies once again affirmed that there must some unseen matter in addition to what we can see in the nearer objects to account for the amount of bending of the light from the more distant objects.

These results taken together meant that there had to be clouds of matter unseen and undetected by us surrounding these galaxies and clusters of galaxies, and to this the name dark matter was given. Today scientists are beginning to map these clouds of dark matter surrounding the large structures in the universe. What is clear is that dark matter must be ubiquitous in our universe and that there must be about six times more of it than the ordinary matter we can directly detect.

But what might this dark matter be? Antimatter was suggested early on as being a good candidate for dark matter, but the idea was soon rejected because, if this were true, the antimatter should be colliding with the matter particles that they draw together, causing matter-antimatter annihilations that would produce radiation we could measure. Since the radiation is not there, dark matter cannot be antimatter.

Another suggestion was that dark matter might be ordinary matter we simply cannot see—for example, the planets, dust, and gas in space we cannot directly observe because they do not shine, like stars and galaxies do, and so cannot be seen against the black backdrop of space. However, planets, dust, and gas do give off radiation and so can be detected. And even when all the planets, gas, and dust—which make up a significant part of the ordinary mass in the universe—are added together along with that locked up in stars and galaxies, there is still much too little of it to account for the gravitational force required to explain the motions of the galaxies and galaxy clusters.

Yet another candidate might be *dark stars*. These can come in the form of *brown dwarfs*, which are star-like entities that never gained sufficient mass to trigger the nuclear reactions that power larger stars and cause them to radiate visible light; *white dwarfs*, which are the remnants of lower mass stars whose nuclear fuel is spent and so have collapsed into a small volume and shine very dimly; and *neutron stars*, which are the remnants of more massive stars whose gravity is so great that, after their nuclear fuel is spent, they collapse into so small a space that their protons and electrons combine to form neutrons, which is all that remains of such stars. Scientists have searched for and found such dark stars in the halos surrounding galaxies—hence their other name, *massive compact halo objects*, or *MACHOs*—but these also appear to account for far too little of the missing mass to explain galactic motions.

Small *black holes* have been proposed as yet another possible candidate for dark matter. Black holes are formed when an unusually massive star collapses in on itself, shrinking into a *singularity*, literally a point in space, whose gravity is so great that even photons of light, if they draw close enough, are sucked into the black hole and can never escape (hence their name). Black holes come in two varieties: smaller ones that form when stars about ten times the mass of our Sun collapse and which can be found anywhere, and those that are a million times or more massive than our Sun that reside in the centers of galaxies. Still, the total mass of both types in the universe is tiny compared to what is required of dark matter, and so this candidate also does not fit. It has been further suggested that miniscule black holes, invisible to us, might have been formed in the earliest universe and these might be the missing matter, but so far no theory explains why or how these would have been made, so this theory too is, at least for the present, not a promising candidate.

If none of these ordinary matter candidates suffice, then what might dark matter be? Theory indicates the dark matter particle should be massive—far heavier than ordinary matter particles. It should also be electrically neutral, like a neutron; otherwise, we would detect it by its charge. It must also be stable, unlike the neutron, or it would have long ago decayed and so not exert the gravitational force it does. In addition to interacting with ordinary matter gravitationally, it might interact by means of the weak nuclear force. Unlike ordinary matter, though, dark matter evidently does not emit electromagnetic radiation. Finally, dark matter particles must be relatively slow-moving—*cold* in physics terminology, as opposed to fast-moving, or *hot*, matter—so that they could clump together under the force of gravity. Thus, this elusive material is sometimes called *cold dark matter*. Another name for these particles is *weakly interacting massive particles*, or *WIMPs*.

One possible candidate that fits these prescriptions is something neutrino-like only much more massive and slower moving. Neutrinos, as we have seen, are stable, electrically neutral, and interact weakly with other matter, but have very little mass, so that even though the universe is literally filled with them, they exert very little gravitational force. Therefore, they cannot be the dark matter physicists seek. But if there were something essentially like them but far more massive, they could well be the dark matter.

The *theory of supersymmetry* (often called *SUSY*) suggests how these WIMPs could have originated. According to supersymmetry theory, for every ordinary matter fermion or boson that exists, a superpartner also exists. That is, for every fermion that exists, a boson will also exist with the same properties, and for every boson that exists, a fermion with identical properties will exist. *Superpartners* would be the same as ordinary fermions and bosons except that their spin would decrease by ½ and their masses would all be some hundreds of times—current estimates predict 300–1,000 times—greater. This means that there would exist the same number of superpartners as there are ordinary fermions and bosons. The names for these *fermion superpartners* were created by simply adding an "s-" onto the beginning of the name of the ordinary particle; thus, the electron superpartner is a *selectron* and the quark superpartner is a *squark*. The names for *boson superpartners* are formed by adding "-ino" onto the end of the ordinary particle; thus the photon superpartner is a *photino* and the graviton superpartner is a *gravitino*.

One appeal of supersymmetry theory is the elegance of its mathematical equations and the symmetrical nature of the particles and forces that constitute our universe. With all the fermions and bosons having partners, the Standard Model chart of particles becomes balanced, which is aesthetically pleasing. Another reason physicists find the theory of supersymmetry appealing is the fact that it helps solve some problems with the energy state of a vacuum in the natural world when considered from the perspective of quantum mechanics. A third is that it would provide support for Superstring Theory, which, as we will see, offers the only widely accepted theory to date to describe *gravity* in a way that reconciles General Relativity and quantum mechanics. Thus, supersymmetry is attractive for many reasons. It is also significant to note again that physicists generally operate with a deep-seated conviction that at its deepest levels nature and the equations that describe it are ultimately simple and beautiful, and supersymmetry fulfills these convictions.

Despite this interest in supersymmetry, to date no superpartner has yet been found, so whether supersymmetry is a true description of the constituents of our universe is yet to be determined. Physicists theorize the existence of a *lightest supersymmetric partner*, or *LSP*, which would be heavy, stable, and would likely feel the weak nuclear force. Power upgrades to the Large Hadron Collider in Switzerland, which are currently going on, should give it the ability to detect at least this lightest superpartner, should it exist, and this would not only provide a first empirical support for the supersymmetry theory, it would also support superstring theory and perhaps help us begin to unravel the dark matter mystery. However, recent evidence has suggested that WIMPs might be significantly lighter than even the lightest superpartners—on the order of a few proton masses. Only time and continued experimentation will finally resolve this mystery.

Although supersymmetrical partner particles provide the most popular current candidates for dark matter, they are not the only ones that have been suggested. *Axions, sterile neutrinos,* and even miniscule, curled-up *other dimensions of space,* as proposed by superstring theory, have been suggested. Further investigation is needed to tell if any of these candidates prove to be the real dark matter. Whatever dark matter proves to be, it was very important in the early universe, contributing gravitationally to the clumping of ordinary matter into the patterns of structure we observe in the universe today. Without it, our universe would have expanded at a far more rapid rate, preventing the formation of most structural elements in the universe and so making the production of any life within it impossible.

If dark matter is a mystery to us, *dark energy* is even more so. It was theorized to exist only in 1998 when two groups of astronomers, one in the United States and another in Australia, both announced that their observations indicated that the universe was not slowing down its expansion, as had been expected because of the ceaseless pull of gravity since the Big Bang, but was in fact expanding at an accelerating rate. The whole fabric of spacetime was being stretched outward at an ever-increasing rate. Because any type of matter, ordinary or dark, acts to slow down the universe's expansion due to its gravitational force, whatever was causing the expansion had to be something totally new and unknown.

Dark energy is believed to have little influence on small scales within the universe; instead, its influence is on the universe as a whole. Its density must have remained essentially constant throughout the history of the universe, not becoming more diluted like matter and radiation as the spacetime fabric stretched. If it did dilute, it would act like these others and increase the gradual deceleration of the universe over time rather than cause the acceleration we have observed. It must also be spread evenly throughout space, because if it weren't and clumped together like matter does, it would have affected the motions of galaxies and galaxy clusters in ways we could observe but do not. Instead, it seems to exert a negative pressure on the universe, sometimes called *tension*, pushing it outward rather than drawing it inward. Thus, we can say that dark energy is essentially constant in density throughout both space and time.

But what exactly is this dark energy? The simplest suggestion and the one most popular currently with physicists is that it is *vacuum energy*, a constant energy inherent in what we normally call empty space and which does not interact with anything in the universe.

From the perspective of *quantum field theory*,[15] one cannot assert that empty space, or *the vacuum*, as we call it, is actually empty; rather, it is permeated by fields that cause *virtual particles* to come into and go out of existence constantly. Virtual particles are matter-antimatter pairs, like quarks and anti-quarks or electrons and positrons, which emerge out of the vacuum and then annihilate immediately, disappearing back into it. Their energies contribute a small mass-energy to the vacuum of space. This energy has in fact been found to exist in space and has been measured. The problem is that the measured energies are far too small to account for those predicted as necessary for dark energy to have the effect it does. The measured value for the vacuum energy we have observed is 10^{-8} ergs per cubic centimeter; the amount predicted for dark energy is 10^{112} ergs, meaning the measured and predicted value differs by a factor of 10^{120}—an inconceivably large amount! Interestingly, if the vacuum energy were significantly different from its measured value, the universe as we know it would not exist and we would not be here to wonder about it. This inconsistency between theory and measurement is termed the "cosmological constant problem," and at present no one has any idea how to resolve it.

This impasse has led to another possible explanation for dark energy, called *quintessence*. The novel idea in this theory is that dark energy may be dynamic rather than constant, as in the vacuum energy model. In this view, dark energy is generated by a dynamical field, bosonic in nature, termed *quintessence*, and being such a field means it could, unlike vacuum energy, interact with other things in the universe. The energy density of this field would remain nearly constant as the universe expands, unlike other fields, yet fluctuate somewhat over time. If true, this would mean that there are not four fundamental forces in nature, but five. Such a field would again account for the expansion of the universe, and because quintessence claims that the

15. The highly complex quantum field theory deals with matters we need not know for our purposes, so I will not attempt to lay out its fundamental teachings here.

rate of universal expansion should vary rather than be constant, as in the vacuum energy model, it should be observationally detectable. Scientists are now beginning to try to determine the means by which this quintessence field might be detected.

Yet another proposal has been suggested to explain both dark matter and dark energy in a single move. Because both have been detected only due to their gravitational effects, some astrophysicists have suggested that perhaps neither dark matter nor dark energy exists, but that the effects we attribute to them are the result of our incomplete understanding of gravity. In other words, Einstein's theory of general relativity might need to be modified. Specifically, it is suggested that perhaps general relativity is very accurate on smaller scales, such as that of the solar system, where it has been tested and confirmed repeatedly, but not on larger cosmological scales, such as those of galaxies, galaxy clusters, and beyond—just as Newton's laws of gravity broke down under certain circumstances. To date, several theories modifying general relativity have been proposed that have sought to eliminate either dark matter or dark energy, but thus far it seems that none has succeeded.

Another intriguing idea builds off the theory of *inflation* in the earliest universe. Postulating this *Inflationary Period*—a brief time of extremely rapid expansion of the universe, occurring around 10^{-36} to 10^{-33} seconds or so after the Big Bang—has been necessary to explain the nature and evolution of the cosmos as we have observed it. It suggests that the large-scale structure we see in the universe today was created by the quantum fluctuations that occurred during this brief period and were greatly expanded by inflation. Since this theory's inception in the early 1980s, evidence from different sources has provided increasing support for it, leading to the suggestion that the force that caused the early universe's rapid inflation and dark energy, both of which are forces causing the universe to expand, might be related. At the present, we simply do not know enough to say whether this is the case or not.

At present, then, dark matter and dark energy remain theories that do a good job explaining the nature of, and large-scale behavior of, the universe we observe today. At the same time, what each is remains unknown. While experiments are underway to try to begin to answer these questions, the quest to understand dark matter and dark energy will likely occupy researchers for decades, if not longer.

THE ANTHROPIC UNIVERSE

Several decades ago certain scientists began asking what the universe might have been like had the laws of nature been somewhat different. They began by varying certain constants, such as the gravitational constant G, increasing and decreasing it slightly to see what the history of the universe would have been, given these changes. To their surprise, what resulted was, in either case, a universe that could not possibly have ever produced living beings. Had the gravitational force been slightly greater, the force of the Big Bang would have been overcome relatively early in the universe's history, and the universe would have collapsed in on itself before life could ever have gotten started. Had the gravitational force been slightly less, the universe would have expanded so rapidly that stars could never have been created, and so again life could never have gotten started.

When these scientists did the same thing for the other three forces of nature, they came up with similar results: vary their strengths slightly up or down and you would not have the life-bearing universe in which we live. Altering other things also in this way, such as the speed of light, electrical charge, Planck's constant, the mass of the universe, the masses of various particles, and more, yielded the same result—a barren universe devoid of life. The lesson was that ours is a very special universe, a "Goldilocks universe," that was neither too much nor too little of anything, but "just right" for the production of life. But given all the seemingly possible values for all of these different physical entities, that they should all be "just right" was a puzzle. Indeed, the odds against a universe having just these life-producing values appear to be one in a trillion or quadrillion—very long odds, to say the least.

When scientists, philosophers of science, and others considered this situation, they came up with only three possible reasons why this should be the case: either these numbers had to be just what they are even though we can today see no reason why that would be the case; or there exist trillions or quadrillions of other universes in addition to ours, most of which are not life- bearing, and we just happen to be in the life-bearing one; or God exists and has set the laws of nature in just this way in order for life, and eventually intelligent life like ours, would come to be. The last position is a modern variation of the classic Argument from Design for the existence of God, in which the universe is seen to have a goal that could have been given to it only by God. What is particularly interesting about this modern variation is that it has emerged from science, not religion or philosophy.

Having now examined the nature of space-time and the particles and forces that constitute the Standard Model of physics, it will be of use to us, as we attempt to recount the history of the universe, to have a somewhat deeper understanding of the most significant scientific theories of the past century and a half that have attempted to explain the nature and function of the universe and its constituents. We have already addressed Einstein's special and general theories of relativity, dealing with the large-scale nature of the universe, and so we will begin our next chapter with a closer look at the microscopic world as understood by quantum mechanics.

5

The Quantum World, Superstrings, Chaos, and Evolution

QUANTUM MECHANICS

CLASSICAL MECHANICS, WHICH IS essentially Newtonian mechanics, did a good job of describing how things behave on typical human scales. Einstein's special relativity theory showed that, when dealing with objects moving at great speed, Newton's laws did not apply and so could not be understood to be universal laws of nature. General relativity showed, likewise, that Newton's theory of gravity did not apply when dealing with massive bodies in space. It was left to *quantum mechanics* to show that things on the atomic and subatomic scales also failed to behave in the ways described by Newtonian mechanics.

The world that quantum mechanics reveals, as with special and general relativity, does not always make sense to us given our "common sense" understandings of reality and so requires a significant rethinking of some our fundamental beliefs about the natural world. Yet exactly what the correct interpretation of quantum mechanics is with regard to its meaning for the nature of reality is a matter of contention. Most physicists agree on the mathematics of quantum mechanics and the accuracy of its predictions, but there are several different interpretations of what quantum mechanics tells us about the nature of physical reality. Here the metaphysical philosophies of the researchers come into play.

The dominant interpretation of quantum mechanics has been the *Copenhagen interpretation*,[1] and that is the one I will use in the description that follows. Once I have laid out some of the important findings of quantum mechanics from that perspective, I will return to this issue of interpretation and discuss the other two major perspectives on quantum mechanics and the way in which each paints a very different picture of reality from that of the Copenhagen understanding. But first we must lay out some of the essential discoveries of quantum mechanics that show how the behavior of atomic and subatomic particles deviates from our commonsense (Newtonian) understanding of how they ought to behave.

One of the first discoveries that led to quantum mechanics was Max Planck's (1858–1947) discovery that radiation, or energy, does not increase or decrease in a continuous way in subatomic particles. It increases or decreases only in discrete, step-like jumps, which he termed *quanta*. For example, an electron as it travels around its nucleus orbits only in certain shells with particular energies, with each higher (further from the nucleus) orbital shell requiring the electron to have a greater, specifically defined, amount of energy. Thus, the electron's *energy* is said to be *quantized*; if the electron gains energy, it stays in its current orbital shell until enough energy is added that it attains the next quantized energy level, in which case the electron jumps to the next higher orbital shell.

Einstein applied this quantization of energy to propose that light also came in quantized packets and that it therefore behaved like a particle, which came to be called the "photon," as well as like a wave, as was the prevailing understanding at the time. This understanding solved the "photoelectric-effect problem," which concerned the question of why electrons were not increasingly emitted from a metal foil when the intensity of the light increased. Einstein said that it was only when the energy of the photons reached the certain required level that they could dislodge electrons from their energy shells. After Einstein's proposal was further tested, most physicists concluded that light behaves *both* like a particle *and* like a wave, though it cannot behave like both at the same time.

This is the concept of *wave-particle duality*, which is applicable to both energy and matter, and it stands at the heart of the quantum-mechanical *principle of complementarity*—that things can have multiple properties but that all of the properties cannot be viewed or measured at the same time. Thus, another understanding of classical mechanics, that something has to be *either* a particle *or* a wave, was shown to be untrue at atomic and

1. It is named "Copenhagen interpretation" because one of the earliest quantum theorists, Niels Bohr, created an institute to study quantum mechanics in the city of Copenhagen, and it was from Bohr and his group that this interpretation arose.

subatomic scales.[2] But how to conceive of, say, light as *both* wave-like radiation *and* a photon particle at the same time remains unclear. We simply have to adjust our normal way of thinking of things to match the experimentally verified results of the nature and behavior of things that exist at the quantum level.

The principle of complementarity is expressed with regard to measurements of a particle's location and momentum (its velocity multiplied by its mass) in the famous *Heisenberg Uncertainty Principle*, which says that the location and momentum of a thing can never be *simultaneously* measured accurately. The more precisely one measures the location of, say, an electron orbiting a nucleus, the less one can know about its momentum, and the more one knows about its momentum, the less one can know about its location. The reason is that to measure a quantum particle, e.g., an electron, to determine its position accurately, you need to use some kind of measuring tool, such as a photon. To measure the electron's position, you need to bounce a photon off it, but in so doing you change the location and momentum of the electron due to the energy transferred from the photon when they collide. To measure the electron's position accurately, you must use a photon of the shortest wavelength; but the shorter the wavelength, as we have seen, the greater its energy, meaning that it will affect the electron's momentum a great deal, making the measurement of momentum more uncertain. Thus, the more we know about the electron's location, the less we know about its momentum. On the other hand, should we use less energetic photons, their longer wavelengths mean that they will affect the electron's position less, and so we can determine its location the more precisely. But the longer the wavelengths, the less accurately we can specify the electron's position, since it could be anywhere within that wavelength. Thus, the more precisely we know the electron's momentum, the less we can know about its location.

For Werner Heisenberg (1901–1976), this uncertainty was not fundamentally a problem of our measuring abilities, such that with better instruments we might be able to determine precisely both the position and the momentum of a particle at the same time; rather, the uncertainty was a condition inherent in nature itself. The particle simply cannot not be described as *having* a precise location and momentum simultaneously, a claim that has been verified repeatedly via experimentation. Thus, nature itself is indeterminate, and so he preferred to call his principle the "Uncertainty Principle."[3]

2. In actuality, this wave-particle duality is also true for things at the human scale and larger, but their effects are extremely difficult to observe at these scales.

3. The Uncertainty Principle also holds true for other pairs of measurements, such as time and energy: the more precisely one knows the time of an event, such as an

If nature is indeterminate, then the traditional scientific belief that everything is determined, and therefore could be predicted if one knew all the initial conditions and influences over time, is incorrect. The behavior of atomic and subatomic particles/waves cannot be known precisely because there is indeterminacy, or freedom, in their behavior; and therefore we can only speak of the *probabilities* of any given thing, such as the location of an electron orbiting a nucleus. Just as the social sciences describe human behavior probabilistically, using a bell curve to describe all the *possible* behaviors a human can exhibit (the horizontal width of the bell) and the *probabilities* of a person acting in each possible way (given by the vertical height of the bell at the point corresponding to any particular behavior along the width of the bell), so too the behavior of quantum particles in a given situation is given by a bell curve, albeit one with an extremely narrow base (range of possibilities).

It was this implication of quantum mechanics that Einstein could not accept, given his deterministic worldview, leading to his famous saying, "God does not play dice with the universe." Einstein (along with others of a like mind) proposed an alternative to this Copenhagen interpretation of the meaning of quantum mechanics, the *Hidden Variables interpretation*. Einstein believed that there must be something else going on that quantum mechanics has so far missed, and that when that unknown something was discovered, it would show that the behavior of quantum particles, along with everything else, was determined. For Einstein there was no such thing as freedom in nature—including, evidently, ourselves as well. Einstein never discovered what these hidden variables might be, nor has anyone else to date. The Hidden Variables interpretation of quantum mechanics, while it continues to be held by a few physicists, is a distinctly minority view.

If nature is inherently indeterminate and therefore our knowledge of the behavior of things, at least on the quantum scale, is only probabilistic, this means that any quantum entity, such as an electron, does not have a precisely defined state, such as its path within its orbital shell around its nucleus, but rather travels all possible paths it could take simultaneously, and each path has an equal chance of being taken by the electron. It therefore *has* no defined path until it is measured; it is the measurement of the electron that causes it to have a definite path. This constitutes one of the fundamental general rules of quantum mechanics: anything that can happen on the subatomic scale must happen. This is yet another finding of quantum mechanics that goes against our commonsense (i.e., classical) understanding of reality.

---

electron jumping to a new orbital shell, the less one can say precisely about its energy.

Another odd feature of quantum mechanics is the notion of *entanglement*, which deals with *composite systems*—those with two or more particles, such as a pair of photons. Each of the two particles can have its own particular state, or they can have a superposition (something like overlapping waves) of their individual states, in which case neither has its own unique quantum state. Having interacted in this way—having become entangled—the two particles are now linked in such a way that what happens to one instantaneously affects the other. For example, if two entangled photons together form a *total spin zero system*, in which the photons have equal but opposite spins (+ ½ and -½), then changing the spin of one from +½ to -½ means the other *instantaneously* changes from -½ to +½ without anyone having done anything to it, and this is true even when the photons have been separated and are now a great distance apart. Einstein referred to this as "spooky action at a distance," because they are not close enough to affect each other physically and the change is instantaneous. Even a massless particle travelling at the speed of light, such as a photon, could not traverse the distance between them instantaneously and so cause the effect. How this can be true continues to baffle physicists, since it goes against all our normal understandings of causality, which are based on causation due to particles being in the same location. Nonetheless, this quantum entanglement has been confirmed experimentally.

Again, *all* these odd features of quantum mechanics have been tested, and there is evidence, though not unanimously accepted, that they are in accord with reality. The world of the very small, of atomic and subatomic scales, is very different from the world of our normal experience. On that all physicists agree. It is a place of frenzied activity, where particles constantly vibrate and spin and speed around, interacting with other particles, with each being affected by all the other particles it encounters. It is also a place where the variety of fields that permeate space, such as the gravitational, electromagnetic, and Higgs fields, also are not smooth and calm, as they appear on larger scales, but undulate feverishly on the quantum level.

Even in what we term "empty space," there is a feverish bustle of unceasing activity, which warps space at this ultramicroscopic level. The activity comes from the fact that, as we have seen, quantum particle pairs, one of matter and the other of antimatter, can pop briefly into existence, having "borrowed" energy from the energy present in the vacuum, before annihilating and thereby returning the borrowed energy to the vacuum. Again, we call these matter-antimatter pairs *virtual particles*.

This quantum understanding of feverishly roiling space is at odds with general relativity's picture of a smooth, quiescent space and is a reason for the fundamental incompatibility of quantum mechanics and general relativity. Either one, the other, or both are inaccurate understandings of nature. What is needed, therefore, is a *quantum theory of gravity* that would somehow reconcile the two, and physicists have been working for decades trying to create such a theory, mostly without success. The best candidate to date for such a theory appears to be superstring theory, which we will examine below.

But what quantum mechanics tells us about the actual nature of physical reality, as we have seen, is debated. As noted, I have been presenting aspects of quantum mechanics as understood by those who follow the Copenhagen interpretation, the first, most influential, and still most widely held understanding of quantum mechanics. This view understands nature at its most fundamental level to be indeterminate, not determined, as in classical physics. I also noted a second, deterministic interpretation of quantum mechanics, Einstein's Hidden Variables theory, and the fact that this theory has not been supported by any solid evidence and so has not persuaded many physicists to accept it.

More recently another deterministic understanding of quantum mechanics has arisen and appears to be gaining a number of adherents. This is typically called the *Many Worlds interpretation* and represents one of a number of multiverse theories that currently exist in physics. The multiple worlds interpretation claims that when any physical event takes place, such as a measurement of the location of an electron as it orbits its nucleus, the location *appears* probabilistic to us because we exist only in this one universe; but in reality *all possible outcomes occur*, though not in our universe (or, as some say, our branch of the universe). Reality splits, as it were, into a whole collection of parallel universes (or branches of this universe), and each particle continues on its own path, splitting again with each new event that occurs and thereby spawning another whole new set of alternate universes. In the history of our universe alone, if one considers all the seemingly probabilistic events that have occurred, the number of new universes that have arisen in which each of these probable outcomes are enacted must be virtually infinite, and still increasing exponentially, moment by moment.

As I have noted, the mathematics of quantum mechanics permits both interpretations, and it is also the case that no empirically testable experiments exist, and may never exist, to enable us to know whether there are such parallel universes. Since science can offer us no help in deciding which interpretation to accept as true, we must decide on the basis of other considerations, and it is here that a person's worldview will shape his or her interpretation of the meaning of quantum mechanics. If, like Einstein and others, you believe there is no such thing as freedom, that everything you and everything else in the universe do in each moment you *had to do*, then you will favor the deterministic Many Worlds interpretation. If, however, you believe that freedom exists in the universe, that you have real options for behavior, such as what you will think, say, and do, you will favor the Copenhagen interpretation.

It seems to me that our whole experience of life and our deepest intuition about ourselves tells us we are free, and that even if we deny it with our words, our every action affirms that we in fact recognize that we do decide, between a number of options, what we will do moment to moment in our lives. To deny this freedom of choice would mean that, for example, if in the midst of teaching a class I suddenly screamed, then did a summersault, raised the pinky finger on my left hand, put a pen in my right ear, took off my belt, turned around four times, stuck out my tongue, dropped to the floor and lay there face down, got to my knees and crowed like a chicken thirteen times, spit on the blackboard, took off my shirt and patted down my chest with an eraser, sang precisely five bars of "Yankee Doodle Dandy," and a thousand other such things in a row, all of these actions were things I *had* to do, that they were somehow programmed into me such that I had to do precisely these things in that order. To me this is totally incredible. Our deep intuition and commonsense understanding that we freely choose among options how we will behave, what we will do moment to moment, tells quite decisively against the deterministic understanding of reality and so also against the many worlds interpretation of quantum mechanics. In order to support their deterministic worldview, these physicists literally have to invent an infinite number of universes, which strikes me as a rather desperate strategy for saving that worldview, comparable to the desperate strategies adopted by fundamentalists in order to maintain their view of biblical inerrancy. For this reason, it seems to me more reasonable to accept the Copenhagen interpretation with its assertion that, at least on the quantum level, freedom exists and therefore the behavior of quantum entities—as with humans—must be understood probabilistically.

SUPERSTRING/M-THEORY

String theory,[4] or, as it has become in the past decade or so, *superstring/M-theory*, is a speculative, evolving theory about the most fundamental

4. It would be more accurate to speak of superstring *theories*, since there are a number of forms of the theory that differ from one another in significant ways, and it is not clear which, if any, is the correct version. Also, it is common shorthand to speak of superstring and superstring/M-theories simply as string theories, a practice I will occasionally use in what follows.

constituents of the universe that unites general relativity and quantum mechanics.[5] It also offers the possibility of being able to answer a host of other questions physicists have regarding the Standard Model, such as why there exist the precise number of fundamental particles and forces that there are, why their masses or strengths are what they are, and the like. And it does so in a fundamentally simple and mathematically elegant way. As early superstring theorist John Schwarz (1941–) said, "the mathematical structure of string theory was so beautiful and had so many miraculous properties that it had to be pointing to something deep."[6]

String theory's core claim is that the ultimate building blocks of the universe are not the various point-like particles, such as quarks, leptons, and bosons, of the Standard Model. They are instead extraordinarily small, one-dimensional strings, or loops, of energy that vibrate. The particular shapes and vibrational patterns of these tiny strings constitute what we perceive to be the particles and force carriers of the Standard Model; each unique type of vibrating string is a different elementary particle. The shapes and vibrational patterns of these different strings are also what give the elementary particles their properties, such as mass, spin, and/or force strength. Thus, all the disparate fundamental particles of nature and their properties are constituted by a single kind of thing: strings that vibrate in different ways. It is an elegantly simple and powerful theory in this respect.[7]

These strings are unimaginably tiny. As we have seen, atoms are about 10^{-10} meters in diameter and have only recently become observable by our most powerful microscopes. The nuclei of atoms are 1,000 to 10,000 times smaller, roughly 10^{-15} meters in diameter, and the quarks that make up the protons and neutrons of the nucleus must logically be smaller still, though given their confinement in nuclei because of the strong force and their quantum mechanical particle-wave duality, physicists are hesitant to speak about their having size. But strings—if they do indeed exist—are

5. *Quantum loop gravity* theory is currently the only main competitor with string theory that attempts to unite general relativity and quantum mechanics. We will not go into that theory here, but if you are interested in pursuing it see, e.g., Smolin, *Three Roads*, or Bojowald, *Once Before Time*. It is possible that these two theories will eventually prove to be the same theory but one that has been approached from the different starting points of the two theories.

6. Quoted in Greene, *Elegant Universe*, 137.

7. Interestingly, however, some recent findings in superstring/M-theory have indicated that even these miniscule strings might be composed of still smaller particles, termed "M-partons," or simply "partons," which are like tiny pearls that are strung together to make up the necklace-like string. In this case, we would be back to particles, not (energy) strings, as the most fundamental constituents of nature. However, this new wrinkle on superstring/M-theory—like the larger theory itself—remains controversial.

estimated to be roughly 10^{-35} meters in size, roughly the Planck length (the smallest length that could possibly exist or be measured, no matter how advanced the measuring instrument), or some 10^{20} times smaller than an atomic nucleus. Being the same size as the Planck length, these strings are the smallest possible things that can exist. One important conclusion, then, is that the way quantum mechanics treats quarks, electrons, and the like as zero-dimensional point particles cannot be correct. They only appear to be such because of their incredibly small size. Rather, each is a vibrating, one-dimensional string that has length. As we will see, this is the key to uniting quantum mechanics and general relativity into a single, coherent theory. Another implication is that space is not smoothly continuous, nor is time; rather, space—at least, space as we conceive it—cannot exist in smaller bits than the Planck/string length, and time cannot be partitioned any more finely than the Planck time, which is the amount of time necessary for light to traverse the Planck length. Space and time are therefore grainy, not smooth and infinitely divisible.

Given the unimaginably small size of these strings—and the fact that only recently have we been able to actually observe atoms, though not their nuclei, quark constituents, or electrons—it should not be surprising that the only possible way currently to study them is through mathematics. The mathematics behind superstring theory indicates that these strings come in two basic forms: open-ended (or "open") strings and closed-ended (or "closed") strings. Open strings are like little, one-dimensional, spaghetti-like filaments. Closed strings are tiny, circular, doughnut-like filaments—like an open string that has had its two ends joined together.

Both types of strings move in two different ways. First, both kinds of strings vibrate. The vibrations are essentially of the same type for both, forming wave-like movements with peaks and troughs, though their different geometries mean that their vibrations are not precisely the same. The tension of a string also has much to say about its vibrational behavior, and tensions vary, much as in musical strings that we are familiar with. Specifically, the stiffness of the string is proportional to its length squared, which means the shorter the string, the stiffer it is. Thus, longer strings are easier to bend, while shorter strings require much more energy to vibrate. Most strings are understood to be of the shorter kind.

Also, the more complex the vibrational pattern is, the greater the amount of energy the string has. Only the simplest strings have relatively small energies. Strings that vibrate faster and more intensely have greater energy and so, per $E=mc^2$, greater masses, while massless particles, like the photon, vibrate as slowly and mildly as is possible for a string. Put enough energy into the string and, beyond vibrating more frantically, it begins to grow longer, and there is no limit to how long it might become. Many of the particles predicted by string theory are far more massive than any we have thus far encountered in nature and may indicate the existence of the superpartners of supersymmetry.[8]

Open strings, with their open ends, vibrate along their length, including at their ends, whereas closed strings vibrate only within the limits imposed by their circular forms. Because the string is closed, its waves can only oscillate in whole integer patterns. Between the waves of both types of strings there are fixed points, called "nodes," separating one half-wave (e.g., peak) from the next (trough). For each full wave there are two nodes, so that a string with a two-wave pattern has four nodes, one with a four-wave pattern has eight nodes, and so on. All these vibrational patterns are similar to those we see in our macro world in any oscillating string, such as a guitar string.[9]

One way of conceptualizing these vibrating open strings is to imagine you have a pencil made of rubber that is not so stiff that the ends stand straight out when you grasp it in the center nor so flexible that the ends fall straight downward, but rather just stiff enough to allow the ends to droop gradually down to about a 45-degree angle. Holding it in the middle and moving your hand up and down, your fingers form a single node that stays stationary relative to the pencil as a whole while both ends of the pencil move upward and downward in unison. This is one possible movement of a string. But you can also grasp the pencil with the fingers and thumbs of both hands, dividing the pencil into three segments, in which case you now have two nodes. When you move your hands up and down in unison, the ends continue to move in unison, but the wave motion of the pencil between the two nodes is opposite to that of its ends. This constitutes a second type of open string. For every node you add, another type of vibrational pattern

8. Interestingly, there is a relationship here between the masses of the vibrating strings and musical tones. Just as in music, in which each fundamental tone (frequency) has harmonics that are integer multiples of that tone, so in string theory the masses of the vibrating strings are integer multiples of the fundamental string mass, which is called the Planck mass. Thus, the massless particles would be strings with masses of zero times the Planck mass, or zero mass, while others would be one, two, three, etc. times the Planck mass. Once again we see a relationship between music and physics, including the fact that beauty is involved with each of them.

9. Because these strings vibrate they can be thought of as analogous to guitar strings, with each different vibrational pattern creating different notes. If these strings do exist, and could we bring these incredibly high frequency vibrations up to our macroscopic level of existence and somehow be enabled to hear them, we would be hearing the "music" of the strings, the music generated naturally by the ultimate constituents of physical reality.

emerges, and so another type of open string. These different types of vibrational patterns constitute the family of possible open strings, with each pattern forming, from our vantage, a distinct particle type. Open strings constitute a more limited number of these fundamental particles, which are photons and other objects closely related to them. They can hold electrical charge if they are photon strings, or color (strong nuclear) charge if they are gluon strings, or the weak nuclear charge if they are W or Z boson strings. With such force-carrying bosonic strings, the longer the string is, the stronger the force.

The vibrational patterns of *closed strings* follow the same general form as those of open strings except that now there are no open ends to vibrate. To image this you must think of a large circle (such as a hula-hoop) made of the same pencil-diameter rubber as in the open string analogy. If you and a friend stand opposite one another with hands together and both push the circle inward at the same time, you will see that the segments of the circle to your right and left will move outward to the same degree your segments moves inward. When you simultaneously pull your sections outward to the same limit you had pushed them inward, the parts of the circle to your right and left will move inward to the same degree. If you continue this back and forth movement regularly you will see that stationary nodes form halfway between the peaks and troughs of the vibration. This motion defines one possible vibration of a closed string.

Each time you double the number of people around the circle, you double the number of nodes and create a different vibrational pattern, and thereby add another kind of vibrating string to the family of closed strings. Because of their far greater set of possible (and highly complex) vibrational patterns, closed strings constitute the majority of fundamental particles. Importantly, closed strings can combine to form a new closed string with different properties or divide into two daughter closed strings with their own unique properties, and it is this joining of closed strings into a new closed string and its subsequent splitting into two closed strings again that explains how particles interact and how the fundamental forces of nature work.

In superstring theory, it is also the case that another kind of motion is permitted in strings: the entire string can and does *spin*. That is, it not only vibrates but it also rotates either clockwise or counterclockwise as it vibrates. Like the spin associated with particles in quantum mechanics, superstring theory states that each kind of string can spin on its axis, with open strings spinning like a long, skinny top and closed strings spinning like a ring standing on one edge. It was this innovation of "spinning strings" that, among other things, led to the rise of superstring theory from the earlier string theory (termed "bosonic string theory").

One of the strangest implications of string theory is the possible existence of more than our perceived three dimensions of space and one of time, the four dimensions of space-time. String theory does not demand that these extra dimensions exist, and there are some string theorists that have approached the theory from this perspective and come up with a consistent mathematical description of superstring theory in the four dimensions that we all experience directly. But the mathematics involved in describing the theory are more complex and are perceived by many physicists as less elegant than those of the other forms of the theory with added dimensions, and so most string theorists think that these added dimensions do, in fact, exist.

In an earlier form, string theory postulated an additional six dimensions of space, which, when added to our normal three dimensions of space and one of time, yielded a total of ten dimensions. They seemed to be needed to describe the motions of the strings. Mathematically, adding these extra dimensions is simple and there is no conceptual difficulty. If you consider what it would take to locate an object is space, you would have to know its position in the three dimensions of space plus the one of time that tells you when it would be there (since all things are in motion relative to one another). Thus, you need 3 + 1 pieces of information to locate the object. For each extra spatial dimension, you would need an additional piece of information to locate it, so that for the ten dimensions of string theory you would need nine pieces of spatial information and one of time to locate the object.]

This seems simple enough in theory, but envisioning what these other spatial dimensions are is another story. Not even mathematicians and physicists can imagine what these extra dimensions look like. Nonetheless, the way most superstring theorists think of these extra dimensions is as extremely tiny and warped structures pervading all of space, curled up upon themselves, as it were, into a variety of possible topologies (shapes) that are so ultramicroscopic in size as to be undetectable even with our most powerful instruments. The equations of superstring theory determine what these topologies can be, and they must be one among a large variety of what are called Calabi-Yau spaces (see the QR code below), though exactly which Calabi-Yau space is the form taken by these extra dimensions remains at present unknown.[10] If this is the way space-time really is, then space-time is much more textured than previously understood, and when we move through space we are really moving through nine spatial dimensions in every moment of our lives. Its relevance to string theory is that the tiny strings that constitute all the various particles in the universe always vibrate not just in our three spatial dimensions but in nine, and the complex shape of the extra Calabi-Yau spatial dimensions affect the vibrational patterns of the strings. Thus, the more spatial dimensions there are and the more complex their shapes, the more complex the vibrational patterns of the strings are and so the more properties of the fundamental particles, such as mass, spin, and charge, they can specify.

For an image of Calabi-Yau space, see http://en.wikipedia.org/wiki/File:Calabi-Yau-alternate.png.

It was quickly realized that adding the theory of supersymmetry to string theory would also resolve some problematic aspects of string theory, and this led to the formulation of *supersymmetric string theory*, or *superstring theory*, in which, as we saw above, for every fermion and boson there exists a *superpartner* with a spin one-half less than itself. Not only did the combination of these two theories resolve some problems in the earlier string theory, it also made the theory more symmetrical and so more aesthetically pleasing.

The nature of the extra dimensions of superstring theory is understood according to the theory of *superspace*. Superspace is a postulated kind of space which includes our three spatial dimensions plus a number of other, *"anticommuting" dimensions*—that is, dimensions in which, when two objects are multiplied in one order one number results, but when multiplied in reverse order the negative of the first number results. Further, any number squared in these dimensions equals zero. Thus, when an ordinary matter particle travels into one of the *superdimensions* (extra dimensions), it becomes its own superpartner. In fact, all the equations describing our universe can be described in the mathematics of superspace and so yield the superpartners.

Part of the impetus for this new superstring theory came from another supersymmetric theory called *supergravity*, in which general relativity is combined with quantum field theory to yield a theory of gravity in which,

10. Part of the difficulty in this regard, as in string theory generally, is that so far string theorists have been working with equations that are only approximations of the exact equations of string theory, and these exact equations remain unknown. This is the great challenge facing string theorists and will likely continue to be well into the future.

besides the graviton, its superpartner, the *gravitino*, also exists. In the theory of supergravity, the mathematics indicate that there must be not ten but eleven dimensions (one additional spatial dimension, for a total of ten spatial dimensions and one time dimension), and this is the dominant view of space-time in superstring theory today.

The mathematics of superstring theory led to the postulation by 1985 of five types of superstring theories, each associated with a different type of string: the Type-I theory, the Type IIA and Type IIB theories, the Heterotic type O (32) theory (or simply Heterotic-O), and the Heterotic type E_8 X E_8 theory (or simply Heterotic-E)—the differences between them involving fine points that we do not need to go into here. In each of these theories, ten dimensions were required. These extra dimensions consist of an ultramicroscopic curled-up, six-dimensional space present throughout all of the three-dimensional space that we experience, and, again, it is the specific geometry of this space that specifies how strings can vibrate and so determines properties such as charge, spin, and mass that the fundamental particles have.

The question then became whether the different superstring theories with their different strings could somehow be combined with supergravity to yield a single, overarching theory, and the result of this attempt was superstring/M-theory. This theory indicated that the five earlier theories were all different ways of analyzing mathematically—that is, different ways of describing—the same one theory. Sometimes the mathematics of one of the theories regarding some question is simpler than that of the others, and therefore what is an extremely complex mathematical answer in some of the five theories can be understood more readily by using another's simpler math. This meant that by combining the insights from each of the five theories, more could be understood about the single superstring/M-theory than one could get from any single theory. Importantly, it was demonstrated that the heterotic-E string could have an eleven-dimensional description and so actually be a two-dimensional membrane-like structure. Thus, space was not ten-dimensional with only one-dimensional strings but eleven-dimensional with strings but also two- (or more-) dimensional membranes, with three- or more-dimensional membranes being called "blobs." This was a major insight that the more approximate equations of the five prior theories failed to reveal but that the superstring/M-theory mathematics clearly showed.

These *membranes*, or *branes*, then, can actually have as many as ten dimensions. These branes are usually called *P-branes*, where P denotes the number of spatial dimensions the membrane occupies, from zero to nine, and *DP branes* refers to the total number of possible branes.[11] All branes can carry charge and are much more massive than individual strings. In brane theory, open-ended strings are affixed to one or more P-branes that hold its endpoints in place. They can vibrate and move about on the P-brane(s) to which they are attached but can never move from that brane or those branes to another or other branes, and for this reason the branes are said to be "sticky" with respect to open-ended strings. The properties of a given P-brane are thought to be determined by the open strings attached to them. Closed-ended strings, however, having no ends by which to be attached to a brane, are not affixed to a particular brane but are free to move from one brane to another.

According to superstring/M-theory, it is possible that our entire universe might be a single P-3 (three-spatial-dimensional) brane, with the time dimension adding the fourth dimension. This is termed the *braneworld scenario*. Nearly all of the fundamental particles of our universe are understood to be open strings with their ends attached to our three-dimensional brane. This is also true of three of the force carriers: the photon that mediates the electromagnetic force, the gluon that mediates the strong force, and the W and Z bosons that mediate the weak force. However, the graviton, mediator of the fourth force, gravity, is understood to be a closed string and therefore able to move from our brane to another. An important implication of the theory is that now the extra spatial dimensions of superstring theory need not be ultramicroscopic but could also be large, like our three dimensions.

If such other branes exist, then our universe/brane would be only one among a myriad of others existing in other dimensions. The lengths of strings in our brane are theorized to be related to the size of the other dimensions, in which case higher other dimensions would mean longer strings—perhaps large enough to be detected in the near future by the Large Hadron Collider or some future-generation particle accelerator. Also, if gravitons can move between different branes, then gravity in our universe might provide the contact point between our P-branes. This might even provide the reason why the force of gravity is so weak relative to the other three forces in our universe: the gravitons that constitute it might be leaking out into other-dimensional branes, taking some of the strength of that force into those other branes.

These various branes are thought to be able to interact with one another, and one cosmological theory that has come from superstring/M-theory

11. A zero-dimensional brane would be a point, a one-dimensional brane would be a line, a two dimensional brane would be a plane, and so on. A nine-dimensional brane would add five spatial dimensions to the four space-time dimensions with which we are familiar.

is that our universe is reconstituted repeatedly over the eons by its collision with another nearby three-dimensional brane. But the number of types of P-branes increases exponentially as the number of dimensions increases, making the mathematical description of them and their interactions extraordinarily difficult. This has led a number of physicists to wonder whether superstring/M-theory constitutes a coherent mathematical theory. Superstring/M-theory has also brought a new problem, in that the mathematics seem to indicate the existence of particle-like entities that are strung together like pearls on a necklace to constitute the strings.[12] Might these be zero-branes? The question whether these partons, rather than strings and branes, are the truly fundamental particles in nature continues to be investigated.

One of the key implications of superstring/M-theory concerns the importance of the shape, or *internal space*, of the possibly curled-up extra dimensions. These apparently can have an enormous variety of these Calabi-Yau shapes, and the precise shapes of these miniscule extra dimensions in our universe are believed to have determined the fundamental laws of nature as we know them. If they were to be curled up in a different way, then the laws of nature in our universe would be different. This means that different shapes of the extra dimensions would result in different universes with different laws of nature. Were these different shapes to be actualized, the result would be an enormous number of other universes—perhaps as many as 10^{500} of them—and our universe would be only one among a vast multiverse. Whether what superstring/M-theory allows has actually occurred cannot be known, and so whether this multiverse actually exists remains a matter of conjecture.

The central and most hotly debated question today regarding superstring/M-theory is whether it is simply a very interesting mathematical system or in fact describes the way the universe is actually constituted. In other words, is it science? The physics community generally as well as the mathematicians and physicists currently researching superstring/M-theory are essentially agreed that for the present, at least, superstring theory is not strictly science. The reason is that there have been no empirical findings yet that would either support or falsify the theory. There are several lines of experimentation that are already providing, or could in the next few years or decades help provide, at least indirect support or refutation of superstring theory. For example, superstring mathematics has recently proved helpful in explaining the apparently liquid behavior of both extraordinarily hot soups of quarks and gluons of the kind that existed in the first small

12. See the discussion of partons on p. 100 n. 7.

fraction of a second after the Big Bang and ultra-cold gases a tiny fraction of a degree above absolute zero. It has also made predictions about certain features of elementary particles that match what we know about them from experimentation. So there are reasons to think superstring/M-theory might describe the universe in which we live. But more evidence is needed.

The Large Hadron Collider experiments taking place in Geneva, Switzerland, might find evidence of the superpartners for ordinary matter predicted in superstring theory, although predictions regarding their mass are so various and indefinite that it would be difficult to know exactly what evidence would falsify them. Still, they might provide some empirical evidence in support of supersymmetry and so of superstring/M-theory. Also, if some strings are relatively large, as noted above, they too might be detected with the Large Hadron Collider or some future-generation, higher-powered particle accelerator. Closer examination of the fine structure of the cosmic microwave background radiation that will soon be possible through a new generation of more sensitive telescopes might also provide evidence for or against the theory. However, as already noted, given the unimaginably small scale at which these strings must exist, we cannot yet imagine the kind of technological advances that would be necessary to detect *directly* the strings, if that is even possible. Given these complexities, it is not surprising that most superstring theorists think it could well be centuries before we can answer the question of whether superstring/M-theory is simply an interesting mathematical system or whether it is in fact science and describes at a deeper level than ever before the fundamental nature of our universe.

Why, then, are some physicists and mathematicians spending their entire careers working on this theory? There are several reasons. First, it is in our human nature to want to *understand*, whether the question is how our minds work, what it is that makes us human, or what the fundamental nature of the universe is. This is what science exists to do—help us understand ourselves and the universe in which we live.

Second, even though superstring/M-theory is now essentially a mathematical theory, it has often been the case, as we have noted, that mathematics has led the way to scientific discoveries that were unexpected and have revolutionized our understanding of the world. Recall how Paul Dirac's mathematical descriptions of the electron indicated that there must also be an anti-electron—and therefore antimatter—which was subsequently confirmed experimentally. In fact, it has often taken a long time before the technology has advanced sufficiently to test the scientific predictions of the mathematicians, for example, finding evidence of the existence of all six quarks once they were proposed theoretically, which took decades). Indeed, in 1916 Einstein's general relativity theory predicted the existence of gravity waves, and today, almost one hundred years later, we are just beginning to have the technological sophistication to test experimentally this mathematical prediction. So to claim that superstring theory is currently "only" mathematics does not necessarily count against it. It may simply be the latest in a long line of cases where mathematical theories have pointed the way to true science even though experimental verification has had to await future technological advances.

In this regard, it must also be noted that superstring/M-theory is mathematically elegant, and this fact prompts string theorists to believe it is true. Thus, in the words of physicists Sheldon Glashow (1932–) and Paul Ginsparg (1955–), "superstring theorists pursue an inner harmony, where elegance, uniqueness, and beauty define truth."[13] In the physics of the past century, as we have seen, the beauty of the equations has often been a guide to the truth of a theory. In superstring/M-theory, a host of symmetries long recognized by physics are consequences of the theory, with the additional symmetry of supersymmetry theory also being added. The elegance, including the power and simplicity of superstring/M-theory's understanding of the ultimate constituents of the universe, leads the physicists investigating it to believe it is true.

In addition, superstring/M-theory has managed to do what, thus far, no other theory has done so convincingly: solve problems that other theories have been unable to solve. In addition to the two problems already noted, superstring theory is the only theory that has been able to offer a mathematical description precisely in line with Jacob Bekenstein (1947–) and Hawking's prediction of the radiation of heat by black holes. Also, its mathematics predicts the existence of the massless, spin-2 graviton and that all known fermions and bosons have superpartners. Still another is that the theory explains why there are precisely three families of fermions and, potentially, why the properties of the particles of the Standard Model are what they are.

But probably most important of all, superstring theory has solved Einstein's problem of the *unified field theory*. It has managed to unify the force of gravity as described by Einstein's general relativity theory with the other three fundamental forces—the strong and weak nuclear forces and electromagnetism—as described by quantum mechanics. It does this by smoothing the quantum mechanical picture of the fabric of space-time, one filled with bustling point-particle undulations, due to the fact that strings have size. Because strings have size and are the smallest things than can exist, the point-particle view utilized in quantum mechanics cannot be correct. The

13. Cited in Greene, *Elegant Universe*, 212.

result is that space-time is smoother than suggested by quantum theory, and this smoother space-time on the quantum level now fits with general relativity's picture of space-time as essentially smooth; therefore, the two theories can be united.[14] This same argument applies to consideration of gravitational fields and the quantum understanding of the jitteriness of the gravitons that constitute it.

These things plus the others already noted give the hundreds of superstring theorists working today the sense that they are on the right track toward a unified theory that explains everything about the fundamental nature of the physical world and does so in an elegantly simple way. But there is still an enormous amount that string theorists do not understand about the mathematics of the theory and its meaning, and it is for this reason that it will likely take a long time before these elements of the theory are worked out.

Superstring/M-theory, then, is clearly a work in progress, with its extraordinarily difficult mathematical structure yet to be fully realized and the problem that experimental data adequate to test the theory is seemingly decades, if not centuries, away. It may turn out to be the long-sought "theory of everything" or it may end up being another dead end, as have other theories in physics before it. Only time will tell.

CHAOS AND COMPLEXITY THEORY

Some have referred to the development of *chaos theory* as the third great revolution in twentieth-century physics, with general relativity and quantum mechanics being the other two. Unlike the others, however, chaos theory does not deal primarily with the extremely small-scale or large-scale features of the universe: it deals largely with nature on the human scale. It is also *non-reductionistic*, in that it views chaotic systems in nature not by tearing them down into their most fundamental constituents, such as quarks and electrons, and looking for the answers there, but rather by viewing the parts of these systems in their interactions with other parts, that is, *holistically*. The claim of chaos theory is that things interact with other things in chaotic systems in unexpected ways and produce something more than the sum of the parts, something seen only from the perspective of the whole. This reemphasizes the essential nature of relationality in the universe

14. Speaking more technically, when one tries to combine the mathematical equations of general relativity with those of quantum mechanics, the result is almost always infinities, which is impossible in the physical world. Therefore the two theories cannot both be correct in their present formulations.

that we saw, for example, in our discussions of relativity theory and quantum mechanics. What chaos shows us is that physics as it has traditionally been practiced has looked at nature almost exclusively from a reductionist perspective and has also failed to describe a large portion of nature in its messier aspects. Chaos theory, though beginning in the world of pure mathematics, describes much of how nature behaves.

The term "chaos," however, is not the most helpful descriptor for this theory, in that the word as normally used implies the absence of order, or randomness, and that is *not* what this theory is concerned with. Rather, chaos theory describes systems in math and in nature, the behavior of which appears erratic and random but actually has a deep inner order. These systems are *dynamical* (that is, they evolve over time according to a certain mathematical rule) and are seemingly *deterministic* (there appearing to be no randomness in their behavior), yet their behavior is so complex that prediction of their long-range states is impossible. This is new, because physics has traditionally focused on the regular and predictable side of nature as defined by deterministic equations that constitute the laws of nature. From the beginnings of modern science, this understanding of nature as orderly and predictable was assumed, leading to Newton's machine-like universe. But chaos theory, applied to the real world, focuses on its more irregular and unpredictable side, such as heartbeat irregularities, earthquake patterns, stock market fluctuations, internet traffic, and a host of other such complex events. Thus, chaos theory studies both mathematical and natural systems that behave in seemingly random ways yet are evidently not random and finds the hidden order inherent in them, a kind of order that science had never before known existed.

The first intimations of chaos theory arose with late-nineteenth-century theoretical physicist and philosopher of science Henri Poincare (1854–1912) and his attempts to solve the intractable "three-body problem," which had bedeviled astronomers and physicists for centuries. The problem concerned how to predict the motions of two bodies revolving around the Sun given the differing gravitational attractions of each on the others. He discovered that if one began with two extremely similar initial states in this system, the subsequent behavioral paths could vary wildly over time—or, more generally, that an apparently deterministic system could be unpredictable regarding its future behavior given tiny disparities in its initial conditions.

The implications of this were unappreciated except by a small cadre of mathematicians and therefore went essentially unnoticed for another seventy years. Then meteorologist Edward Lorenz (1917–2008) discovered in 1961, through his work on the simple artificial weather models he studied with the aid of the recently invented computer, that very small changes in the initial conditions specified in weather systems could lead to vastly different outcomes over time. Eventually this came to be known as *the butterfly effect*—that a butterfly flapping its wings in Brazil could eventually cause a tornado in Texas. This was essentially Poincare's discovery restated: that, even in relatively simple dynamical systems, subsequent behavior is highly sensitive to initial conditions. Small differences in initial conditions accumulate exponentially, thereby making prediction of that behavior impossible in the long run. This is why, Lorenz realized, weather will never be predictable beyond the short term.

The key to this behavior is that chaotic systems are also *non-periodic*, that is, they never settle down to a periodic pattern but change endlessly through time, never repeating exactly the same behavior. Its opposite, *periodic behavior*, can be visualized by thinking of a swinging pendulum. If there were no friction at the pivot point, the pendulum would swing back and forth in exactly the same way forever. Add friction, however, and the pendulum will gradually slow down, tracing ever-smaller swings centered on the bottommost point of its arc, until finally it comes to rest there. That point is its *attractor*, in that it always seems to draw the pendulum's motion toward itself.

Non-periodic behavior, however, never does settle into a repeating pattern, and its cumulative, non-repeating behavior is pictured by what is called its *strange attractor*. This type of attractor is called "strange" in that the shape of the attractor is *fractal*. *Fractals* are geometrical forms that have the character of being *self-similar* on all scales—that is, parts of the fractal are repeated in similar (though not identical) forms at ever smaller scales. The more you magnify any point on a fractal, the richer the structure appears, and the structure of the whole repeats as the magnification increases, though with slight variations from one scale to the next, and so on down to the infinitely small. This latter characteristic is termed *scale symmetry*, indicating that, no matter what the scale is as you look at the fractal, from the large to the infinitesimal, you see the same general patterns. This holds true absolutely in mathematical fractals, while in nature this scale symmetry eventually ends at some point as the scale decreases. Our blood vessels, nervous system, and lungs, for example, all express this same branching, self-repeating fractal structure, and each ends with the smallest physical entities of the system—in the case of blood vessels, the capillaries, or in the case of Romanesque broccoli, the smallest parts of its flowers. Fractals also have an incredibly rich structure, as is evident from the Romanesque broccoli example below.

For a photo of Romanesque broccoli by Jon Sullivan, see http:// pdphoto.org/PictureDetail.php?mat=&pg=8214.

The reason fractals have this richness and self-similarity—and, indeed, that chaotic systems exist—is that they are the product of *non-linear equations*. In mathematics, *linear equations* are those in which the output of the equation is proportional to the input, such as x = 2y. Thus, when x = 1, y = ½, when x = 2, y = 1, when x = 4, y = 2, and so on. If you were to mark each of these points on a two-dimensional graph, where the horizontal axis is x and the vertical axis is y, you would see that the sum of these points yields a straight line on the graph. Chaotic systems, however, are described by nonlinear equations, in which the output of the equation is not proportional to the input, as in Newton's inverse square equation for the strength of gravity, $F = m_1 m_2 G / r^2$, where the force of gravity between two bodies, F, is equal to the mass of one body, m_1, times the mass of the other body, m_2, times the gravitational constant G, all divided by the distance between the centers of the two bodies squared.

A strange attractor, then, is the product of *all* the solutions to a given nonlinear equation mapped in *state space* (also termed *phase space*). State space is a multidimensional graphing system—for example, a three-dimensional graph—in which each point represents the total state of the system at a given time. As time moves forward, the different state of the system in each moment is represented by another dot, so that eventually the collection of dots on the graph represents all possible states the system can have. This total picture of the system is its strange attractor—that state toward which the chaotic system tends regardless of the state of the system when the mapping begins.

The significance of the strange attractor is that it shows graphically the fact that the system behaves in a relatively orderly way, even though that way is non-repeating. Lorenz's strange attractor (see the QR code below), is flattened into two dimensions and so fails to convey the fact that the pattern he found actually exists in three dimensions. Thus, on the two-dimensional image, lines appear to intersect when in reality they never do. The line describing the system over time moves in all three dimensions, creating a series of layers, like stacked sheets of paper, which, like the points and lines that constitute them, never touch. Thus a chaotic system's behavior never repeats itself. The strange attractor, then, is the general spatial shape the system assumes over an infinite amount of time—in Lorenz's weather-modeling case, forming a pattern like a butterfly wing.

For an image of "Lorenz's strange attractor," see http://en.wikipedia.org/wiki/File:Lorenz_system_r28_s10_b2-6666.png.

Probably the most famous of all fractals is the Mandelbrot set (shown in its large-scale totality and in two successive magnifications in the three QR codes below).[15] The fractal image indicates that the really interesting behavior takes place at the boundaries between the points that appear within the fractal image and the outside area where the points never appear. These boundaries represent the behavior of the chaotic state as it exists between ordered (the black spherical objects seen here) and random (the blue background) behavior. This kind of boundary is often seen in nature, such as in phase transitions of substances like water as it changes from a gaseous state to a liquid state to a solid state, or when a metal changes from being nonmagnetic to magnetic.

For an image of the Mandelbrot set, created by Wolfgang Beyer, showing the large pattern repeating on ever-smaller scales, see http://en.wikipedia.org/wiki/File:Mandel_zoom_00_mandelbrot_set.jpg.

For a partial view of the Mandelbrot set, created by Wolfgang Beyer, showing showing the "gap between the 'head' and the 'body,' also called the 'seahorse valley,'" see http://en.wikipedia.org/wiki/File:Mandel_zoom_01_head_and_shoulder.jpg.

For a partial view of the Mandelbrot set, created by Wolfgang Beyer, showing "double-spirals" (on the left) and "seahorses" (on the right), see http://en.wikipedia.org/wiki/File:Mandel_zoom_02_seehorse_valley.jpg.

15. For those of you who might be interested in the mathematics of the Mandelbrot set, the non-linear mathematical equation that gives rise to the Mandelbrot fractal is $Z_{n+1} = Z_n^2 + C$, where n is any number, beginning with 0, and Z and C are complex numbers of the form $a + ib$, where a and b are normal numbers and i = the square root of -1 (an *imaginary number,* in the sense that it violates the standard laws of mathematics, in which no possible number squared can be a negative number) and the result of the equation is plotted with a along the x-axis of a graph and b along the y-axis. The equation is repeated to infinity, giving a new number with each repetition, and so generating the image above.

At bottom, then, fractals show the hidden, complex order in chaotic (mathematically nonlinear) systems and also the transitional boundaries between states. That order, as seen in the Mandelbrot set or any other such fractals, is quite beautiful. Once again, we have truth in science and mathematics intimately tied to beauty.[16] Since these equations typically describe natural systems, and fractals are simply mathematics graphically presented, one might well suggest that these natural systems are inherently beautiful, too. Perhaps this is one of the elements undergirding the typical human aesthetic appreciation of nature.

Another way to image the boundaries between orderly and chaotic behavior is the *orbit* or *bifurcation diagram*, in which the system's behavior is viewed not in space, as with fractals, but rather through time. The vertical axis of the orbit diagram represents some variable in the system, such as an insect population in a wheat field, and the horizontal axis represents another system variable, such as the growth rate of the insect population. The diagram allows one to see how the first variable (the insect population) changes as the second parameter (the growth rate of the insect population) rises. At first the population of insects rises with the increased growth rate because this results in more insects, and since there is plenty to eat, the insect mortality rate is low. So far it is a very orderly system, behaving predictably in the short run—a higher growth rate results in more insects for a given food supply. There are no cycles in the insect population, only a steady increase over time. At some point, however, the amount of food is no longer adequate to supply the increasing growth rate and thus the number of insects in the field. Suddenly there is a bifurcation in the diagram, yielding two lines instead of just one. This indicates that a two-cycle system has now begun in which there is a year when the population increases followed by a year in which it decreases. The resulting diagram is said to have a *periodicity* (or *Feigenbaum number*) of one—referring to the spacing between bifurcations.

See http://en.wikipedia.org/wiki/File:Logistic_Bifurcation_map_High_Resolution.png/ for Jordan Pierce's diagram showing the movement of a system as it bifurcates and then bifurcates several times more before falling into chaos, then finds islands of order before moving into chaos again, and so on forever, repeating the cycle on ever smaller scales.

16. Fractals may also tell us something about why we consider things beautiful: they are neither too regular nor too irregular but sit somewhere between boring regularity and chaos.

Increasing the insect growth rate further soon results in another split, or bifurcation, in which each of the two lines splits again, resulting in a total of four lines. This indicates that the period has doubled again to yield a four-year cycle of insect population booms and busts. The diagram now is said to have a periodicity of two, since it now takes twice as long to complete one cycle. A yet greater increase in the growth rate soon causes another period-doubling, now yielding eight lines and, meaning an eight-year cycle of insect population increases and decreases, and so a periodicity of four. This period doubling could theoretically continue to infinity, but at some point the diagram reaches what is called the *accumulation point*, which is where the system's behavior suddenly turns chaotic (as seen in the dark regions of the orbit diagram [see the QR code above]). The graph changes dramatically, yet it indicates that even in the chaotic state there is still order, albeit of a far more complex sort. This more subtle order can be seen in the various lines that appear up and down (much clearer when the diagram is magnified) but also in straight and curved lines that run through parts of the diagram. This is order, but it is order of an incredibly complex type.

The three vertical alleys that appear between the larger dark structures are islands of relative order. They reveal that periodicity in the cycles of insect populations continue, but these no longer double; rather, they have a seemingly random periodicity of, from left to right, 6, 5, and 3. What is even more fascinating about the diagram is that when you magnify these lined alleys, you find that the general structure of the entire orbit diagram is repeated within each of them, only at a much smaller scale (this is most easily seen is what appears to be a rounded white structure at about $r = 3.82$ and $x = 0.5$). But more vertical alleys also appear as the growth rate continues to increase, each with its own periodicity, until it reaches growth rate 4. The final periodicity pattern for the entire diagram ends up being 1,2,4,6,5,3,6,5,6,4,6,5,6.

Several things are interesting about this sequence of periods. First, the entire pattern repeats continually as you magnify any of these regions farther and farther, all the way down to infinity. Here we see the fractal structure of the diagram, analogous to the fractals we have seen of such nonlinear systems plotted in space, only now seen with regard to the dynamical system's behavior over time. But even more startling is the *universality* of this sequence that is seen across a wide range of nonlinear systems: this exact same sequence is seen in *every* orbit diagram of a nonlinear system. Thus, it is called the *universal series* (or the *Feigenbaum sequence*). These nonlinear (chaotic) systems are found throughout nature, such as in irregularly beating hearts, oscillating chemical reactions, and even in humanly-created complex systems, such as the stock market.

The universality seen in the orbit diagram, however, is not confined solely to the periodicity pattern of the bifurcation diagram. If one measures the heights and widths of the wishbone-shaped lines as they approach the accumulation point, each successive height of a wishbone is about 2.5 times smaller than the one that preceded it, and its width is about 4.7 times smaller. Near the accumulation point they converge on 2.5029 and 4.6692 respectively—and again, this no matter what the nonlinear equation or chaotic system is. The overall shapes of the orbit diagrams describing these different systems also display quite similar shapes. Clearly, then, chaos theory is not just about interesting mathematics but in fact describes the way many dynamical systems in our world actually behave. In this regard, it can be said to have uncovered new laws in biology.

At the same time, however, chaos researchers thus far have investigated only relatively simple chaotic systems, systems that involve only a limited number of parameters. Current research is extending the number of variables in more complex chaotic systems (hence its name, *complexity theory*) and is finding surprising tendencies toward self-organization, such as synchronization, within them. One example here is the often-noted tendency of women living in close proximity to have their menstrual cycles synchronize. Others include the tendencies of fireflies, pacemaker cells in the heart, and even pendulum clocks to synchronize when some form of information is permitted to be exchanged between them. In the future we will need to extend that knowledge into yet more complex chaotic systems with far more variables that are constantly changing and interacting with each other from moment to moment, such as the incredibly complex way in which all the different, interrelating parts of the brain function to produce thought, or the millions of different things that interact in various ways to make up our national economy. For such complex systems, our current understanding is quite inadequate. Still, since most of the great, unresolved problems in the natural sciences deal with nonlinear, dynamic, and complex systems, it will be the further developments in chaos theory regarding increasingly complex systems that will increasingly move our understanding of nature forward.

THE THEORY OF EVOLUTION

When discussing evolution, one must make a distinction between the evolution of nonliving things and the evolution of living things. We understand today that the universe as a whole evolves, as well as all the things in it; thus everything that exists changes over time, which is the general meaning of evolution. Stars are born, live for a (typically long) time, changing all the while, and then die. Earth, like all planets, has also evolved over time, changing constantly as its gigantic plates slowly slide around the planet's surface, making and unmaking continents, oceans, mountains and every other geological and geographical feature. But the *theory of evolution* in its proper and specific sense deals with only one kind of evolution: *organic evolution*, or the evolution of *life on Earth* (or on any other planet on which we might someday find life), particularly as described historically by Charles Darwin in his theory of evolution by means of *natural selection* and the subsequent elaboration of that theory by others. It should also be noted that, while the theory of evolution speaks primarily about the development of life once it had begun, it is also relevant to how life originated in that it theorizes about how chemical evolution might have occurred that led to the first living organisms.

In the study of the earth sciences, such as geology and geochemistry, it is the theory of plate tectonics that forms the unifying concept of the entire field (see chapter 11). In the biological sciences, it is Darwinian evolution that constitutes the central unifying concept. While some people today continue to choose not to accept the theory of evolution for their own ideological reasons, many, if not most, biologists regard it as a natural law, akin to Einstein's theory of gravity, quantum mechanics, and the like. In fact, the evidence for the evolution of life over billions of years on earth is so overwhelming, based on findings in biology, geology, paleontology, anatomy, physiology, and a host of other fields, that, like gravity, scientists consider it to be a fact. Some points of evolutionary theory continue to be debated, just as they are in any active field of research, but *that* evolution has occurred is essentially a fact, i.e., something no rational and objective person could deny given the evidence. The evolution of species can even be witnessed firsthand in the mutations of disease-causing organisms in the laboratory, where new generations arise with great rapidity due to the incredibly small size and short life spans of the organisms. Thus, it is that evolution is the central concept in all of biology, the one that ties all the varied aspects of biological science into a single, coherent whole.

Darwin did not invent the idea of evolution—another common misconception. Rather, that idea, though appearing under different names (such as *transmutation*) had been championed by certain geologists and biologists well before Darwin's time. What Darwin contributed to the discussion was a theory of a primary force that drives evolution: *natural selection*. In his *On the Origin of Species by Means of Natural Selection*, published in 1859, Darwin made several commonsense observations about nature. First, all populations of a *species*[17] produce more offspring than can survive because of the limited supply of available resources, such as food and space, and, in many cases, predators that eat them. Second, within each species one notes differences between individuals, such as in size, coloration, speed, and the like. Third, the traits any individual has tend to resemble those of its parents. Fourth, some traits confer advantages in the competition for survival, such as greater speed or better eyesight in a hunting animal.[18] Fifth, the individuals that have the advantageous trait tend to have a greater chance of reproducing successfully and thereby passing on this new trait to their offspring. Possessing this advantageous trait, their offspring also tend to have more reproductive success and pass the trait on to their offspring, and so on.

In this way, those having the survival-enhancing trait gradually increase in numbers while those that do not tend to decrease. Eventually, those with the new trait come to dominate the population, and at some point the whole population of that species comes to have this advantageous trait. If a sufficient number of these new, survival-enhancing (or, at least, survival-neutral) traits accumulate in the species over relatively long periods of time, the later entities can come to be so different from their ancestors that a new species is said to have emerged.[19] This process of new species arising is called *speciation* and typically occurs when populations of the same species have become geographically separated from each other, after which each tends to go its own evolutionary way depending on the mutations that appear in each population.

17. *Species* is actually a rather fuzzy term in that what constitutes a species or separates one species from another is not clearly defined. It is generally said that what distinguishes two closely related species is their ability to have sex and produce fertile offspring. However, various types of animals currently defined as different species, such as lions and tigers or zebras and horses, can reproduce fertile offspring. Genotypical (genetic) and phenotypical (physical form) factors have also been used to define species, but in general what it is that constitutes a species remains poorly defined.

18. These four points together constitute the idea that evolution is about the "survival of the fittest." As the subsequent discussion will reveal, this is an overly simplistic summation of Darwin's theory of natural selection, even though he himself did use the phrase. More properly construed, the phrase should be something like "the better reproduction of the fitter," since it is not ultimate fitness but rather only superior fitness that enables the greater relative ability to reproduce that is involved in natural selection.

19. Helpful or harmful, of course, are relative terms. Helpful genetic mutations will generally be passed along to the offspring of those having the mutations and retained in the population. Other genetic mutations can be mildly harmful yet may persist for a number of generations. Other changes might be very harmful, in which case the individual having such a change will be more likely to die before reproducing. Other changes may have no effect on the survival of the creature and so are said to be *survival neutral*.

In formulating this theory of natural selection, Darwin was inspired by a practice employed by farmers and animal breeders called *artificial selection*. Whether seeking more productive crops or breeding animals for certain traits or tasks, people had for centuries been taking individuals within a species that had the desired attributes (or at least something closer to them) and breeding these individuals with other such individuals to produce offspring having, or more closely approximating, those traits. This process was continued over many generations until it resulted in animals specialized in the ways their breeders had desired. Today we know that all species of dogs, from Chihuahuas to St. Bernards and everything in between, are descended from wolves that lived only about 10,000 years ago, with each type of dog being the result of this artificial selection. What Darwin realized was that something similar might have happened *naturally* over the course of long periods of time in the earth's history, occurring accidentally rather than intentionally. Thus, it was the *artificial selection* practiced by humans that prompted him to term this natural process *natural selection*.

The major factor determining which traits are favored is the environment in which a species lives. Environments change over time, sometimes slowly and in minor ways, but at other times rapidly, such as in the case of severe drought, when a food source disappears, or when a river abruptly changes its course. Such changes mean that those individuals of a species that have traits adapted to the changed environment will tend to be the ones that survive. A classic example of this was the effect of changes in England's air quality on the survival of differently colored pepper moths over the past four hundred years. Pepper moths, some of whose wings are light in color and others whose wings are dark, spend much of their time clinging to the trunks and branches of trees. Prior to the Industrial Revolution in England most of the trees had bark that was light in color, and so it was the lighter colored moths that predominated. Darker moths alighting on light-colored trees were easily spotted by their prey and eaten, while the lighter moths were less discernible and so survived in greater numbers, reproducing and creating more light-colored moths like themselves.

But once the Industrial Revolution was well under way, factories were belching out black smoke which, over time, coated the bark of trees, turning it darker. Then it was the light-colored moths that were the easy prey, and so in time the darker moths became predominant. Eventually, in the twentieth century, smoke-mitigation technologies were employed at factories across England to clean up the air, and since then the bark of trees has begun to become lighter-colored again, with the result that now the lighter-colored pepper moth is once again the dominant form. As we will see later, environments around the earth have been in a process of change, both on land and in the seas, throughout the planet's history, and it has been the individuals and species that have been able to adapt to these environmental changes that have tended to survive to reproduce, and so over time have caused the genetic makeup of the population to change.

It should be noted, however, that differential reproductive success is not the only important factor in the process of evolution. *Sexual selection* is another and is vitally important for the passing on of genes to subsequent generations. Sexual selection has to do with the process of mating, which includes especially the attractiveness of males to females, which typically have the final say regarding with whom they will mate, or, in some species, the competition between males to possess females for mating. The former competition, termed *intersexual* (between the sexes) *selection*, explains why, for example, male birds tend to be more beautiful than females of the same species: evolution has caused them to develop plumage ever more attractive to females, and the more attractive the plumage, the more likely the male will be chosen by the female and so pass along his genetic makeup to the next generation. This also explains the complex courting behavior of certain bird species, seen in the sometimes quite elaborate "dances" males perform in the presence of females.

In the second form of competition, termed *intrasexual* (within the sexes) *selection*, the competition in some species between males to possess females explains certain features of their anatomy, such as body size and armaments (teeth and horns, for example), which often dictate which male will intimidate and so deter other males from fighting or, when a fight between males does ensue, determine which male will likely win. What is interesting is that not only random mutation but also *the choices of animals*—for example, which males are chosen by females for mating, or which males choose not to challenge the better-equipped males but instead attempt other means of inseminating females, such as by using their intelligence to approach them when the alpha males are not around—can affect the path of evolution.[20] It is also the case that these choices create *feedback loops*, in which preference for, say, more elaborately showy tails by females leads to mating with showier-tailed males, and this preference is passed along to female offspring, and the possession of showier tails is passed on to male offspring. This passing along of such traits can continue on into the distant future, generating ever-stronger female preference for longer tails and males with increasingly showier tails—as seen, for example,

20. It must be noted that not all species have hierarchical social structures, as in the mammalian example of fighting for procurement of females, nor that it is the females that always choose the mate, as in the bird example.

in the large and beautiful tails of male peacocks. Sexual selection, then, is another determinant of what genes will end up in a population over time and which will not.

Darwin knew that parents passed along traits to their children, but he did not know specifically what it was that caused this inheritance or how the process was accomplished. It was the pioneering work in genetics of Gregor Mendel (1822–1884) that eventually answered Darwin's questions. Mendel learned that it was by means of *genes* (clustered together on *chromosomes*) that the traits of parents are passed on to their offspring, and that the particular way in which the genes from each parent are mixed and combine to form the offspring's chromosomes is what constitutes the uniqueness of its children—not only *genotypically* (in terms of their genome, or total genetic makeup), but also *phenotypically* (the way their bodies are structured and appear). The term used for the total genetic makeup of an individual is its *genome*. In the 1960s, James D. Watson (1928–) and Francis Crick (1916–2004) discovered that it is *DNA* (short for *deoxyribonucleic acid*, a double-helix structured molecular chain in the nucleus of the cells of most life-forms) that carries the full genetic message of how a creature is made and how its body chemistry will work. It is this combination of Darwin's natural selection with Mendel's genetics that forms the core of the *neo-Darwinian synthesis* that is the heart of the theory of evolution today.

According to mainline neo-Darwinian theory, then, evolution is caused primarily by the interplay of two things: chance and necessity. It involves *chance*, or randomness,[21] in several ways. For one, it is by chance (as we normally use the word) that certain things happen, such as an otherwise fit individual dying of some accidental cause, which prevents it from reproducing and passing its genome on to the next generation, or an avalanche separating a population into two groups that can no longer interact with each other, so that each tends to go its own evolutionary way. Thus, chance accidents can play a significant part in the course of evolution.[22]

21. The claim for randomness in events is an interesting case. Mathematicians know that there is no way of proving something to be a random occurrence; randomness, in the way we use the term, reflects how something appears to us, not necessarily the way it really is.

22. It has been argued by some scientists that in fact there is no such thing as chance in evolution. They claim that what we consider chance—such as a cosmic ray striking a gene in the body, causing it to mutate—isn't really a chance occurrence at all. Rather, if we could go back in time we would see that the path that particle took was perfectly predictable if we had known its precise position and momentum at a given time and the interactions it would subsequently have with other particles and forces in the universe, these also being deterministic, as it sped toward Earth. Thus, given these factors, the determinist says, the particle *had to* strike the gene at the time and in just the way that it

In addition, the way the genome of an offspring of a male and a female is constituted is, as far as we can tell, a matter of chance. In sexual reproduction, each sperm contains the entire genome of the father within it, and each egg the mother's entire genome. When these cells combine, the genes from each parent are shuffled together, some being passed along to the offspring and others not, and out of this mixing, the new genome of the offspring is created. This process is called *recombination*.

It is also essentially by chance that something, such as a cosmic-ray strike or damage from radiation or imprecise genetic copying, causes a specific gene to mutate and so change the role or roles that gene plays in the body. Most such genetic changes will be harmful, since there are seemingly few survival enhancements that can be added to the genome due to evolution's having already tuned the creature's genome toward survival for a very long time. Evolution then weeds out such harmful changes over time in the population. Occasionally, however, a genetic change proves useful by somehow aiding an individual's survival and/or its reproductive success and therefore, over the generations, the long-term genetic makeup of the rest of the population of that species will change to include that helpful new form of the gene. Hence, it is at the population level that evolution ultimately acts.

But if chance is what causes genetic mutations in the individual, it is *necessity*, or law, that determines what happens to the individual possessing that mutation. Natural selection is that law, and it says that what happens to the individual with a genetic mutation depends on whether the mutation helps or harms the individual's chances of reproduction. If the mutation aids the individual's reproduction, that individual will likely produce more offspring than those who lack it. If it decreases the individual's relative reproductive success, it will probably not be found in future generations. If it neither aids nor hampers the individual's chances of survival, it may or may not continue to be passed along into subsequent generations. Natural selection, then, constitutes one of the fundamental laws of nature.

It is another central tenet of neo-Darwinian evolutionary theory that evolution has no goal. Evolution is based on random events, such as genetic mutations, and the effects of these in the individual are then acted upon by the law of natural selection. What evolves over time cannot therefore be predicted beforehand and so bears no indication of having been predetermined by the evolutionary process. Ultimately, what exist in the world are simply

---

did. However, this assumes that the movements of the person in whom the gene resides were determined too, a view only determinists would accept. If humans really are free, then the person carrying that gene need not have been at this particular place when the cosmic ray intersected the Earth, and therefore chance, based on that which is unpredictable (the normal way we use the term), would still hold as a factor in evolution.

those species that have survived due to their fortunate genetic inheritance relative to the environment in which they live.

While essentially every evolutionary biologist agrees that evolution has occurred in this neo-Darwinian way, there is significant debate regarding the adequacy of neo-Darwinian theory *alone* to explain *all* the causes of evolution. As we have seen, the choices of animals, such as in sexual selection, also seem to be a factor in which genes are passed on to future generations. In addition, the development of certain phenotypical traits have been shown to depend on the environmental conditions in which the being lives (this is called the *Baldwin effect*). For example, in many turtle and lizard species, including alligators, the sex of the individual depends on what the environment was when it was developing from a fertilized egg. Thus, in some species of alligators, the individuals develop either ovaries or testes depending solely on the temperature around them during their early development. If these conditions continue over time, the alterations become permanent. Thus, the relationship between creatures and their environments are mutual, each affecting the other. This Baldwin effect has also been observed in the determination of castes within honeybee populations and other social insect species, which then determines which bees and insects get to mate, and this also seems to be the case in a variety of other instances. Not all evolutionary biologists, however, seem to share the view that this constitutes a part of what shapes evolution.

Another non-Darwinian evolutionary mechanism appears to be a new form of *Lamarckianism*, the view held by the early nineteenth-century evolutionary theorist Jean-Baptiste de Lamarck (1744–1829) that traits acquired during the life of a creature could be passed along genetically to the next generation. Lamarckianism had long been rejected by mainstream Darwinian thought but has recently appeared in a somewhat modified form. According to these *neo-Lamarckians*, it is the stress of the environment that can cause changes to certain beings, which are then passed along to the offspring—some for the short term and, more controversially, others for the long term. For example, it is well known that water fleas grow large helmet-like structures when confronted by predatory midge larvae. Not only do these helmets cut the water fleas' mortality risk in half, but they are passed on to their offspring for several generations. Bacteria also evidently alter their genomes based on what their environment is like. This view also remains controversial among evolutionary biologists today and constitutes an ongoing area of research.

Still other evolutionary biologists argue that *symbiogenesis*—"the appearance in the *symbiotic*[23] partners of new behaviors, new metabolism, new tissues, new organs or organelles, and new gene products"[24] due to their intimate relationship—is another evolutionary force in addition to genetic mutation and natural selection. Symbiotic creatures may conjoin their bodies, genes, and cells in varying degrees. Such symbiogenesis has been documented in, for example, the lateral transfer of genes that occurs between various symbiotic partners, indicating that evolution has been caused in such cases by these processes and not random gene mutations.

Another example of symbiogenesis seems to have been the origin of the *eukaryotic cell* (a cell with complex internal structures, including a nucleus, which is enclosed by a membrane). It is now almost universally believed among biologists that this kind of cell was created as the result of one organism invading or having been engulfed by another organism. For example, mitochondria within cells are thought be the evolutionary descendants of free-living bacteria that invaded the cell billions of years ago. The benefits that resulted from this merger caused the arrangement to become permanent. According to these theorists, some 250 of the 30,000 genes in humans probably come from bacteria. The green chloroplasts in plants were also evidently free-living bacteria before entering into a symbiotic relation with plant cells.

Other such understandings of the evolutionary process in addition to, or sometimes even in place of, the neo-Darwinian theory have also been proposed (including quantum evolution, chaotic evolution, thermodynamic evolution, and others), but we need not go into them here. The point of noting these various amendments or alternatives to neo-Darwinian theory is that they demonstrate how evolutionary theory remains a work in progress, and that we can expect to see a fuller understanding of the number of facets of the evolutionary process in the future.

Another issue concerns the fundamental unit on which evolution acts. For example, while evolutionary scientists traditionally focused on the phenotype of the individual as the locus of evolution, most now agree that it is at the level of the gene that natural selection ultimately operates.

23. Symbiosis refers to two different species living together in an intimate relationship. This relationship can be one in which the relationship is mutually beneficial (mutualism)—the classic example of which is lichens, which consist of a permanent joining of fungi and algae that helps each to survive. However, the relationship can also be one in which one benefits and the other is unaffected (commensalism) or one in which one species benefits while the other is harmed (parasitism).

24. I use here the concise definition of symbiogenesis given by Lynn Margulis (1938–2011) and Dorian Sagan (1959–) in Cobb, ed., *Back to Darwin*, 178.

Richard Dawkins (1941–) has popularized this notion in *The Selfish Gene* (2006). It is the mutation of genes that causes changes in the appearance and functioning of the entity, and so it is ultimately the genes that determine the survival and reproductive success of the individual.

Yet another debate concerns whether at least some features of evolution are predictable. Stephen Jay Gould famously said that if we went back in time to any particular point in the earth's evolutionary history and started evolution over again from that point, the result, in terms of the kinds of organisms that appeared, would not be the same as exist today—including ourselves.[25] Other theorists, however, point out that nature seems to have a propensity toward certain basic forms that are seen in creatures of different phyla that have evolved from different forebears but under similar selective pressure. This phenomenon is called *convergence* or *convergent evolution*. For example, large, fast-swimming predators in the oceans, such as dolphins, killer whales, sharks, and tuna, appear largely alike in body shape and fins yet have evolved from very different creatures. The reason seems to be that this body pattern is the most effective type given the survival needs of these predatory animals.

A similar pattern is seen in marsupials in of Australia and mammals elsewhere in the world: evolution has caused certain animals that live with similar selective pressures to look and function in very much the same way. There are carnivorous marsupials with very similar appearances to other carnivorous mammals that live in the same kinds of environments, just as there are similar large herbivores of each type. Also, camera-like eyes have evolved in many different animals, including snails, jellyfish, squid, and vertebrates.

Given the existence of such examples of convergent evolution, Gould is at least somewhat wrong in his claim that we could not guess which kinds of organisms would exist if we rewound the evolutionary clock. We might not know the precise animals that would evolve, yet we could predict the general appearance of different animals in different environments. Simon Conway Morris (1951–), a champion of the importance of convergence in evolution, believes that once life starts, given the convergence seen repeatedly in nature, some sort of intelligent life, for example—though not necessarily one resembling human beings—is essentially inevitable, whether on this or another Earth-like planet. This seems to imply a general directionality to evolution and therefore a degree of predictability, although one in which the prediction of the *precise* nature of beings remains impossible due to historical contingencies.

25. Gould, *Evolutionary Theory.*

Another debated topic is the *theory of emergence*. Emergence theory rejects the claim of reductionists that everything that exists can finally be understood by reducing them to their constituent particles. That claim implies that if we knew physics sufficiently well, we could predict the nature and behavior of complex life-forms, including humans. Clearly this view reflects a deterministic understanding of reality, one in which all living beings are essentially only sophisticated robots.

Emergence theory opposes this view, arguing instead that, at certain points in the evolution of the universe, new levels of complexity have arisen and new properties have emerged that could not have been predicted on the basis of the properties of the lower-level entities from which they emerged. Thus, in complex entities, the whole is always greater than the sum of its parts. For example, molecules exist on a higher level than atoms, and therefore the properties of molecules are emergent: that is, they are something new in nature and could not have been predicted based on the knowledge of the properties of the atoms that constitute them. An example of this is the feeling of "wetness" we experience with water. The emergentist claims that this characteristic of large groups of water molecules could not have been predicted from our knowledge of hydrogen and oxygen atoms alone. Wetness is thus an emergent property that arises when one moves from the level of atoms (H, O) to the higher level of molecules (H_2O).

This fact holds true, according to emergence theory, all the way up to intelligent creatures such as humans. The properties of living cells could not be predicted from their constituent molecules any more than human self-consciousness could have been predicted from the molecules of which we are made. If this is true, then the reductionist and deterministic understandings do not tell us the truth about reality.

This also relates to the debate about causality in creatures, particularly whether the whole creature influences the parts that constitute it (called *top-down causality*) in addition to the parts influencing the whole (*bottom-up causality*). The ability of the human mind to control to some degree the behavior of its parts—such as deciding to raise an arm and the arm subsequently rising—is an example of top-down causality. The influence of parts on the whole in humans is indicated by the way ingesting certain drugs such as alcohol, or contracting diseases such as cancer, influences the mind (as well as the rest of the body). Reductionists and determinists (often one and the same) reject top-down causality, insisting, for example, that the brain causes and determines the mind, with the brain in turn being determined by molecular and atomic events, with the result that humans are not at all free despite their near universal belief that they are.

One final factor in the evolution of life that we must mention is the role of extinctions. As we will see below, there have been multiple major extinctions in the history of the earth, caused by such things as asteroid strikes from space, huge amounts of volcanism within a relatively short time, or massive and long-lasting weather changes. The most devastating of these events have wiped out a significant portion of life on earth, which has then opened the ecological niches vacated by the extinct species to new forms of life better suited to the changed conditions. This is a form of natural selection, if it can rightly be termed as such, that largely negates what we usually think of with natural selection in that it acts on much larger scales than the population level at which we generally think selection occurs.

What should be clear from all of this is that, while evolution is understood to be a fact by virtually all scientists, the proper description of evolutionary theory remains an area of vibrant continuing research. It will likely be decades at least before some common consensus on the evidently multiple causes of evolution arises, and it is probable that this consensus will have a multifaceted understanding of how evolution happens in the world.

6

The Origin of the Universe

THE BIRTH OF THE UNIVERSE

THE SCIENTIFIC UNDERSTANDING OF the origin of the universe in its major features is reasonably well established, supported by a significant amount of experimental and observational evidence. The theory that summarizes this science is called the *Inflationary Hot Big Bang Theory*. Theorists continue to study and debate certain details of this understanding, particularly the very earliest moments in the universe's history and what, if anything, preceded it. Nevertheless, the general understanding that, very early on, our universe experienced some sort of explosive "Big Bang" is at present so well supported that it is the single reigning theory, accepted by virtually all contemporary physicists.

The current reigning theory is that our universe came into being at a theoretical "time zero," originating in some sort of "Big Bang," an explosive creation of the space-time fabric and its contents in an unimaginably hot, dense, and small space some 13.75 bya.[1]

The origin of the theory was the discovery in the twentieth century that the universe is expanding outward in all directions. That means that in the past the universe must have been smaller; and if one reversed its current

1. As we will see, not all physicists agree with this dating. The age of the universe has recently been determined to be 13.73 billion years old, plus or minus 0.12 billion years. Thus I will use the rounded-up number, 13.75 billion years, as the approximate age of the universe.

expansion—that is, reversed the flow of time—then everything would move closer together. At some time in the distant past, then, the entire universe would have been only a single point, called a *singularity*, and this would be the "time zero" when the Big Bang occurred.

At this time, all of the matter and energy in our present universe—one so enormous that we can only guess at its actual size—was jammed together in a space so small that neither matter nor energy as we know them could exist, since the size of even the smallest subatomic particle we are aware of or the shortest wavelength of a photon would have exceeded the size of the entire universe at that time. It was a state of such enormous energy and density in such an infinitesimally small space that our minds simply cannot conceive it. Imagine taking the entire Earth—something whose size we can reasonably well comprehend—and shrinking it gradually smaller and smaller without changing its *mass*[2] until it was so tiny as to be invisible. Then imagine its mass to be a trillion times greater and its size a trillion times smaller, and then another trillion times more massive and smaller, and yet another trillion times more massive and smaller, and you begin to get some sense of the initial staggeringly massive yet miniscule size of the universe at its origin.

Cosmologists do not know what, if anything, preceded the Big Bang and what caused it to occur. Prior to what is termed the *Planck time*, about 10^{-43} (.0000000000000000000000000000000000000000001) seconds, none of the laws of nature as we currently know them existed. In addition, we cannot say anything about something smaller than the *Planck length*—about 10^{-35} (.00000000000000000000000000000000001) meters. Therefore, what did exist prior to this time—called the *Planck Epoch*[3]—and at less than the Planck length, no one can say with any significant scientific support. Much speculation exists, with some theories, such as the Stephen Hawking-James Hartle theory, claiming that our universe had no beginning but has existed forever, while others, such as quantum loop cosmology, assert that our universe is the rebound resulting from the collapse of a prior universe. M-theory claims that such cycles are the product of repeated collisions between the three-dimensional membrane, which is our universe, and another such membrane existing nearby. Yet another theory suggests that our universe

2. In physics, *mass* denotes a property inherent in matter itself, while *weight* refers to mass existing within a gravitational field. The weight of an object of a given mass would be greater in a stronger gravitational field or less in a weaker gravitational. In space, away from the pull of gravity, an object, no matter what its mass, is weightless.

3. There is no one universally recognized schema for dividing the early history of the universe into epochs or eras. I have simply chosen one that makes sense to me based on the changing circumstances of the early universe.

was born from a black hole in some other, antecedent universe. Whether science will ever be able to provide the kind of evidence that would indicate which of these theories, or another as yet unformulated, is true, no one can presently say. All we can say is that, at present, we do not know what existed prior to the Big Bang. The true origin of the universe, if it had one, remains a mystery for the present and likely will remain one far into the future.

We come closer to saying something meaningful about the early universe only after the Planck time and when it was the Planck length in diameter, at which point it is estimated to have been roughly 10^{32} (100,000,000,000,000,000,000,000,000,000,000) degrees Kelvin.[4] In dealing with such enormous temperatures, physicists often speak in terms of *gigaelectronvolts*, or *GeV*, where one GeV equals about ten trillion (10^{13}) degrees; thus at 10^{-43} seconds, the temperature of the universe was about 10^{19} GeV and had a density of approximately 10^{92} (one followed by 92 zeros!) grams per cubic centimeter. Because the four forces of nature as we know them were all combined into a single force, this period is often called the *Grand Unification Epoch*.

The extraordinarily large temperature and density scales, and small time and space scales, we are talking about here are truly beyond our comprehension. To make any sense out of these numbers, and particularly the enormous changes in the universe that take place over the infinitesimal time scales of succeeding periods within the first second of the history of the universe, we need to address the "*scale problem*."

Everything in the universe operates at its own scale of time and space. As a general rule, things on a larger scale of being appear to things on a smaller scale to move slowly, and things on a smaller scale of being appear to those on the larger scale to move faster. For example, to us, the Moon and other planets seem to move slowly, while electrons appear to move at a blinding speed. But to a planet, its motion (if it were able to observe and think) would seem swift in comparison with the rotation of the galaxy it inhabits, while humans would seem to move with astonishing speed. Likewise, to an electron (again, if it were able to observe and think), the motion of humans would appear incredibly slow, while the vibrations of a string (if such a thing exists) would seem unimaginably fast. Understanding this scale

4. The Kelvin scale of heat measurement differs from the Centigrade and Fahrenheit scales in that it starts from absolute zero, the lowest possible temperature, in which there is no energy whatsoever present. The Centigrade scale assigns 0° to the temperature at which water freezes and 100° to the temperature at which it boils, while in the Fahrenheit scale water freezes at 32° and boils at 212°. The Kelvin scale otherwise follows the centigrade scale. Absolute zero is -273.15° C and -459.67° F; water freezes at 273.15° K and boils at 373.15° K.

relationship will allow us to understand that, although the time and space scales discussed in the first few seconds of the history of the universe seem infinitesimally small to us, on the scale of the subatomic particles we are discussing, the pace is appreciably more leisurely, which is why so much can happen in such a small time (from our perspective). Thus, from a quark's perspective, the first second in the history of the universe is an enormous amount of time, perhaps proportional to how we would view the entire history of the universe.

Somewhere perhaps around 10^{-41} and 10^{-39} seconds after time zero, when its temperature had dropped to somewhere between 10^{31} to 10^{28} degrees, the rapidly expanding and wildly chaotic quantum universe evidently underwent what to our knowledge was its first important change—a *symmetry breaking*, in which the unity of the single force of nature that existed at the time was lost as the *gravitational force* separated to become its own unique force, leaving the other three forces we know today—the strong and weak nuclear forces and the electromagnetic force—still together as a single force. Evidently, the energy/temperature had dropped below some minimum required to hold the four forces together.

Given the extreme temperature and density of the miniscule universe, no matter as we know it could have existed, except perhaps only the strings of superstring theory (again, if that theory is correct). According to our present understanding, the high-energy, very short wavelength photons that we call gamma rays overwhelmingly exceeded whatever particles of matter might have existed during this time in terms of its mass-energy density.[5] Some physicists refer to this as the *Radiation Era*.

As the space-time fabric continued its expansion, the highly energetic, short-wavelength photons within it were stretched along with the space-time fabric within which they existed, which meant their wavelengths became longer. Longer wavelengths meant that the photons had lost some of their energy. Consequently, they and the universe in which they existed grew cooler. During this early phase of the universe's history, they were still bunched so closely together (that is, their density was so great) that these photons were in immediate contact with one another, exchanging heat with one another and, therefore, cooling slowly. Accordingly, the universe at this

5. Because mass and energy are really only different forms of the same thing, as expressed in Einstein's formula $E=mc^2$, we can explain the nature of the universe at any given time in terms of its mass density or its energy density. Therefore, we can compare the energy (or mass) density of the *radiation* in the universe with the energy (or mass) density of the *matter* in the universe to determine which predominated at any given time: radiation or matter.

time existed in a perfect thermal equilibrium, in which all areas of the expanding universe were connected and had precisely the same temperature.

With the continuing explosive expansion of the microscopic early universe, the wavelengths, temperature, and density of its photons decreased proportionally. Once the density dropped to the point where the rate of heat exchange between photons could no longer keep up with the rate of their cooling, the temperature of the radiation ceased to be perfectly uniform. Precisely when this occurred remains unknown. Due to random quantum fluctuations within this roiling sea of radiation, some areas had ever so slightly greater energy density than others. These areas of microscopically greater density exerted a slightly greater gravitational attraction than the other areas around them, thereby drawing more photons to themselves and thereby creating a very slight clumping within the otherwise perfectly homogeneous universe. These small areas of increased density were the ultimate origins, the seeds, as it were, of the larger structures that would eventually develop in the universe—the stars, star clusters, galaxies, galaxy clusters, and galaxy super-clusters and "walls" of these that we observe today.

Some physicists and cosmologists believe that the real Big Bang actually occurred somewhere around 10^{-36} to 10^{-32} seconds (the precise time varies in the different types of inflationary theories that currently exist), when space-time underwent a rapidly accelerated expansion, termed *inflation*. The period within which this occurred is called the *Inflationary Period*. Suddenly, the roiling quantum universe (or some small part of it) began to expand at an exponential rate, doubling in size about every 10^{-36} seconds, which is faster than the speed of light.

Because the universe expanded at super-light speed, all the photons traveling within the universe at "only" the speed of light could no longer interact with one another. As a result, parts of the universe were cut off from one another, with the result that the universe could no longer maintain absolute thermal and density equilibrium. In this way, the quantum fluctuations in the dense sea of radiation became larger scale fluctuations, stretched along with the expanding space-time fabric and so furthering the process of structure formation in the early universe. Nonetheless, these fluctuations were still very small relative to the average temperature and density of the universe, varying only by approximately one part in 100,000. Overall, the universe remained *almost* homogeneous.

This inflationary doubling of the size of the universe took place some 50 to 100 times, and possibly many more, before it suddenly stopped between perhaps 10^{-34} and 10^{-32} seconds, bringing the Inflationary Period to a close. Overall, the universe grew during this brief era from approximately a miniscule 6×10^{-28} centimeters to perhaps the size of a basketball, increasing the volume of the universe by some 10^{78} times what it had previously been. Scientists' estimates vary here, as our current lack of understanding of this period makes it impossible to determine the true size of the universe and the duration of the Inflationary Period.

Why this inflation happened we do not know, though there are several theories that might explain it. One is that gravity, normally an attractive force, can become a repulsive force under the kind of conditions that existed at this stage of the early universe's evolution. Another suggests that inflation occurred when the strong nuclear force decoupled from the electroweak force at approximately 10^{-36} seconds, when the universe had cooled to about 10^{28} degrees/10^{15} GeV, or perhaps slightly after that. This is generally termed the *Grand Unification Transition*, or *GUT*, which was followed by the *Electroweak Epoch*. This decoupling could have caused a phase transition within the tiny universe not unlike what happens when steam cools to become water, releasing heat energy in the process. This second case of symmetry breaking of the forces of nature would thus have released an enormous amount of energy, overwhelming the counteracting force of gravity and so powering the rapid outward expansion of the space-time fabric. Currently, uncertainty about the relative beginnings of the GUT transition and the Inflationary Period prevents any certain tie between the two.

For the NASA/WMAP Science Team's illustration of the rapid expansion of the universe during the inflationary period compared with the rest of the history of the universe, see http://map.gsfc.nasa.gov/media/060915/060915_CMB_Timeline75.jpg.

Alternatively, the cause of the universe's sudden inflation may have been something else, such as the repulsive vacuum energy of space, or *dark energy*, suddenly exerting its expansive force with great energy, pushing the space-time fabric and its contents outward by countering the force of gravity while supplementing the outward thrust initiated by the Big Bang. After that, it has been suggested, the dark energy decayed into ordinary matter and radiation. Whatever the cause, though, cosmologists are generally confident that this Inflationary Period did occur.

At the end of this brief period, the universe's expansion rate dropped precipitously. The universe continued to expand at a great rate from the initial force of the Big Bang. However, the enormous gravitational attraction of mass and energy within it caused the rate of that expansion to slow progressively throughout the remainder of the early universe's Radiation Era. That slowing of the universe's expansion would continue, though at a decelerating rate, throughout the subsequent *Matter Era*. During that era, the mass of matter came to dominate that of radiation. After that, around 9.4 billion years aBB, the universe's expansion began to accelerate under the force of dark energy. The rate of this expansion has continued to accelerate ever since, and, as far as we know, the universe will continue to expand ever more rapidly forever. This event inaugurated what some have called the *Dark Energy Era*, the era in the history of the universe in which we now live.

THE QUARK-LEPTON OR ELECTROWEAK EPOCH

By about 10^{-32} seconds or so—in other words, after the inflationary stretching of space-time and cooling of the universe—the temperature of the universe had dropped enough for the first matter and antimatter particles to form as distinct entities from the collisions of gamma-energy photons in the intense radiation soup. These were the relatively heavy quarks and anti-quarks and relatively light leptons and anti-leptons (i.e., electrons, positrons, neutrinos, and anti-neutrinos) and gluons that cause the strong nuclear force. Their appearance initiated what some cosmologists call the *Quark-Lepton (*or *Quark* or *Quark Gluon) Epoch* of the early universe. These elementary particles moved at near light speed through the dense inferno that was the universe at that time, carrying such enormous energy that, when they collided with other matter particles, the strong and electroweak nuclear forces were not strong enough to bind them together. Therefore, they collided and scattered, rather than forming larger and more complex particles of matter, such as protons and neutrons. (Where dark matter particles fit into this schema is currently unknown due to our ignorance of their nature and properties.)

At the same time, all these quarks and leptons themselves existed only briefly. They collided almost instantaneously after their birth with their particle-pair opposites (quarks with anti-quarks, electrons with positrons, and neutrinos with anti-neutrinos). The opposites annihilated each other (physicists simply say "annihilated" for this process), thereby creating new gamma-energy photons. These photons in turn collided rapidly with one another to create more matter-antimatter particle pairs, and so on, repeating this conversion process from energy to matter to energy again with astonishing rapidity, all according to Einstein's $E=mc^2$. Importantly, however, these energetic interactions altered slightly the conditions within the nascent universe from one microscopic place to another, creating, as we have already noted, areas of slightly greater and lesser temperature and density. Nevertheless, on the whole the universe continued to be essentially homogeneous.

In addition, during this period, and of crucial importance for the future of the universe, the symmetry between ordinary matter particles and their anti-particles was broken. We would expect that this relationship would have been symmetrical—that is, that the gamma photon collisions would have formed equal numbers of matter and antimatter particles, which would just as quickly have annihilated, releasing gamma photons back into the universe. But somehow, during the Radiation Era, this did not always happen. Instead, for reasons we do not yet understand, for every billion matter-antimatter particles that annihilated, one excess matter particle remained. This was fortunate for us, because everything we see in our universe today is made up of these excess ordinary matter particles. Were it not for this slight favoring of matter over antimatter, the universe as we know it, with its stars and planets—and ourselves—would not exist.

Still, throughout the Quark-Lepton Epoch, the number of photons dwarfed the number of matter particles: roughly a billion photons existed for every matter particle. The mass-energy of radiation still vastly exceeded the mass-energy of matter in the universe, though from this time forward the degree of its domination would decrease as the radiation lost increasingly more of its energy, due to the stretching of the wavelengths of its photons, which, as we have seen, occurred along with the continuing expansion of the space-time fabric of the universe.

If the theory of supersymmetry is correct, the heavy superpartners to the particles in the Standard Model were created during this Quark-Lepton Epoch, perhaps around 10^{-17} to 10^{-15} seconds. As we noted in chapter 4, the lightest of these superpartner particles would have been a *weakly interacting massive particle*, or WIMP, which is perhaps the mysterious *dark matter* particle for which physicists today are searching. Since dark matter does not interact with ordinary matter and feels only the force of gravity, little more will be said about it until later in the universe's history, when it contributes significantly to the creation of the large-scale structure of the universe. Meanwhile, the universe continued to expand, though ever more slowly, because the force of the Big Bang was being countered by the inward pull of gravity. The universe remained a superheated plasma of matter and antimatter and photons coming into and going out of existence as they collided with enormous energies.

Then, at about 10^{-12} seconds, when the universe had cooled to about 10^{15} degrees, the final symmetry breaking of the fundamental forces of nature occurred as the weak nuclear force decoupled from the electromagnetic force. Thus, what had originally been a single force had now completed its division into the four fundamental forces that govern the interactions of ordinary matter particles in the universe today: the strong and weak nuclear forces, electromagnetism, and gravity. Also, around this time, the *Higgs transition energy* was reached and so the *Higgs mechanism* began to function, giving mass to all matter particles as well as the W and Z bosons that governed the newly liberated weak nuclear force. This initiates what many physicists refer to as the *Quark Epoch*. From this time forward, the fundamental laws of physics as we know them today would apply, making scientists' reconstruction of the nature of the events in the subsequent evolution of the early universe far more certain.

THE HADRON, LEPTON, AND PHOTON EPOCHS

Once the continuing expansion of the universe had cooled its temperature to around 10^{13} or 10^{12} degrees (1 GeV), which occurred at around 10^{-6} seconds, the energy of the quark collisions had dropped sufficiently so that the strong nuclear force could bind quarks together into *hadrons*, the protons and neutrons that are the basis of all the ordinary matter that exists in our universe today. This is termed the *quark-gluon phase transition*. This marked the end of the Quark Epoch, with its quark-gluon plasma universe, and ushered in the new *Hadron Epoch* of the early universe. At this point, the universe was perhaps a hundred times larger than the size of Earth's orbit around the Sun.

These newly created protons and neutrons were unstable, however, because the universe was still too hot for them to remain permanently combined. Therefore, protons turned into neutrons either by adding an electron and releasing a neutrino or adding an anti-neutrino and releasing a positron. Neutrons turned into protons by the reverse processes. In addition, when a proton and neutron collided with each another, they might combine to form a *deuterium* nucleus composed of one proton and one neutron; however, in the dense, hot, cosmic soup of the universe, these deuterium nuclei were destroyed almost as soon as they were created by energetic collisions with other protons, neutrons, or gamma-energy photons. Most of the protons ended up colliding with anti-protons, annihilating to release more energetic photon radiation. While the protons were disappearing, neutrons continued forming protons, thereby resupplying the stock of protons in the universe but diminishing the number of neutrons. By the time the Hadron Epoch had ended—at about a hundred-thousandth of a second (10^{-5} sec)—only a relatively small number of protons and neutrons had survived.

Meanwhile, gamma photons continued to collide with one another to create electron-positron pairs, which in turn annihilated back into photons. With relatively few protons and neutrons left, it was the lighter leptons—the electrons, positrons, neutrinos, and anti-neutrinos—that dominated all other particles in terms of their total energy densities for the next ten minutes in the early life of the universe. Because these lightweight particles are collectively termed *leptons*, this period is called the *Lepton Epoch*.

By .01 second, the universe's temperature had dropped to about 100 billion (10^{11}) degrees. The universe was still very dense, with a mass density of approximately 3.8 billion kilograms per liter of volume, and so all matter particles continued to interact rapidly with one another. The most abundant particles remained electrons and positrons of such great energy that they behaved essentially as radiation, plus neutrinos, anti-neutrinos, and photons. The universe continued to exist at near thermal equilibrium, though the areas of slightly higher or lower temperature and density grew as the fabric of the universe continued to stretch outward. The number of protons and neutrons was about one for every billion photons, electrons, and neutrinos.

At 0.12 second, the temperature of the universe had dropped to some thirty billion degrees. At this lower temperature, it was easier for the slightly heavier neutrons to decay into protons and electrons than for protons to combine with electrons into neutrons, and so the proportion of protons to neutrons increased to 62 percent to 38 percent. It was still too hot for these protons and neutrons to combine into stable nuclei under the strong nuclear force.

At 1.1 seconds, the temperature of the universe had dropped to about ten billion degrees (about the temperatures reached inside a hydrogen bomb explosion) and its density to about one ton per cubic centimeter. The decrease in temperature and density allowed neutrinos and anti-neutrinos to decouple from electrons, positrons, and photons and so behave like free particles. Therefore, they began speeding through the universe. From this time forward, their wavelengths would gradually lengthen, their energy decrease, and their temperature cool with the continued expansion of the universe. Because neutrinos interact very weakly with one another and with other matter, those primordial neutrinos are still around today, filling the universe like a very fine mist, but with energies so low that they cannot be detected directly. Scientists refer to this as the *Cosmic Neutrino Background*. At this point, the universe was perhaps three light-years across.

With the temperature in the universe continuing to drop as it expanded, photons became less and less energetic, making it harder for electron-positron pairs to form. As a result, electrons and positrons began to annihilate faster than they could be created by photon collisions, so the total number of electrons and positrons in the universe declined.

These two changes—the decoupling of neutrinos from matter and the loss of electron-positron production—meant that the two mechanisms for creating neutrons had come to an end. Because free neutrons decay into protons and electrons by the weak nuclear force, from this time forward the supply of neutrons in the universe would steadily decrease until the temperature had cooled sufficiently for them to combine with protons under the strong nuclear force to become stable *deuterium* nuclei. Due to this decreasing number of neutrons in the universe, the proton-neutron ratio continued to rise. For every proton that existed, there were still about a billion photons that, along with neutrinos, electrons, and their antiparticles, constituted the majority of the contents of the universe. In this way, radiation continued its dominance over matter in the early universe.

By ten seconds after the Big Bang, the expansion of the space-time fabric had caused the density of radiation in the universe to drop to about that of lead, or roughly five kilograms per cubic centimeter. Electron-positron annihilations accelerated as the temperature decreased until finally there was about one extra electron left over for every billion electron-positron annihilations. The result of this was that essentially the same total number of electrons existed as protons. After this decline in electrons and positrons, the dominant component in the universe was photons, whose energies ranged from the gamma end of the spectrum down through X-ray to ultraviolet wavelengths, plus neutrinos. This continued domination of photons over matter in the universe would last for an unimaginably long time, given the time scales we have so far been dealing with: to approximately 380,000 years. Hence its name: the *Photon Epoch*.

At about this time, the universe's temperature had dropped to approximately three billion degrees. Electrons and positrons continued rapidly disappearing, their annihilations producing heat that was absorbed by photons that were now more energetic/hotter than neutrinos. Some deuterium nuclei, or *deuterons*, began to form but were quickly broken apart by energetic collisions with other particles. Were it not for the formation of deuterons, the remainder of the neutrons in the universe would have disappeared within another ten minutes, leaving only protons and so populating the universe ever after with only hydrogen nuclei and, later, hydrogen atoms. However, this reaction preserved neutrons, and this preservation would permit the creation of heavier elements—the kind we are mostly made of—later in the history of the universe. This process froze the proton to neutron ratio of the universe at 83 percent protons to 17 percent neutrons. By three minutes after the Big Bang (aBB), the universe had grown to perhaps fifty light-years across.

THE PERIOD OF PRIMORDIAL NUCLEOSYNTHESIS

The next period in the history of the early universe is called the *Period of Primordial Nucleosynthesis*, or nucleus building. By around 3 minutes 46 seconds aBB, the temperature of the universe had dropped to about 900 million degrees and its density to around ten grams per cubic centimeter. Under these conditions deuterons could survive, permitting other reactions that, through a variety of means, could make nuclei of *tritium* (or ^{3}H, the heaviest *isotope* or kind of hydrogen, made up of one proton and two neutrons), *helium-3* (or ^{3}He, composed to two protons and one neutron), the more stable nucleon *helium-4* (or ^{4}He, composed of two protons and two neutrons), small amounts of the heavier *lithium-6* and *lithium-7* nuclei (or ^{6}Li and ^{7}Li, composed of three protons and three or four neutrons), and perhaps a trace of *beryllium*—the fourth element on the periodic table after hydrogen, helium and lithium).

Given the initial proton-neutron ratio of 83% protons to 17% neutrons, or about five to one, this meant that once these reactions had stopped, around ten minutes aBB, they would leave the universe with roughly 75% hydrogen nuclei and 25% helium nuclei by mass, with a trace of deuterium, helium-3, lithium, and perhaps some beryllium nuclei as well. This marked the end of the primordial creation of nuclei in the universe, with ratio of primordial hydrogen to helium set for all time. Only a relatively small additional amount of the helium and all other elements would be created throughout the entire subsequent history of the universe by thermonuclear fusion occurring within stars and by the explosive deaths of massive stars (as we will see in chapter 8). Today the universe is composed of about 73% H, 26% He, and 1% all the other elements of the periodic table combined.

At about 34 minutes and 40 seconds aBB, the universe's temperature had dropped to some 300 million degrees. Electrons and positrons had finished annihilating one another, leaving about one extra electron for every billion annihilations, or about the same number of excess electrons as excess protons.[6] The energy released from these annihilations left the universe with a mass density in which about 69 percent existed in the form of photons and 31 percent in the form of neutrinos and anti-neutrinos. Thus the universe was still dominated by radiation compared with matter, and that by a large margin. All the ordinary matter nuclei and other particles continued to travel at extremely high velocities and with such great energies that they could not capture and hold the electrons that were whizzing through the universe. Whenever an electron was captured, it was quickly knocked out

6. The excess of electrons over positrons is produced by two anti-up quarks annihilating into an electron and a down quark.

of its orbit by some other particle, meaning that atoms of hydrogen, helium, and the rest could not yet exist for more than a tiny fraction of a second.

The matter in the universe at that time consisted of a dense plasma of ionized atomic nuclei and electrons flying through the expanding universe at enormous energies, colliding constantly. This state continued for the next several hundred millennia, with the only real change being the constant decline in the temperature and density of the universe as it expanded and the consequent diminishing energies of the particles within it. By about 1 year aBB, the universe's temperature had dropped to around three million degrees. By the year 10,000, it had become a "mere" 20,000 degrees, and its density had declined to significantly less than the density of air on our planet today. However, the nucleons and electrons were still too energetic to unite to form the first stable atoms. Photons continued to outnumber protons and electrons by a ratio of about a billion to one. All the while, the expansion rate of the universe continued to drop at a rapid pace under the influence of gravity. By 100 years aBB, it was expanding "only" at a rate of some 22,000 kilometers/second (km/sec), or about 13,670 miles per second (mi/sec).

As the universe continued to expand and cool over the next several hundred millennia, the nuclear particles within it lost energy and slowed down, so that the universe gradually became a less violent place. Photon wavelengths continued to increase, stretched along with the expanding space-time fabric, thereby losing energy. Radiation was still coupled with matter, so no photon or other particle could travel far without hitting another particle. Photons, because they could interact only with charged particles, would be absorbed if they collided with electrons only to be released again as other photons. The conditions in the universe at this time were essentially like those in the interior of our Sun today. Because of the great density and heat within the Sun, photons created by the fusion of hydrogen into helium at its core constantly interact with other particles, so that each photon takes some 100,000 years finally to reach the surface of the Sun, where it is radiated away into space. Photons of the proper wavelength can be seen by us as the light of the Sun. Just as we cannot see through the Sun, so the universe at this early stage was opaque.

As had been the case since the earliest moments of the universe's history, the matter and radiation within this opaque universe continued to be not quite evenly distributed. Areas of slightly greater radiation and matter density drew other material to themselves under the influence of gravity, clumping into patches of various sizes. These areas of greater density, however, continued to be only very slightly denser than their surroundings. Because of the constant pressure caused by the much greater amount of radiation energy, they were unable to grow into significantly larger structures. In addition, as the matter and radiation clumped, it heated up somewhat, due to the more frequent collisions of particles with one another in these denser regions, creating another outward pressure, making them less dense again. These oscillations between greater and lesser densities of radiation and matter produced low frequency waves within the dense primordial soup, creating sound waves of such a low frequency that we could not have heard them even if we had been around to do so.

Dark matter,[7] however, does not interact with radiation in the way ordinary matter does, since it does not absorb or emit photons. For this reason, it was unaffected by the outward radiation pressure that kept ordinary matter from remaining in clumps. The result was that dark matter continued its gravitational contraction, providing the seeds for the stars and galaxies and larger structures that would eventually form in the universe. Had it not been for this dark matter, it would have taken billions of years longer for stars and galaxies to form, and we would not be here now to observe them. By 10,000 years aBB, the universe had grown to perhaps two million light-years in size.

THE MATTER ERA

Somewhere around 70,000 years aBB, when the universe had cooled to about 9,000° K, the energy density of matter and radiation finally became equal. This was the first time in the history of the universe when radiation/photons were not the dominant material in the cosmos. After this, matter would continue to increase in mass energy relative to that of radiation, signaling the beginning of the *Matter Era*. The Matter Era would last for more than nine billion years, finally giving way to the last major era in the history of the universe, the *Dark Energy Era*, in which we live today.

By approximately 370,000 years aBB, the universe had cooled to about 3,000° K, slightly cooler than the surface of our Sun. This meant that the energy of photons had dropped to the point where they could no longer knock electrons out of their orbits around nuclei once they had been captured by the electromagnetic force. Thus the first stable atoms were formed. This is called the period of *recombination*. This binding of free electrons caused the supply of free electrons in the universe to vanish in a relatively short amount of time and so with them the pervasive fog throughout the universe they had created by their collisions with the photons. Since photons do not interact with particles that are not electrically charged, and all the charged

7. It is called *dark* matter because it is invisible to us.

particles—protons and electrons—were suddenly confined within atoms, these photons were suddenly able to travel freely throughout the universe just as the neutrinos had done earlier. For the first time in its history the universe became transparent. The dissipation of the fog also brought to an end the subsonic sound waves that had previously permeated the universe—including not only primary notes, but their second and third harmonic overtones as well—and the cosmos went silent. Cosmologists refer to this sudden and important event the period of the *decoupling of radiation from matter*.

This recombination period also saw the *origins of chemistry*, in which the newly created atoms began to interact with one another to form molecules. Thus, for example, two hydrogen atoms combined to form hydrogen gas, or H_2. Such molecular combinations occur, not because of anything going on in the nucleus, but because of the electrons travelling in their shells around the nucleus. As the nested shells move out from their innermost locations relative to the nucleus, they can hold successively more electrons. Thus, the innermost shell can hold two electrons, the next two shells can hold up to eight electrons each, the next two up to 16 electrons, and so on. Molecules are most stable when their electron shells are filled. Thus, H_2 is stable, because its electron shell is filled with two electrons; He (helium) is very stable, because its shell is filled by its two electrons, and H_2O (water) is stable, because its first shell is filled with two electrons and its second shell is filled with eight electrons (because oxygen atoms contain eight protons and eight electrons). At this early stage, where only hydrogen and helium existed in significant amounts, hydrogen gas and helium atoms were the main constituents of the universe. Only later, when in the hearts of stars heavier atoms were created and eventually blasted into space in the death of larger stars, could more complex chemical molecules arise.

This dissipation of the initial kind of cosmic fog caused by electrons and nuclei flying freely around the early universe, however, ended up giving rise to another, if less dense, fog, created as hydrogen gas (H_2) filled the universe. Then, as the space-time fabric continued to stretch apart, the photons streaming throughout the universe gradually cooled. The energies of electromagnetic waves travelling in the universe, formerly in the visible light range, dropped into the infrared range and, eventually, into the microwave range. Once this occurred, the universe was plunged into darkness. Consequently, this era in the history of the universe is called the cosmic *Dark Age*.

The primordial photons, created by the Big Bang and now freed by their decoupling from matter more than 13 billion years ago (bya), still exist today in the form of the cosmic microwave background (or CMB) radiation that pervades the universe. Scientists see it everywhere they look when they turn their microwave telescopes to the sky; they are literally seeing the universe as it was only 370,000 years aBB. We nonscientists also see it—or used to see it—on our analog television screens, as it constituted some of the static we would see when the channel selector was tuned to channels where no station was broadcasting.

Because these photons ceased their interaction with matter, the two would no longer exist at the same temperature, and the essential temperature homogeneity that had existed in the universe from its beginning was thus finally at an end. Radiation would cool faster than matter as its wavelengths grew, stretched along with and at the same rate as the expanding space-time fabric of the universe and so losing energy. Today, that radiation has cooled to just 2.725 degrees Kelvin—more than a thousand times colder than at the moment it began to stream freely through space. The extremely low energy of this radiation is the reason space is so cold.

See http://map.gsfc.nasa.gov/media/121238/ilc_9yr_moll4096.png for the NASA/WMAP Science Team's illustration of the cosmic microwave background of the sky showing warmer and cooler temperature variations. In reality, the differences are much less than the diagram implies, varying by only about one part in 100,000.

Even though the CMB appears nearly homogeneous to us, mapping it with ever greater precision has allowed us to see that there are indeed very small differences in temperature and density within it. Denser regions had a slightly greater than average temperature and less dense areas had a slightly lower than average temperature, differences amounting to no more than about one part in 100,000, or about 200 microkelvins (10^{-6} ° K) above or below the average temperature. Today this CMB temperature variation measures 2.725 +/- .0001 degrees Kelvin, or a range from about 2.7249 to 2.7251 degrees above absolute zero (roughly -270° C or -454.7° F).

From this point forward, matter and radiation would interact with each other only through gravitational attraction, per Einstein's theory of general relativity. By the end of this brief period of decoupling, the universe was made up of about 12% atomic material (versus 4.6% today), 15% photons (versus about .005% today), 10% neutrinos (versus about .0034% today), 63% dark matter (versus about 23% today), and a negligible amount of dark energy (versus about 72.4% today).

With radiation no longer interacting with ordinary matter, other than gravitationally, it became possible for this matter to begin to fall into the slightly hotter and denser primordial regions caused by dark matter's gravitational collapse. With the presence now of both dark and ordinary matter, the gravitational force grew stronger, and these dense regions grew larger and denser. At this point, dark matter performed another crucial service that aided the process of structure formation in the early universe. When ordinary matter was collapsing gravitationally, it was aided by dark matter, which was also collapsing. When ordinary matter began to rebound and push outward under the increasing pressure caused by its condensation, the dark matter did not; instead, it continued to collapse. The overall effect was that the presence of dark matter blunted the rebounds of ordinary matter, thereby hastening the gradual collapse of the total system. The fact that there was more than five times more dark matter than ordinary matter further aided the rate at which this collapse took place, speeding thereby the formation of stars and galaxies in the nascent universe.

After the decoupling passed, the radiation continued to cool, the wavelengths of its photons stretched by the expanding universe, leaving only a fading glow throughout the universe. With no stars yet existing to provide light after the decoupling, the universe was gradually plunged into an ever-deeper darkness. By the time the universe turned half a million years old, it had expanded to the point where it was roughly one-thousandth the size of the visible universe today, or about 40 million light-years across, with only the faintest glow standing between it and total darkness. This expansion of space created ever more dark energy, though at this point its amount and strength was still negligible compared to that of the gravity caused by the combination of ordinary and dark matter.

By around one million years aBB, the temperature of the universe had fallen to about 2,000 degrees. By five million years, most photons had been shifted to infrared wavelengths by the stretching of the space-time fabric, so that the universe would have appeared completely black had any human been around to view it. The Dark Age persisted until the first stars ignited around 200 million years aBB and sent visible light streaming out again through the universe.

Throughout this period, the universe continued to grow, though ever more slowly, as gravity continued to counter the outward push from the Big Bang. By 15 million years, the visible universe had grown to about a hundredth of its size today and had cooled to a temperature about equal to that of a hot summer day. Gravity continued to enhance the denser regions of gas as more and more hydrogen, helium, and trace amounts of lithium were drawn into them, setting the stage for the formation of the first stars and an end to the universe's Dark Age.

7

The Christian Understanding of God as Creator

IT SEEMS APPROPRIATE TO follow a chapter on the origin of the universe with a theological consideration of the *Christian Doctrine of Creation*. This doctrine has traditionally comprised several assertions, though not necessarily those most people think of. First, and most fundamental, is the claim that all physical reality exists only because God causes it to exist. God is thus understood to be the *Creator* and *Sustainer* of the universe and of everything within it. The doctrine further asserts that this distinction between Creator and creation is *ontological* (having to do with the nature of being): God is a *necessary being*, one for whom existence is part of God's essence, and therefore one who necessarily exists, who could not *not* exist. This is distinguished from *contingent being*, the kind of being that the universe and everything within it has, meaning that it might not have existed and comes into and eventually goes out of existence. In the Christian tradition, this distinction has sometimes been made by defining God as *Being* and creatures as *beings*. This distinction between kinds of being is also meant to reject any notion of the creation being divine.

This fundamental distinction between Creator and creation lies at the heart of *the cosmological argument*, one of the classical arguments for the existence of God noted in chapter 1, which is used not only in the Christian tradition but in the Jewish and Islamic traditions as well. This argument seeks to answer the question, "Why is there something and not nothing?" The cosmological argument is grounded in the *Principle of Sufficient Reason*, fundamental to both science and philosophy, which states that "nothing comes from nothing," that no effect exists without a cause. If the universe is an effect, there must be a cause for its existence.

Here we must note that, in classical philosophy and the theologies grounded in it, several types of causation exist. When we today think of causation, we think only of *efficient causation*—that type of sequential, historical causation in which something is the cause of some subsequent effect, which then becomes the cause of some other subsequent effect, and so on throughout time. In classical thought, however, there are other kinds of causation. One of these is *metaphysical causation*.[1] It is this type of causation that we are concerned with here as constituting the foundation of the Christian doctrine of creation, as well as one form of the cosmological argument for the existence of God. Metaphysical causation is the kind of causation that undergirds the entire chain of efficient (temporal) causation of contingent beings, holding all contingent beings and the universe in which they live in being and so preventing them from slipping into nonbeing. (A third type of causation, *final causation*, is relevant to another type of argument for the existence of God, *the teleological argument*, which we will discuss later.)

Thus, the first—and what I understand to be the foundational—form of the cosmological argument claims that the metaphysical cause of the universe cannot be something contingent, since any contingent thing must itself have a metaphysical cause of its being and that of all other contingent beings. Also, no contingent being (creature) could hold all physical reality in being, because it is only a part of that reality, nor could it cause the laws of nature to be operative, since it was these laws that governed its coming into and remaining in being. In the end, then, the existence of all such contingent things is in need of a metaphysical cause that explains why they exist and continue to exist through time. If this cause is not contingent, something that exists *necessarily*—something that cannot *not* exist—must be the ontological or metaphysical cause of all contingent beings. This necessary being is what the Christian, Jewish, and Islamic traditions understand God to be. Thus, the argument concludes, the existence of our universe of contingent beings means that a necessary being, God, must exist. This is the kind of argument found in the first and third of Thomas Aquinas's "Five Ways," which are his five arguments for the existence of God.[2]

1. Metaphysics, a branch of philosophy, deals, in its most general definition, with the nature of being, with its universal and necessary features. As such, the terms ontological causation and metaphysical causation are essentially synonymous.

2. Aquinas, *Summa Theologica* I a q. 2 art. 3. That the first way concerns ontological or metaphysical, rather than efficient, causation is not obvious to those not versed in the terminology of classical Greek metaphysics.

There are several interesting things to note about this form of the cosmological argument. First, it did not originate within Christianity or any other religious tradition. It originated (in the West, at least) within ancient Greek philosophy and is found, for example, in Aristotle's writings. It is thus based solely on the logic of being, not on any religious *apologetics* (attempts to defend the rationality of religious belief).

Second, it says nothing about *how* God created the universe, or what conditions, if any, preceded God's creation, nor the historical process of such creation. These are separate considerations and relate in part to the second form of the cosmological argument, one based on efficient causation as seen in Aquinas's second way and in the similar argument of the medieval Muslim scholar al-Kindi. We will address this argument below in conjunction with another aspect of the traditional doctrine of creation, the doctrine of *creatio ex nihilo*, or "creation out of nothing."

Third, the cosmological argument does not say anything about the *nature* of the Creator. It only claims that a necessary being must exist who causes the universe to exist, and without whose power all ordered physical reality would disintegrate into nothingness. David Hume (1711–1776), the skeptical English philosopher, made this same point in his assessment of this argument for the existence for God.

In my opinion, the cosmological argument is the most powerful *philosophical* argument for the existence of God (for most religious people, it is their own religious *experience* that is the fundamental reason for their belief in God's existence), though it by no means settles the case. Still, an example of its logical power may be seen in the case of Antony Flew, who, as we noted earlier, was arguably the staunchest proponent of atheism among twentieth-century philosophers. In a striking turnabout in 2004, Flew announced that he could no longer deny the logic of the cosmological argument and that he was now a theist, though not necessarily one whose concept of God accorded with any particular religion.[3] Several scientists have sought to refute the cosmological argument on the basis of scientific theories that suggest the universe might have arisen without a cause. Their common weakness, however, has been that they have assumed that something existed prior to the appearance of the universe, especially physical laws that govern how the universe might have arisen.

The second aspect of the doctrine of creation to consider is the question of the state of things when God began to create. The traditional Christian answer to this question has been, as noted above, the doctrine of

3. Habermas, "My Pilgrimage," 200–201. See also Flew and Varghese, *There Is a God*, Introduction and Section II.

"*creatio ex nihilo*." This doctrine claims that no physical reality existed prior to God's creative act. There was nothing, then God created, and suddenly there was a universe. The original intention of this doctrine was theological, not scientific. It sought to argue for God's absolute and complete power over the creation, and so, against Aristotle, denied that anything existed prior to God's creative work. Those who have elaborated this doctrine have claimed that it accurately reflects the biblical understanding.

As we have seen, the *creatio ex nihilo* doctrine undergirds the second form of the cosmological argument for the existence of God. This argument is grounded in the notion of efficient causation, not metaphysical causation, arguing that every effect in the universe is the result of a prior historical cause, which itself is the effect of a prior historical cause, and so on back through time. However, such a chain of cause and effect cannot go on infinitely into the past, they argue; it must have had a beginning, a first, uncaused cause that started the universe's causal chain in motion, and that cause must be God.

It is interesting that contemporary biblical scholarship has challenged the claim that this *ex nihilo* notion is what the Bible teaches. Here again we encounter a case where manuscript and translation difficulties bear pivotally on a doctrinal matter. The most ancient biblical manuscripts we that have create problems for us, because their texts consist of long, uninterrupted strings of letters, with no spaces between words, sentences, or paragraphs, nor any punctuation or other indications of where words, sentences, and paragraphs begin and end. (Chapter and verse divisions were added to the biblical text only much later, in the Medieval Era.) Biblical scholars must use their own judgment as to what letters to group into which words, and what words should be combined into sentences, or clauses within sentences, and so on. In most cases, this is not difficult to do, but occasionally it is quite unclear what the best text is and how to translate it.

The primary biblical texts cited in support of the doctrine of *creatio ex nihilo* are from the first two chapters of the book of Genesis. There, mainline biblical scholars discern two separate creation stories, as we have noted. The first of these stories, Gen 1:1–2:4a, has been the more important in developing the *ex nihilo* doctrine. The chief translational problem bearing on this doctrine appears in the first two verses of this story, Gen 1:1–2. These verses have been translated traditionally as two independent sentences, as seen, for example, in the Revised Standard Version of 1952 (RSV): "In the beginning God created the heavens and the earth. The earth was without form, and void; and darkness was upon the face of the deep. And the Spirit of God was hovering over the face of the waters." However, more recently a number of scholars, both Christian and Jewish, have championed another translation, equally valid grammatically, which rejects the division of the initial part of the text into two sentences. These read, "In the beginning, when God began creating the world, the earth was a formless void . . . ," or some variation thereof.[4]

While both readings are grammatically possible, the latter is to be preferred, for two main reasons. First, God's creative activity occurs, according to this first creation story, over a period of seven days: on day two God creates the heavens, and on day three God creates the earth. This makes no sense if God had already created the heavens and the earth prior to the first day, as in the traditional translation of Gen 1:1.

Second, in the traditional translation, there appears to be a logical inconsistency between Gen 1:1 and 1:2. In that reading, God makes the heavens and the earth; then, for some reason not given, there exists a formless, dark, watery void, and from this God begins to make the world as recounted in the six days of creation that follow. Where did this watery void come from? Did God create it? If so, why create this watery void rather than simply creating the creations of the six days out of nothing if that were what God did? Perhaps, in the traditional interpretation, the opening sentence should be taken as a summary of what will be explained in greater detail in the verses that follow. If so, however, it is still the case that God creates the heavens and earth from the preexisting void, turning chaos into an ordered cosmos.

The newer translation avoids these problems and makes better sense of the Hebrew text. In this view, when God began to create the world, waters in absolute darkness already existed. Several features of this first creation account hearken back to the Babylonian creation story. This is significant since, according to Genesis 11, the founders of the Hebrew people, Abraham and Sarah, were Babylonians who, with their families, emigrated from Babylon and eventually settled in the land of Canaan, which later would be called Israel. Moreover, many centuries later, the Babylonians conquered Israel and took its leaders to Babylon as captives, where they remained in exile for more than sixty years (2 Kings 24–25; 2 Chr 36:5–23). Scholars believe it was while in Babylon that Judean priests developed the Genesis 1 creation account in its present form, in contrast to and rejection of the theology expressed in the Babylonian story. This is relevant to our concerns here, because the watery darkness that precedes God's creative activity is described using the Hebrew term *tehom*, translated as "the deep" (just as today we sometimes refer to oceans as "the deep"). *Tehom* is a Hebrew

4. Better English translations of the Bible, such as the NRSV, note these alternate possible readings in footnotes to the text.

cognate of the Babylonian word *Tiamat*, which was the name of the evil dragon representing the destructive, chaotic power of waters/floods in the Babylonian creation story. In this way, it becomes clear that the Genesis 1 account pictures God's creative activity, not as creation out of nothing, but rather as creation out of chaos, the bringing of an ordered cosmos out of primeval chaos—in other words, as God's victory over the Babylonian evil dragon, *Tiamat*.

The conclusion that we should draw from all this is that the traditional doctrine of creation *ex nihilo* is not supported by the first Genesis creation story, which instead asserts a creation out of a preexisting chaos. The same can be said of the second Genesis creation account (beginning at 2:4b), where a barren desert already exists when God begins to create. Discussion of all the other biblical passages that are relevant to the *ex nihilo* doctrine, including a third creation story found in Ps 74:12–17 (God creates by defeating the chaos/water dragon, which also refers to the Babylonian creation story) would take us well beyond the scope of our present concern. The conclusion we can draw from our analysis of these Genesis creation stories is that they do *not* offer support for the *ex nihilo* doctrine; instead, they suggest that God created by bringing order out of chaos. It is also important that the doctrine is not religiously necessary, which is the most fundamental point that we should emphasize here. Once again, the Bible is concerned with religion, and is always misconstrued when it is taken to be about science.

Given these considerations, and that God's *ontological/metaphysical* causation of the universe is the true foundation of the doctrine of creation, the view expressed by some scientists—that if the universe did not have a beginning, then there is no need for God and, therefore, no reason to believe that God exists—is a weak argument. For example, Steven Hawking, in laying out his theory of an unbounded universe, concluded, "So long as the universe had a beginning, we could suppose it had a creator. But if the universe is really completely self-contained, having no boundary or edge, it would have neither beginning nor end; it would simply be. What place then for a creator?"[5] A more adequate understanding of the Christian doctrine of creation would have enabled Professor Hawking to answer his own question.

Yet another question that has arisen within the Christian doctrine of creation is *how* God created the universe. This question has garnered by far the most public attention over the past century and a half, leading many to think—mistakenly, as we have seen—that it constitutes the central question addressed in the doctrine of creation. This issue *not religiously* important.

5. Hawking, *A Brief History of Time*, 141 (both editions).

The existence of this debate only reflects the confused belief that the Bible is a science book. Nevertheless, three Christian positions on the question of how God created the universe are proffered today. They are *Creationism*, *Intelligent Design*, and one we might call *Scientific Theism* (which includes the *Theistic Evolution* position). We will examine each of these options in turn.

Creationism is the view typically held by biblical inerrantist Christians who take the Genesis 1–2 creation accounts as literally, scientifically true statements of how God created the world (using some creative, ad hoc ways of attempting to avoid the contradictory features of those two stories, which never hold up under scrutiny). Actually, creationists come in two varieties: the "young earthers," who believe that God created the universe in six twenty-four-hour days (which is what they believe Genesis 1 says), usually some 6,000–10,000 years ago, and the "old earthers," who take the six days (of Genesis 1) to represent six relatively long periods of time and so date the creation to some millions of years ago.

However, although they can be distinguished in this way, both share a rejection of any science that contradicts their respective views—which in essence means they must reject *all* science, because every natural science, from physics to astronomy to chemistry to geology and biology, provides evidence that refutes the creationist claim. For example, according to the Genesis 1 creation account, God's first creation was light, which, following the preexisting darkness, made the first day.[6] By day three the earth and its vegetation have been created, and then, on day four, the Sun is created. Yet virtually everyone knows that the Sun causes the light of day and causes plants to grow. Thus, the Genesis 1 account, with the appearance of light and vegetation before the creation of the Sun, cannot be correct. Creationists will attempt to counter this problem by saying, for example, that the Sun was created on the first day but that it was hidden by clouds so that it could not be seen on earth until those clouds were dispersed, which happened on the fourth day. The obvious major difficulty for this view is that the text simply does not say this, and if this was what the author, whom they presume to be God, meant to say in Genesis 1, then why did not God simply say so? On every other day, God creates something new, so why would this day be the exception?

In addition, if you examine the six days of creation more closely, you will see a pattern: on the first three days God creates what we might call

6. This pattern of darkness, then light in the first day is also seen in the phrase that concludes each of the other days of creation, "And there was evening and morning, the (second sixth) day." This reflects the Jewish understanding of time, according to which days begin and end at sundown, so that the first half of the day is darkness and the second half is light.

three distinct environments, and on the last three days the creatures that inhabit those environments. Thus, on the first day God begins with darkness and then creates light and, on the corresponding day four, creates the creatures that "inhabit" the darkness and light: the Sun, Moon, and stars.[7] On day two God creates a dome filled with air and uses it to separate the preexisting waters, so that now there is water both above and below the atmosphere (literally, *heavens*) and air in between; on the corresponding day five God creates the various creatures that inhabit the waters and air. Finally, on day three, in two separate creative acts, God creates the dry land and the vegetation that covers it, and on the corresponding day six, in another pair of creative acts, God creates the animals and the humans that inhabit the land and eat the vegetation. The creationist "cloud cover" explanation would thus violate this clear pattern in the six days of creation, so it must be rejected for this reason as well.

I also want to argue that, apart from any scientific considerations, the very Genesis 1–2 creation accounts themselves, upon which creationists base their position, decisively deny the validity of their view. Creationists typically insist that these two creation stories are complementary accounts that tell the same story, with the first account telling how God created the universe and the second account focusing on how God created animals and humankind. However, based on a careful—and literal—reading of the stories themselves, this view is demonstrably false. In such a reading, a number of important differences and several irreconcilable contradictions appear between the accounts (several of which we noted in chapter 3 above).

First, the Genesis 1 story is written in the form of Hebrew poetry, which emphasizes repetition of phrases or ideas. Thus, in the description of each day's creation event, we find typically repeated these phrases (or some variation thereon): "Then/And God said . . . ," "let there be . . . ," "and it was so," "And God saw that it was good," and "and there was evening and morning, the (first, second, etc.) day." On the other hand, the second creation story is written in prose. Not only do these two different genres of writing suggest that we are dealing with two different stories, but also one might well want to ask why creationists would take the *poetic* story in Genesis 1—or any of the other myriad poetic writings in the Bible as well—literally! Does any educated person take poetry literally? Do creationists take poetry as expressing literal truth anywhere other than in the Bible? At any rate, the presence of two different literary genres in the two Genesis creation stories

7. This verse was also intended to repudiate the theology of the Babylonians, according to which the heavenly bodies were divine beings. In Genesis 1 they are merely things that were created by the one and only God that exists.

certainly suggests they are not simply a single story with slightly different foci.

The second difference between the two stories is that they name the Creator differently: God is always referred to as *Elohim* in the first creation story, but God is called *Yahweh Elohim* in the second.[8] Third, as we have noted, in the first Genesis story the creation takes place over the span of six days (plus a seventh on which God rests); but in the second story, there is no indication of the passage of days, so that the commonsense reading would understand that God's creation of the plants, animals, and humans took place in a single day. Fourth, as we have seen, the initial condition when God began to create in the first story was a dark, watery void; but in the second story, the initial condition prior to God's creating was a barren desert.

Fifth, where God creates humans, animals, and everything else in the first story simply by speaking a word, in the second story God creates these creatures by molding them from the moistened soil, like a potter, into their respective forms, and then breathing on them, bringing them to life.[9] Sixth, the order of the creation of plants, animals, and humans differs in the two stories. The first account has them created in the order just given, but the second story has the human being created first, then the plants, and finally all other animals. Finally, the first creation story has the male and female humans created at the same time, while the second story either has the male created first and the female created second, or, more in line with the Hebrew text, an evidently asexual human being is created first and only later is divided into a male and a female (recall here our discussion in chapter 3). Taking all these seven differences into account, we are driven to the conclusion that the only reasonable understanding is that Genesis 1–2 contains two distinct and contradictory stories of how God created the world.

8. Editor's note: *Yahweh* is one of the vocalizations of the Latin transliteration of YHWH, the four Hebrew letters (the *Tetragrammaton*) that represented one of the ancient Hebrew's names for God. The King James Version (KJV, 1611) vocalizes it with the name *Jehovah* in four places (Exod 6:3; Ps 83:18; Isa 12:2; and 26:4). In the American Standard Version (ASV, 1901), an update of the KJV, *Jehovah* appears 6773 times. Ancient Israelites, and religiously observant Jews up to the present, avoid saying the name of God and, instead, vocalize the *Tetragrammaton* with the Hebrew word *Adonai*, which means "the Lord." The Greek version of the HB/OT (the Septuagint, middle third to late second century BCE) translates YHWH as *kyrios*, "Lord." The RSV (1952) introduced the convention of translating the *Tetragrammaton* as "the Lord" (with small capital letters), which other standard Christian Bible translations have followed ever since.

9. The root Hebrew word here is *ruah*, which means "wind," "breath," or "spirit." Thus, by *breathing* on the creatures, God endows them with *spirit* and they become alive.

This, then, raises the obvious question: Why would the Bible contain two conflicting creation stories (not to mention a third in Psalm 74, which contradicts both of these Genesis stories), if its purpose were to tell *how* God created the universe? One would have to conclude either that the ancient Jewish thinkers who put these two stories next to each other when they were assembling the book of Genesis missed the fact that they contradicted one another—a possibility I find to be incredible—or that they understood that the purpose of these stories was *not* to tell *how* God created. However, if the purpose of the two creation stories was not to tell how God created the world, then what was their purpose?

The answer becomes clear when we understand, again, that the Bible contains a wide variety of literature, including histories (in the sense discussed in chapter 3), prophetic utterances, legal material, wisdom sayings, hymns, devotional literature, apocalyptic literature, and myths; and that the reader needs to understand what type of literature he or she is reading in order to interpret it properly. The key to understanding Genesis 1–2, as we noted in chapter 3, is to recognize that what we are reading here is myth, not history—that is, *theology in story form*. Because of this theological intent in Genesis 1–2, creationists, by interpreting these chapters as science, is—on the basis of the very texts upon which they make their case—misguided. The Bible, in fact, never addresses the question of *how* God created the universe. The main reason it does not is probably that the Bible's primary concern is trying to teach and motivate the Hebrew people to live their lives in loving relationships with God and all other people and the earth. Given this concern, *how* God created the world really does not matter.

The case for Intelligent Design is essentially the same as the case for Creationism, in that Intelligent Design is at bottom another form of Creationism that was developed after the United States court system ruled against the teaching of Creationism as science in public schools, because it was based on religious belief, not science. Leaders of the Intelligent Design movement thought they could get Creationism in a new form taught in public-school science classes if they did not refer to the Bible or God but, instead, invoked an unnamed Designer as the creator of the universe.

Intelligent Design advocates thought that, by slightly revamping William Paley's (1743–1805) classical presentation of the *Argument from Design* (also called the *teleological argument*) for the existence of God,[10] and making it scientific, they would succeed where Creationism failed in their common objective of getting their views taught as science in the schools. Although they did reiterate Paley's claim that the eye was a complex organ

10. Paley, *Natural Theology*.

with a specific purpose and so must have been the creation of a Designer,[11] they left out Paley's conclusion that this meant there must be a God. Instead, they emphasized the complexity of the eye—along with a number of biochemical process that they claimed were also "irreducibly complex" (i.e., so interconnected and complicated in functioning that they could not have evolved)—to argue that evolution was a dubious theory, and that intelligent design "science" should have equal time in science classrooms in public schools.

In the case of "Kitzmiller v. the Dover Area School District," the court ruled against permitting Intelligent Design to be taught as science in public schools for two primary reasons. First, no scientist promoting Intelligent Design had ever published an article or book that presented scientific evidence in support of Intelligent Design, nor had its supporters ever submitted their work for the standard peer review process prior to publication. Second, biologists working in the field had shown that the claim that "irreducibly complex" biological systems could not have evolved was false. Third, and perhaps most damaging, was the leak of an internal document of the Discovery Institute, the supposedly scientific organization promoting Intelligent Design, which stated that their goal was to counter the teaching of Godless evolution and promote an understanding of the world as God's creation. The judge in the case, himself a Christian, not only rejected Intelligent Design as valid science, but also castigated its proponents for their dishonesty in attempting to hide their true beliefs and motives. This has done nothing, however, to lessen the enthusiasm of its followers in the Intelligent Design argument.

The third option regarding how God created the world is the one I have referred to above as Scientific Theism. This view affirms the belief that any theist must affirm—that God is the Creator, the ultimate reason why a law-governed universe exists—and, at the same time, acknowledges that science is a wholly valid source of knowledge and provides our best means of discovering the way God chose to create the universe. This is, of course, the position I espouse in this book. In my understanding, God created the universe by establishing and holding in continuous operation (as metaphysical cause) the laws of nature that govern the nature and behavior of the universe and everything in it, so that the universe does not dissolve into a lawless chaos. However, precisely what those laws are, and how they have operated throughout the history of our universe and brought us to where we are today, is what science is best able to discover.

11. Ibid., ch. 3.

We now know that the universe is not static, as virtually everybody, scientists included, believed until the early twentieth century. Rather, it is dynamic and continuing to evolve in novel ways. Therefore, most theologians today (other than those subscribing to biblical inerrantism) no longer separate the doctrine of God's creation of the world from the doctrine of God's *providence*, or continuing action in the world, as has been the norm within traditional Christian theology. Rather, they essentially fuse these two doctrines into one doctrine of *creatio continua* (continuing creation). This doctrine affirms that God continues to be active in creating our universe. It also affirms that *all* God's action in the world has always been fundamentally creative and always will be, and that the aim of God's ongoing, creative action is to uphold and keep in operation the laws that govern the universe, and to bring novel beings and more ideal modes of existence into the world. It also means that the creatures in the world, like the creator, are creative. By means of their behavior, they also contribute to the advancement of the world in new directions. God seeks to use this creaturely creativity to contribute to God's aims in the world, which is above all to foster harmonious and satisfying relationships with and among all creatures capable of entering into such relationships.

The Scientific Theism position affirms that science helps us understand the means God has used to create the universe. Therefore, it also affirms that evolution is the means by which God has created the unimaginably varied life forms on our planet, ourselves included. This specific aspect of the larger scientific theism view is called Theistic Evolution. There are varying views within this position on whether and, if so, to what degree and in what way God acts in shaping or influencing the direction of evolution (this will be addressed in chapter 17). Nevertheless, all are agreed that evolution and Christianity are not at odds. In this way, they reject the thesis that proponents of both Creationism and atheism advocate—namely, that theology and science are necessarily in conflict with one another.

One more question is religiously and theologically relevant in our discussion of the doctrine of creation, and that is *why* God would create the universe. To address why God would *do* anything, it is essential to determine who God *is*, what God is like, since a person's actions reflect the kind of person one is. The answer to this logically prior question constitutes the heart of Christian thought and is the one absolutely fundamental affirmation for any and all authentically Christian theologies: namely, that God is in God's very essence *Perfect Agapic Love*.

The English language is impoverished in its words for love. We say we love our car, we love our job, we love dogs and cats, we love our friends, we love our boyfriends, girlfriends, and spouses, and we love God. Surely, there are vast differences among these kinds of love. Among all these loves, we might distinguish three types, depending on our concerns in each. There is the kind of love in which we are solely concerned with fulfilling our own desires, such as having a great car or, sadly, in some cases, getting what we want from other people without really caring about them in themselves. This is approximately what the ancient Greeks called *eros*: desirous love. Then there is the kind of love in which we are concerned that both we and the other we love get something good out of that relationship, such as in friendship. This the ancient Greeks called *philia*. A third kind of love is one in which we are so concerned with the welfare of the one we love that our own wants and needs are forgotten. This is what the Greeks termed *agape*.[12] Agapic love is completely other-centered. We see it only rarely in this life, it seems, in instances such as the soldier who falls on a grenade and dies willingly to save his friends, or in the person who runs out in front of a speeding car to push another to safety though it costs her life.

To claim that God is Perfect Agapic Love is to claim that God's Love is *not self*-concerned, but rather wholly concerned with the welfare of all creatures. Being *agapic*, the kind of love Jesus and NT writers attributed to God, God's Love is so wholly concerned with the well-being of all creatures that God's own well-being is not a concern.[13] In fact, in choosing to create a universe of creatures, God has guaranteed that God will suffer, because real love suffers when the beloved suffers, and suffering is inevitable in this world. In this, God's Love is like that of loving parents, who also suffer with their children and risk being rejected by the very ones who gave them life. God suffers *all* the suffering of *every* creature just as that creature feels it, throughout this and all other universes that might exist, a suffering we cannot begin to conceive. That is the magnitude of God's Love. However, God also does not let this suffering stand, but tries always to bring some goodness out of such evil. This puts an enormous responsibility on humans, in particular, with our self-awareness and freedom, in that, when we cause others to suffer, we are also causing God to suffer. And yet, even when we cause God to suffer, God never ceases to love us and to bring us into agapic love for others, for God's Love for us is not conditional, based on what we do, but unconditional, given to us simply because we are God's creations. This ceaseless and perfect Agapic Love is the reason God, and only God, is worthy of worship.

12. The word used for this kind of divine love in the HB/OT is *hesed*.

13. This is the reason Christianity rejects the God of the Deists, who creates the world and then sits back like some great cosmic voyeur, watching what happens but never interacting with creatures.

On the other hand, that God's Love is wholly altruistic does not mean that God gets nothing positive from it. As creatures return God's Love, God experiences it and enjoys it. God also experiences and enjoys the love shown to creatures by other creatures, their enjoyment of it, and every other enjoyment or feeling of satisfaction that each creature experiences. The key point here, however, is the *motive* in God's Love for the creatures, which is always centered on the welfare of the creatures, not on God's own well-being.

This means that those theologians, past and present, who have claimed such things as that God is concerned that people worship God and that this is their primary duty in life have utterly failed to understand the fundamental teaching that, as the author of the first letter of John in the NT said, "God *is* Agape" (1 John 4:8 and 16, emphasis added). God does not need our worship or prayers. It is we who need these things to help us center our lives on God, in whom we can find our greatest possible fulfillment and peace in this life by, like God, loving others agapically and experiencing such love returned to us. This is God's intention for the creation: that, through agapic love, every creature should find fulfillment and peace in whatever way is possible for it.

It is out of this profound, otherworldly Love, which is who God is, that God acts. As we know, real love involves both feeling and action. We can claim we love someone, but if we never act to try to make that person's life better, it would not really be love. Therefore, that must be true for God as well. This is the answer to why God created the universe. God expresses the infinite divine love by creating a universe and laws that govern it in such a way that creatures can come into existence and experience whatever it is that brings them enjoyment and satisfaction, and God does this solely out of Agapic Love. This also explains the continued existence of the universe. Should God decide to withdraw support of it by letting the natural laws go out of existence, the universe would dissolve into a sea of chaos; but it does not because of God's Love. God's Love initiated the creation by means of the natural laws God established, and God's Love is what upholds those laws in such a way that the universe continues to exist as a cosmos.

There is more. God not only creates (via evolution) all creatures but also works in every moment to bring them enjoyment in every positive way possible. If creatures cooperate with God and God's will, to whatever extent that is possible for them, they will love others in the form they are able to love and so experience the greatest satisfaction possible for them as creatures. Because we humans, like all other creatures, are essentially relational, maximum satisfaction can come only when those whom we love also experience maximum satisfaction; and thus it is that God's aim is to attempt to elicit love in all for all, so that a harmony exists among us all.

This love between all creatures is, therefore, God's ultimate goal, expressed, as Jesus put it, that we love one another as we love ourselves—because, by loving others, we are helping ourselves to experience love in return. Again, in agapic love, everyone wins.

Unfortunately, this fundamental Christian truth of God as Agape has been obscured in certain strains of Christian theology over the past two thousand years, particularly those most influenced by the ancient Greek philosophical tradition, in which feelings of any kind were thought to be a defect, and so they were denied of God. I think that most, if not all, theologians over the past two thousand years would agree with the claim that God is Perfect Agapic Love, in at least some sense, but I have not found many who in practice, in the theologies they have created, have really taken God as Infinite Agapic Love to heart or have recognized how it must necessarily determine all other claims about what God is like and how God acts. We will explore this problem more fully in chapter 10.

In concluding this chapter, I want to address, from the point of view of a theologian, two questions that have emerged from our current scientific understanding of the world and are relevant to the discussion of God as creator. The first is the question of whether life exists elsewhere in the universe. Just as scientists are divided in their answer to this question, so too, I would expect, are theologians. The same is likely true of the second question: whether our universe is the only one to exist and to have ever existed, or whether a multitude, perhaps an infinity, of other universes have existed in the past, exist in the present, and will come to exist in the future.

In answer to the question of whether there is other life, including intelligent life, in our universe, it is on the basis of the central Christian claim about the nature of God as Perfect and Infinite Agape that I answer this question in the affirmative. The logic is simple: God is the perfection of Agapic Love and, because God is infinite in nature,[14] God desires to be in loving relationship with an infinite number of creatures. Out of profound, self-sacrificial Love, God gives the gift of existence to the universe and its creatures and showers them with Agapic Love. Because God's Love is infinite, I do not believe God would establish the laws that govern our universe in such a way that they would restrict the gift of life to the inhabitants of a single planet. Rather, I believe those laws would be established that would make the existence of life, including intelligent beings, virtually inevitable throughout the universe anywhere the conditions for life exist and there has been sufficient time for it to evolve. Scientists have sometimes accused theologians that they make no testable predictions. This is an empirical

14. We will explore the notion of God's infinity in chapter 10.

prediction that we should be able to test someday, perhaps in the not-too-distant future.

By the same logic, I do not believe a single universe would suffice for God's Infinite Agapic Love. Therefore, I would expect that an infinite number of universes bearing an infinite number of creatures have existed in the past; that many, if not an infinite number of such universes, exist at present; and that an infinite number of other universes will exist into the endless future. This, to me, is simply a logical implication of belief in God as Infinite and Perfect Agapic Love. Empirically testing this hypothesis is impossible, given our current technological ability and, indeed, may forever be beyond our capabilities, but it is another empirical prediction I would make based solely on theological premises.

8

From the First Stars to the Creation of Our Galaxy

DURING THE COSMIC DARK Age, from about five to two hundred million years aBB, the slightly denser clouds of hydrogen and helium gas[1] that were distributed throughout the expanding universe continued to draw in more gas, aided significantly by the gravitational clumping of dark matter within and around them. By about one hundred million years aBB, when the universe was some 80 degrees Kelvin (about -193° C, -320° F), the strength of gravity in the densest clouds caused these regions to stop expanding outward along with the space-time fabric and to begin their long, slow collapse inward toward their centers of gravity. By about two hundred million years aBB, the most massive and large-scale clumps of gas and dark matter had condensed sufficiently to form the broad outlines of what would become galaxy clusters, with smaller concentrations of matter within them drawing together into individual *protogalaxies*, and even smaller clumps within these forming *protostellar clouds*. Eventually, these latter clouds would collapse under the relentless pull of gravity to become the first stars.

The gravitational pull from matter outside these protostellar clouds caused them to begin rotating slightly. As they continued to shrink, due to the pull of gravity, their rotational speed increased, just as spinning ice

1. These early gas clouds also contained trace amounts of lithium and perhaps even beryllium, as we noted above. However, because hydrogen and helium were the overwhelmingly predominant gases in these clouds, and for the sake of brevity, I will hereafter refer to these clouds as being composed simply of these two elements.

skaters rotate faster when they pull their outstretched arms in toward their bodies. The spinning of these protostellar clouds introduced a new, outward or *centrifugal force*, which worked in opposition to the inward pull of gravity, slowing the protostars' collapse.

Yet another dynamic was at work in these gas clouds. As they shrank, they became denser, so that the hydrogen and helium atoms that made them up drew closer together. Because they were closer together, they collided more frequently with each other, causing them to become more energetic, which meant that the clouds grew hotter. This heat created pressure within the gas clouds, causing them to expand and, in this way, a second outwardly-directed force was created to counter the inward pull of gravity.

The collapse of the protostellar gas cloud would halt until this heat could be dissipated, which could only occur as the energized gas atoms dissipated their energy out into the colder space surrounding the cloud. How much energy these gases could dissipate was related to how hot their surroundings were. The hotter the protostellar cloud and the space around it, the slower the rate of energy release by the gaseous atoms. For hot clouds of insufficient mass, the collapse would have halted altogether or, if the outward push of the energetic gases and the centrifugal force of the rotating gas cloud were great enough, the cloud might even have dissipated. The result of this interplay of forces and temperature was that only the most massive protostellar clouds in the early universe, those with the greatest gravitational force, would eventually collapse to form the first generation of stars.

However, even the largest protostellar clouds had to deal with this dynamic of gravitational collapse halted by pressure from the heating up of the increasingly dense gas. Once the hydrogen and helium that composed the cloud dissipated enough energy, the force of gravity could again dominate, causing the further collapse of the cloud. This long, slow dance of compression, followed by expansion and then by compression again, was repeated countless times, so that the eventual evolution of these dense protostellar clouds into stars took place only after many hundreds of thousands of years.

The process of the final collapse of these clouds and resultant creation of the first stars is now well understood. The massive protostellar clouds of hydrogen and helium continued to compress into ever smaller areas of space, until finally the heat became so great—that is, the velocities of the molecules were so great, and their collisions in the densely packed cloud were so violent—that the electrons were stripped from the hydrogen and helium atoms, ionizing them into single-proton nuclei, proton-neutron isotopes of hydrogen, and the two-proton-and-neutron nuclei of helium and its isotopes. This, in effect, returned the central regions of these protostellar clouds to the condition in which the whole universe had existed before the decoupling of radiation and matter and resultant transparency of the cosmos that occurred around 380,000 years aBB.

With their electrons stripped away, the ionized protons (hydrogen nuclei) collided with other ionized protons and, through a complex pathway of reaction, fused into helium nuclei.[2] In this fusion process, large quantities of energy in the form of gamma-ray (extremely high-energy) photons were given off. This energy gradually worked its way to the outer edge of the infant star, from which point it was radiated into space as heat and light. In this way, the first generation of stars in the universe was born. As they began to stream photons in the visible wavelengths out into the surrounding space, light came once again into the universe, and the cosmic Dark Age came to an end. From that point on, all the light within the universe would come from stars.

As these first stars ignited and began sending enormous numbers of photons out into the surrounding gas clouds, the photons knocked electrons from their orbits around the hydrogen and helium gas molecules, leaving bare nuclei in their wake. In this way, the first massive stars ionized the gas around them, creating huge bubbles of ionized hydrogen and helium up to tens of thousands of light-years across. As they expanded, these bubbles collided with ionized bubbles from other newly-created stars, forming yet more regions of greater gas density, which could then begin their own collapse to form still more first-generation stars. These new stars in turn spawned more expanding, ionized clouds of gas, which collided with yet other such gas clouds to produce still more stars. In this way, more than 99 percent of the hydrogen and helium gas in the entire universe was eventually ionized, returning the universe to a state similar to that which had existed before the formation of the first atoms 370,000 years aBB or in the central regions of the protostellar clouds, though lacking the far greater density of both earlier forms. For this reason, this era in cosmic history is termed the *Reionization Period.* In all, this reionizing process took some half a billion years to

2. A proton hitting another proton will not fuse with it for very long, due to the repulsive electromagnetic force acting on the two positively-charged particles. For a *diproton nucleus* to last, one of its protons must convert into a neutron, which is created when a proton decays into a neutron and gives off a positron and a neutrino and so forms a deuterium nucleus, or deuteron. Because this occurs only rarely, the more common outcome is that the two protons split apart again. The rarity of the proton-to-neutron conversion is the main reason medium-sized stars like our own last so long—on the order of billions of years—before all their hydrogen has fused into helium in their cores. Once a deuteron has been created, however, it quickly combines with another proton to form a three-particle helium (^{3}He) nucleus. Eventually two ^{3}He nuclei collide to form a highly stable ^{4}He nucleus, giving off two protons, which are then free to begin the cycle anew. This typical way of forming helium nuclei from hydrogen nuclei is termed the *proton-proton cycle.*

complete, concluding around eight hundred million years aBB and leaving in its wake a thin cosmic fog of hydrogen and helium ions.

Because the first stars were extremely large and massive, they dwarfed the kind of stars typically found in the universe today, such as our Sun. They were at least one hundred, and perhaps up to one thousand, times more massive than the Sun and were extremely dense and hot, appearing blue in color. Because of the enormous pressure and heat at their cores, these first stars burned their nuclear fuel at a much faster rate than the smaller and cooler later-generation stars eventually would. This meant that these first stars lasted a relatively short time before their fuel was spent—perhaps a million years for the lighter of these first stars, and only ten thousand years for the most massive.

When most of the hydrogen in the cores of these first stars had been fused into helium, the rate of reactions dropped precipitously, reducing the outward pressure created by the energy released in the fusion process. As a result, the star's massive outer layers collapsed under their own enormous weight, compressing the core and raising its temperature to some 20 billion degrees. This allowed the helium that had previously been created there to fuse into yet heavier elements deeper in the core, including large quantities of lithium (3 protons) and beryllium (4 protons). After that the helium, lithium, and beryllium could combine in various ways to form boron (5 protons), carbon (6 protons), nitrogen (7 protons), and oxygen (8 protons), all of which are essential to life as we know it. This fusion process proceeded, with heavier nuclei continuing to fuse deeper in the star's core to create yet heavier nuclei, all the while releasing energy as a by-product. This fusion process finally ended with the creation of iron nuclei, which contain 26 protons (plus neutrons) and are so stable that energy is required, not given off, for any further fusion to occur. Because of this stability, the energy created by collisions of these iron nuclei with each other was absorbed, rather than expelled. With fusion and, therefore, energy production having essentially ended in the core, the outward pressure that had previously existed ceased.

At this point, the outer layers of these early stars collapsed once again, sending shock waves rushing through the iron core in all directions. This caused the inward-falling solar material to explode violently outward, sending iron along with the nuclei of yet heavier elements forged in this explosion—including gold, mercury, uranium, and all the other naturally occurring elements of the periodic table—and vast numbers of high-energy photons into space at nearly the speed of light. This whole explosive star-death process evidently took only about thirty seconds. Astronomers term the explosive deaths of the relatively smaller of these first stars *supernovae* (plural of *supernova*) and of the most massive stars *hypernovae*.

After the supernovae, the remnants of the less massive of these first stars collapsed to form *neutron stars*, objects whose density is so great that the electrons in their core atoms are crushed into their nuclei, where the electrons combine with protons to form neutrons compressed to unbelievable densities. Such neutron stars typically are only some 24 kilometers (14.4 mi) across, and yet a tiny speck of such a star would weigh more than Mt. Everest.

The remnants of the most massive of these early stars, following their hypernova explosions, collapsed into even denser objects called *black holes*, objects whose gravity is so great that nothing, not even light, can escape once it has been drawn into them. Super-massive black holes of up to several million solar masses formed the cores of at least some of the earliest galaxies. Their immense gravitational attraction drew in huge amounts of gas and dust, which were energized as they swirled in toward the black hole and their ultimate death. The massive amount of energy released in this death dance near the black hole was so enormous that astronomers coined a new term to describe these unique energy sources: *quasars* (short for "quasi-stellar radio sources"). Scientists have observed these quasars, millions of times more energetic than typical black holes today, only in galaxies from this earliest period in the history of the universe. The black holes at their centers might be billions of times the mass of our Sun. Eventually the gas supply feeding these black holes was exhausted, and so the feeding frenzy that had caused the quasars ended. As a result, these super-massive black holes at the centers of the first galaxies became relatively quiescent.

The shock waves created by the supernova and hypernova explosions of the first stars carried the mix of elements created in their cores and in their explosive deaths out into space, where they provided new material, in addition to the preexisting hydrogen and helium gas, for the creation of the second generation of stars in the universe. The pressure of the shock waves, as we have noted, heated the nearby frozen gases and compressed them, while also adding to them the newly expelled heavier elements, collectively referred to by astronomers as *dust*. These interstellar dust particles were typically the size of grains of sand and were predominantly composed of silicon and carbon compounds. The temperatures within these clouds dropped as they encountered the frigid space around them, causing gas to freeze onto these dust particles. Such relatively dense clouds of gas and dust, called *nebulae* (plural of *nebula*), were the nurseries within which all subsequent stars have formed, as they still are forming.

For a zoomable image of "A Giant Hubble Mosaic of the Crab Nebula," the result of a supernova in the year 1054, see http://hubblesite.org/newscenter/archive/releases/2005/37/image/a. Credit: NASA, ESA, J. Hester and A. Loll (Arizona State University).

The ultracold conditions within these nebulae, and the resultant slowness with which the gas and dust that composed them moved, combined with the relatively large distances between them, meant that these particles rarely collided. When they did, however, the gas molecules on the dust could thaw and so interact with one another, sometimes combining into more complex molecules. Some of these complex molecules were *organic*—that is, carbon-based, molecules that constitute the basis of all life on earth—and would become part of the mixture of molecules that formed the planets that revolved around the second-, third-, and possibly even later-generation stars.

Due to the novel presence of this dust in the nebulae, when protostellar clouds began to constrict under the pull of gravity, they were significantly denser than the first-generation clouds. As they continued to collapse and heat up, these dust-filled clouds were able to radiate the heat away with greater efficiency than could the earlier-generation hydrogen-and-helium clouds. This meant that smaller clouds could now collapse to form less massive stars, the kind most typical in the universe today. These second-generation stars ranged from about one-tenth to about one hundred times the size of our Sun, with the lightest stars releasing only about one ten-thousandth the amount of energy as the Sun today and the largest stars some hundred thousand times as much. Because these stars were smaller than the first-generation stars, their interiors were less pressurized and therefore relatively cooler, so they burned their hydrogen more slowly, giving them much longer lives. The lifespan of these stars, as with all stars in the history of the universe, was inversely proportional to their size. Therefore, the largest second-generation stars lived relatively short lives. Astronomers have actually observed the supernova of one such star (named GRB 181913) that occurred less than 900 million years aBB, or about 12.8 bya. The smaller of these second-generation stars lasted many times longer than the largest stars, and it is possible that the smallest of them are still burning today, some 12 billion years or more after they formed.

Once these second-generation stars had turned on, they heated the dust and gas-ice around them, liquefying and gasifying the compounds that had formed on the dust grains, enabling them to react with one another. This phase typically lasted less than 10,000 years after a star's solar furnace ignited. Within these regions, however, many more complex carbon compounds arose, including *ethanol*, *methanol*, and perhaps even certain *amino acids*, the latter of which can combine to form the *proteins* that are the building blocks of all life as we know it.

By about one billion years aBB, most of the early stars in any given region of the universe had drawn close enough to each other to form clusters of stars, with large numbers of these *star clusters* (and the dark matter permeating them) drawing together under the force of gravity to create the infant, irregular galaxies that we see today through our most powerful telescopes. In these first galaxies, most of the star birth occurred in their central regions, unlike the later spiral galaxies, in which most of the star birth has occurred in their outer arms. At this point, the universe had grown to perhaps 10 billion light-years across, and its temperature had dropped to a hyper-frigid 10° K (-260° C, -440° F).

Because the universe was still relatively compact, galaxies were much closer together then than they are today. Consequently, their massive gravitational fields frequently drew them toward one another. In some cases, this resulted in head-on galactic collisions. In other cases, the galaxies passed close to one another. In both cases, a complex dance was initiated, in which the gravitational pull of the two galaxies and the stars within them distorted each other's shapes significantly. Clusters of galaxies were the typical result of this early gravitational dance.

For an image of the two interacting galaxies of Arp 87, see http://heritage.stsci.edu/2007/36/big.html. Credit: NASA, ESA, and the Hubble Heritage Team (STScI/AURA).

Because all galaxies were (and are) composed mostly of empty space, the stars within colliding galaxies very rarely collided. Nevertheless, the interaction of the massive gravitational forces between and within the galaxies changed the orbital patterns of their constituent stars and star clusters, creating shock waves that caused the free gas and dust clouds within them to collide. These collisions produced countless new stars, many of which were blue giants whose short lives ended in supernovae or hypernovae. These explosions sent still more matter and shock waves out into the free gas-and-dust mix in the slowly-merging galaxies, triggering yet more new star formation. The galaxies continued these complex interactions over hundreds of millions to several billions of years, typically resulting in the creation of larger, more mature, *elliptical galaxies*. Probably already in existence by about 3 billion years aBB, these elliptical galaxies ranged in shape from the roughly spherical to significantly elongated ovoid varieties.

Thus, as stars were born, lived, and died in the early universe, the galaxies within which they lived also were born and evolved over time, adding gas and dust, freely roaming stars, and sometimes other galaxies to form still larger "island universes," all under the relentless pull of gravity (including that created by dark matter). Meanwhile, as ever, the universe continued to expand, stretching the space-time fabric, thereby creating larger spaces between structures too far separated to be drawn together by each other's gravitational tug, while simultaneously creating ever more dark energy to fill the expanding volume of space. The more dark energy there was, the more it countered gravity's tendency to slow down the universe's expansion by pulling space-time and the objects within it outward. Thus, three large-scale forces continued to be at work determining the expansion of the universe: the forces of the Big Bang and dark energy pushing and pulling the universe outward, and the force of gravity pulling it inward. Throughout the Matter Age, which lasted until some nine billion years aBB, gravity was the predominant force, continually slowing the expansion rate of the universe and causing relatively nearby structures within it to draw inexorably closer together.

The period of frequent galaxy collisions reached its peak around 2–3 billion years aBB, giving rise, as we have seen, to the earliest elliptical galaxies. Because of the relentless stretching of the space-time fabric and resultant greater distances between galaxies, the pace of collisions has slowed ever since. Nevertheless, such collisions still occur today, even if less frequently, and will continue to occur in the future, though ever more rarely, before eventually ceasing altogether, when gravity can no longer meaningfully counteract the universe's increasing expansion under the pull of dark energy.

As the second-generation stars formed, they, like the first-generation stars, began to spin along with the gas and dust around them. The centrifugal force created by the rotation of the further-out matter and dark matter at some point countered the inward force of gravity, creating an orbital equilibrium that prevented this outlying matter from falling into the protostar, as did the gas and dust closer to it. Gravity gradually flattened this gas and dust cloud, drawing it toward the central plane of the star's equator and causing it to rotate in sync with the star's rotation. As gravity continued to act, this *accretion disk* around the forming star continued to flatten until it finally formed a relatively thin disk. Within this disk, areas of greater density drew ever more matter to themselves, gradually creating larger structures. The end product of these gradual mergers of gas and dust over hundreds of thousands of years was the creation of the first planets, moons, and other smaller objects that, along with their parent stars, constituted the first solar systems.

The nature and abundance of these first planets are unknown. Nevertheless, as already noted, current research suggests that many, if not most, stars have planets orbiting them. The planets' small size relative to that of the stars they orbit, and the enormous glare caused by those stars, make the direct observation of planets around other suns, even those closest to us, enormously difficult. Nevertheless, technological advances in recent decades have enabled astronomers to locate a number of these extra-solar planets, or *exoplanets*. This is typically done either by noting the dimming in their star's light that occurs as they pass between the star and us, or by observing a slight wobbling in the star, caused by the gravitational tug of the planet or planets as they move in their orbits around the star. By these means, in just the past decade and a half, nearly a thousand exoplanets have already been identified, revolving around stars relatively close to us, along with several thousand candidate bodies that require further analysis to determine whether they too are exoplanets. We can expect many thousands more to be discovered in the coming decades; indeed, astronomers now generally believe that our Milky Way galaxy alone contains hundreds of billions of planets, or about one for every star, and that something like seventeen billion of them are Earth-sized.

The finding of yet more exoplanets will help us answer questions, not only about the nature and abundance of these extra-solar planetary systems themselves, but also about which of them might have environmental conditions comparable to those on Earth. The current "holy grail" of exoplanet astronomers is finding roughly Earth-sized rocky planets orbiting in what is called the *habitable* or *Goldilocks zone*—close enough to a star for water to remain liquid, but not so close as to have lost their protective atmosphere to the star's solar wind. Such planets almost certainly exist, awaiting only new technological advances to enable us to detect them.[3]

3. On July 23, 2015, NASA announced that the "Kepler exoplanet explorer" confirmed the discovery of "the first near-Earth-size planet [Kepler-452b] in the 'habitable zone' around a sun-like star. This discovery and the introduction of 11 other new small habitable-zone candidate planets mark another milestone in the journey to finding another "'Earth.' . . . 'We can think of Kepler-452b as an older, bigger cousin to Earth, providing an opportunity to understand and reflect upon Earth's evolving environment,' said Jon Jenkins [leader of the team at NASA's Ames Research Center in Moffett Field, California that discovered Kepler-452b]. 'It's awe-inspiring to consider that this planet has spent 6 billion years in the habitable zone of its star; longer than Earth. That's substantial opportunity for life to arise, should all the necessary ingredients and conditions for life exist on this planet.'" NASA's announcement goes on to report, "the Kepler team has increased the number of new exoplanet candidates by 521 from their analysis of observations conducted from May 2009 to May 2013, raising the number

Because of the organic molecules that probably rained down on these planets along with the rock and ice bodies on which they formed, it is possible that some of these first planets generated life, with some perhaps having evolved intelligent creatures more or less like ourselves. Perhaps some of these inhabitants attained intellectual and technological achievements far surpassing our own. Eventually, we might even come to know some of them. They might even now be aware of our existence and intend to contact us, once we have matured as a species beyond our current violent adolescence. Perhaps some of these intelligent species have died out, as ours too may someday. Given the magnitude of our galaxy alone, with its estimated 200–400 billion or more stars, these seem reasonable, even highly likely, hypotheses. Because the visible universe contains perhaps 10^{23} stars, the likelihood of life evolving on planets surrounding these stars becomes virtually certain.

As the early-generation stars and star clusters were drawing together to become the first galaxies and galaxy clusters—the largest of which, termed *superclusters,* would come to contain as many as a thousand galaxies in addition to enormous quantities of hot intergalactic gas and dust—gravity was also acting to shape the still-larger-scale structure of the universe. On the largest scale, strings and sheets of galaxy clusters and superclusters slowly formed, with huge voids of essentially empty space between them, giving a three-dimensional, web-like appearance to the universe. These largest-scale features of the universe were probably fairly well established by about six billion years aBB, though they would continue to become more sharply defined as gravity continued to pull the galaxies and galaxy clusters in these filaments and sheets closer together.

By five billion years aBB, the universe had grown to perhaps twenty billion light-years in size, and its temperature had dropped to about 5° K, or nearly as cold as it is today (about 2.75° K). The first spiral galaxies were beginning to form then, and by about eight billion years aBB, numerous

---

of planet candidates detected by the Kepler mission to 4,696 . . . Twelve of the new planet candidates have diameters between one to two times that of Earth, and orbit in their star's habitable zone. Of these, nine orbit stars that are similar to our sun in size and temperature." NASA, "NASA's Kepler Mission Discovers Bigger, Older Cousin to Earth." According to another July 23, 2015 NASA press release, "The discovery of a super-Earth-sized planet orbiting a sun-like star brings us closer than ever to finding a twin of our own watery world. But NASA's Kepler space telescope has captured evidence of other potentially habitable planets amid the sea of stars in the Milky Way galaxy . . . The new discovery, Kepler-452b, fires the planet hunter's imagination because it is the most similar to the Earth-sun system found yet: a planet at the right temperature within the habitable zone, and only about one-and-a-half times the diameter of Earth, circling a star very much like our own sun. The planet also has a good chance of being rocky, like Earth, its discoverers say." NASA, "Finding Another Earth."

spiral galaxies, including our Milky Way, were in existence. How these spiral galaxies formed is still not well understood. Today it is widely believed that spiral arms form in galaxies when some faster-moving matter, perhaps riding a pressure wave caused by the supernova of a star in the region, collides with matter revolving more slowly around the outer regions of the galaxy. The faster-moving matter crashes into the slower material, triggering the formation of new stars, often the hot blue stars that give the spiral arms their characteristic color and brilliance. The pressure wave moves on until it collides with more slower-moving gas, triggering another round of star-formation. Thus, it seems to be pressure waves moving relatively rapidly through the well-populated outer regions of some galaxies that are the ultimate cause of the beautiful curving arms we observe in spiral galaxies today. These galaxies are estimated to constitute some 75 percent or more of all galaxies in the universe.

For the NASA, ESA, and the Hubble Heritage (STScI/AURA)-ESA/Hubble Collaboration image of the spiral galaxy M74 by R. Chandar (University of Toledo) and J. Miller (University of Michigan), see http://heritage.stsci.edu/2007/41/index.html.

At about nine billion years aBB, the force of contraction in the universe due to the inward gravitational pull of ordinary and dark matter finally became about equal with the force of expansion created by the force of the Big Bang and the outward pull of dark energy. Until this time, the universe had continued to expand, but did so ever more slowly under the force of gravity. Then, with the continued expansion of space and concomitant increase in the amount of dark energy, the universe began to accelerate its expansion. This point marks the end of the Matter Age in the history of the universe and the beginning of the *Dark Energy Age*. From that time forward, the rate of expansion of the universe has steadily increased. The widely separated structural elements within it—stars, galaxies, clusters, and superclusters of galaxies, filaments and sheets of galaxies—have been receding ever farther away from one another, even while the more proximate stars, galaxies, and larger structures continued to draw more closely together, under the pull of gravity. Coincidentally, it was at about this same time, some 9.15 billion years aBB, or 4.6 bya, that our Sun and its planets, including Earth, were beginning to form.

OUR GALAXY, THE MILKY WAY

The region of space that would eventually, by about half a million years aBB, give rise to our galaxy was already perhaps 0.5 percent denser than the surrounding regions. Over time, under the force of gravity, its density increased, so that by about fifteen million years aBB this region had grown to be some 5 percent denser than the surrounding space. By around 1.2 billion years aBB, its density had doubled to about 10 percent. Astronomers believe that the inner portions of our Milky Way galaxy were already being created at this time, with stars in the outer regions being created or drawn into the galaxy more recently. By four billion years aBB the galactic core was largely formed, perhaps already revolving around a massive central black hole.

For a view of the Milky Way from above, with the Sun's location, by NASA/JPL-Caltech/R. Hurt (SSC/Caltech), see http://www.spitzer.caltech.edu/images/1925-ssc2008-10b-A-Roadmap-to-the-Milky-Way-Annotated-. Note the bar-shaped central galactic core.

The continuing evolution of the Milky Way perhaps included collisions with other galaxies, with one of the earlier-generation galaxies possibly being its ultimate precursor, and one or more other galaxies merging with it during the early universe's period of more active galaxy interactions.[4] While things seem relatively serene in our galaxy today, its strong gravitational pull is slowly drawing the Large and Small Magellanic Clouds, two relatively small and nearby galactic companions, toward it. Many millions of years from now these galaxies will collide with our Milky Way, deforming and eventually being absorbed into it to form a single, larger, and significantly reconfigured galaxy. Some evidence suggests they have collided with our galaxy one or more times in the past, with the Milky Way drawing away stars from these smaller galaxies with each interaction.

Like all spiral galaxies, ours has a relatively flat disc revolving around a central bulge, at the middle of which lies a super-massive black hole named Sagittarius A* (pronounced "A-star"). Residing in the constellation Sagittarius, this black hole is some 2.6 million times the mass of our Sun, yet only some fifteen times its diameter. As we have noted, the Milky Way contains an estimated two hundred to four hundred billion stars, and many of these are continuing to draw closer together into clusters of various sizes. There is also, as in some other spiral galaxies, a massive, dense, tube-shaped

4. Some recent computer models of the formation of the Milky Way, however, have challenged this notion.

collection of stars, which astronomers call "the bar." The stars within this cylindrical structure rotate around its center of gravity, much like a roll of paper towels rotates around the wood or metal rod that holds it.

The two major spiral arms in the galaxy, named Perseus and Scutum Centaurus, begin at the ends of the bar, while other, thinner arms occupy the space on either side of them. A thin halo of diffuse ordinary matter—star clusters and other stray stars, gas, and dust—and an unknown amount of dark matter surround the galaxy, exerting a modest outward pull that keeps the arms from migrating in toward the galaxy's core under gravity's pull. The dark matter also keeps the stars in the outer regions of the Milky Way revolving around the galactic center at about the same rate as the stars in the central bulge, rather than lagging behind as they otherwise would. The diameter of this central bulge is some eight thousand light-years. The overall diameter of our galaxy is one hundred thousand light-years or more.

Our star, the Sun—with its Solar System of planets, dwarf planets, asteroids, meteors, comets, dust, and gas—lies in one of the spiral arms that formed on the outskirts of the Milky Way, some 27,400 light-years from the galaxy's center. The spiral arms are where almost all of the new star formation takes place within the galaxy. Our Solar System has been gradually moving in a relatively open area near the Orion Spur's inner edge (see image above). It is travelling at some 83,500 km/hr (51,900 mi/hr) relative to the rest of the galaxy. Its nearest neighbor, Proxima Centauri, lies some 4.24 light-years (40 trillion kilometers/24.9 trillion miles) away. It takes our Solar System about 230 million years to orbit once around the galaxy's center (referred to as one *galactic year*), travelling at approximately 250 km/sec (560,000 mi/hr). This means that our Solar System has orbited the galaxy about 20 times since its creation, and that the last time it was in its current position relative to the center of the galaxy, amphibians were the predominant form of terrestrial animal life on earth.

Within the wider realm of the visible universe, our Milky Way is part of the Local Group, a relatively nearby collection of galaxies that includes a total of fifty-four galaxies, some of them dwarf galaxies. In addition to the nearby, but smallish, Large and Small Magellanic Clouds, it contains three notably large galaxies, all spirals. Andromeda Galaxy, which recent studies indicate is about the same size as the Milky Way, is also the closest, lying between two and three million light-years away. Andromeda and the Milky Way are caught in each other's gravity, approaching one another at a rate of about 300 km/sec (190 mi/sec), so that the two will collide in approximately three billion years. The next largest of the major local galaxies is the Triangulum Galaxy. Relative to the rest of the universe, the Local Group is moving at a speed of some 600 km/sec (373 mi/sec). It is, in turn, part of the larger Virgo Supercluster, a group of more than a hundred galaxy clusters and groups stretching some 110 million light-years across.

9

The Formation of Our Solar System

As we have seen, some 9.15 billion years after the Big Bang, or about 4.6 bya, our Solar System formed. It started when a massive cloud of gas and dust several light-years across began to collapse, forming a nebula in the outer regions of the inner rim of the Orion arm of the Milky Way. This nebula's creation, and probably that of a number of stars within it, may have been triggered by the explosion of a nearby star, which sent massive amounts of gas and dust outward at enormous velocity. Eventually, this gas-and-dust cloud collided with another such cloud in the nebula, creating an area of increased density, which then began to constrict under the action of gravity, triggering the formation of our *solar nebula*. This smaller nebular cloud was composed of something like 75% hydrogen, 23% helium, and 2% or less of other gases, along with dust that the deaths of earlier stars had forged in their interiors. As this solar nebula slowly shrank, it began to rotate counterclockwise around its gravitational center and flatten toward its rotational equator in the same way that earlier solar systems had formed before. Also, like previous star systems, as this cloud continued to contract it rotated with ever-increasing speed as the cloud grew smaller. Gradually the central bulge shrank and became more defined, and the accretion disk revolving around it flattened, until it became relatively thin and dense as it revolved around the proto-Sun.

It is thought that, once the central bulge had contracted sufficiently, our proto-Sun entered what is called a *T Tauri stage*, which is typical of stars with masses similar to that of our Sun and has been observed in other such stars. During this phase, before nuclear fusion had yet begun in the core of the cloud, gravitational energy generated by the gas and dust's gradual collapse began to power a very strong *solar wind* that blew out into the nebula from the proto-Sun's poles, energizing and pushing nebular gas and dust away from the star. This T Tauri stage was relatively short-lived, lasting probably between 50,000 and 100,000 years.

Eventually—perhaps 100 million years after the solar nebula first formed—the Sun collapsed to a point at which the density at its core became so great that its temperature reached 10,000,000° C (18,000,000° F). At this point, hydrogen nuclei (protons) began to fuse into helium nuclei in its core (according to the proton-proton cycle previously discussed), and our star was born. Of all the matter in the original solar nebula, about 99.86 percent of it ended up in the Sun, with the remaining 0.14 percent forming all the planets, dwarf planets, moons, and other smaller bodies in our Solar System. The Sun's enormous gravity meant that, throughout its roughly ten-billion-year lifetime, it would continue to draw more matter, ranging from gas and dust particles to asteroids and comets, into itself, though in relatively negligible amounts, and with far less frequency, once the planets around it had formed and their orbits stabilized.

For the birth of several stars, perhaps surrounded by protoplanetary disks, in the Orion Nebula, see http://hubblesite.org/newscenter/archive/releases/1994/24/image/b. Credit: C. R. O'Dell, Vanderbilt University.

The turning on of the star's nuclear furnace sent out the energy released in the fusion process, creating a counter-pressure to gravity, which eventually—once the gravitational and radiation pressures equalized—halted any further solar collapse. The rate at which fusion has occurred since the process was fully under way has remained fairly constant throughout the Sun's history. However, as more hydrogen has been converted to helium, the core has grown heavier, and this in turn has gradually increased its luminosity, so that the Sun has grown brighter by about 6 percent every billion years. Today, the Sun converts its hydrogen into helium at a rate of some seven hundred million tons each second.

The formation of our star was somewhat unusual, in that most of the time, when nebulae collapse to form Sun-like stars, they create two or more stars in close proximity. These begin to revolve around one another, forming binary or, occasionally, even triple or more multiple-star systems. Otherwise, the Sun is a somewhat larger than average star, thought to be brighter than probably 85 percent of the stars in the Milky Way. Of the fifty stars nearest the Sun, only three are more massive. In fact, as astronomers now understand it, most of the universe's stars are not like our hydrogen-fusing Sun at all. About half are "brown dwarfs," stars with masses about 8 percent or less that of our Sun. That means that they cannot sustain hydrogen fusion in their cores, and so can last for trillions of years, doing essentially nothing interesting. Of the other half of stars in the universe—those that do fuse hydrogen into helium in their cores—the large majority of these are "red dwarfs," stars about half the mass of the Sun but larger than brown dwarfs. These fuse their hydrogen into helium at a very slow pace. As a result, they can remain active for trillions of years. Most stars, including red dwarfs and larger stars like our Sun—some 97 percent of all hydrogen-fusing stars—end their lives as "white dwarfs." These are very dense stars with huge masses condensed into small spaces, on the order of the size of Earth, which radiate away their heat slowly over billions, if not trillions, of years until it is all gone, at which point they become "black dwarfs."

At 1,392,000 km (870,000 mi), our Sun's diameter is approximately 109 times larger than Earth's. Its mass is about 2×10^{30} kg (2.2×10^{27} tons), or about 330,000 times that of Earth, and it is some million times Earth's volume. The fact that it is a million times the volume of Earth, yet only 330,000 times as massive, means that the density of the Sun is actually less than the density of Earth. This is because of the Sun's internal structure.

At the Sun's center lies the "core," the nuclear furnace where hydrogen is fused to become helium, sending massive amounts of energy outward.[1] The core extends out from the center of the Sun to roughly 25 percent of its diameter. Ninety-nine percent of the core's fusion occurs inside 24 percent of the Sun's diameter, and at 30 percent there is almost no fusion at all. The density in the core is some 150 times that of water, and the temperature there is nearly 15,700,000° K (28,259,540° F). The core apparently rotates somewhat faster than the Sun's upper layers.

Heat from the core is transferred out into the next layer, termed the "radiative layer," by simple heat radiation. Hydrogen and helium ions in the core lose energy/heat by ejecting photons, which travel short distances before being absorbed by other ions. In this way, the heat travels slowly out toward the Sun's surface. The radiative layer reaches from the core boundary, at about 25 percent of the distance to the surface, to about 70 percent of that distance. At that point its density has dropped by a factor of one hundred.

1. The Sun's core also contains heavier elements, such as carbon, oxygen, neon, and iron, remnants of past super/hypernovae that were drawn gravitationally down into its core from the original solar nebula. While these other elements make up only some 1.69 percent of the Sun, their combined mass is some 5,628 times that of the Earth!

For a photo of the graininess of the Sun's surface, with sunspots (dark patches), observed with the Swedish 1-m Solar Telescope, by Göran Scharmer and Mats Löfdahl, Institute for Solar Physics, Stockholm University, see http://www.isf.astro.su.se/gallery/images/2002/24Jul2002_4877.195-196_color.jpg.

The Sun's outer, or "convection layer," which reaches the final 200,000 km (125,000 mi) from the top of the radiative layer to the Sun's surface, lacks the density required for radiative heat transfer, and so "convection cells" carry heat through this layer to the surface. These columns of "plasma" (ionized gas) carry heat upward, with the temperature dropping gradually until the plasma reaches the Sun's surface. At that point, it is a "mere" 5,700°K (9,800° F). This cooled plasma then sinks back down toward the top of the radiative layer, where it is reheated, and then begins its journey back to the Sun's surface. In the convection layer, the density of the Sun is only about 1/6,000 that of air on earth at sea level; still, it is opaque, its convection cells giving the Sun its grainy appearance when viewed through telescopes.

The Sun's surface is called the "photosphere." This very thin layer, comprising only a few tens to hundreds of kilometers, is where solar radiation/heat begins to stream freely into space. By the time the energetic photons created in the core have reached the photosphere, they have traveled for about a million years! The photosphere is also the layer in which sunspots appear. Sunspots are dark patches of lower temperature, relative to the convection cells that surround them (approximately 3,000–4,500° K as compared to 5700° K), that occasionally form and move across the Sun's surface. Caused by powerful magnetic activity, they can range up to 80,000 kilometers (50,000 mi) in diameter and travel at speeds of several hundred meters (yards) per second.

Solar flares eject solar plasma out through the upper layers of the Sun and far out into space. These plasma clouds travel at enormous speeds, reaching Earth in about two days. Because most of the radiation contained within the flares is of high energies well beyond the visible light spectrum, they are invisible to the naked eye. Solar flares occur at differing rates, depending on the level of activity on the solar surface, typically from one per week to several per day.

For an image of a solar prominence compared to Jupiter and Earth, by the Solar Dynamics Observatory, NASA Goddard Space Flight Center, see https://commons.wikimedia.org/wiki/File:Sun_earth_jupiter_whole_600.jpg.

Solar prominences are dense clouds of relatively cool solar plasma that are ejected from the photosphere and reach out into the "corona" as far as 700,000 km (430,000 mi) from the Sun's surface. Some divide, sending massive quantities of this plasma out into space at speeds of up to 1,000 km/sec (620 mi/sec). Readily visible from Earth, prominences typically form gigantic reddish loops under the influence of the Sun's powerful magnetic fields, though some prominences, called "filaments," stay relatively close to the photosphere. Prominences typically form in a day and continue to exist from a few days to several months.

Above the photosphere lies the Sun's "solar atmosphere," which comprises five distinct layers. Closest to the surface is the "temperature minimum," which, as its name indicates, is the coolest place on or near the Sun. At about 500 km (310 mi), the temperature in this layer drops to "only" 4,100° K (6,920° F). Above this is an irregular, roughly 2,000 km (1,250 mi) thick layer, the "chromosphere," in which the temperature rises to some 20,000° C (36,000° F).[2] At such temperatures, hydrogen gives off light in the red spectrum, and calcium ions give off light in the violet, hence the name of this layer (*chromo*, meaning "color").

The "transition layer" is thin, perhaps 200 km (120 mi) deep, notable for its temperature rising to about 1,000,000° K (1,800,000° F). Above this is the "corona," which extends into space to some ten solar diameters, its volume totaling more than that of the Sun itself. In its hottest places, the corona reaches as much as 20,000,000° K (36,000,000° F), and it is here that the solar wind is born. The solar wind consists of massive numbers of charged particles, primarily electrons and protons of very high energies, which stream out from the corona far into space, creating a bubble within which the Solar System exists.

The "heliosphere," the outermost layer of the solar atmosphere, stretches from the corona out some 100 astronomical units (AU – the distance from the Sun to Earth, on average 149.6 million km/93 million mi) to the edge of the Solar System—though this is a very uncertain estimate. The heliosphere is filled with the charged particles of the solar wind, which for the first 10 billion km (6.2 billion mi) travel at over 1 million km/hr (621,000 mi/hr). At about 75–90 AUs, the solar wind begins to slow to subsonic speed as it encounters the "interstellar medium," the mixture of predominantly hydrogen and helium gases that occupies the space between stars. This is called the "termination shock point." The solar wind continues to slow through another region, the "heliosheath," until it reaches the

2. It is not known precisely why the temperature of the chromospheres increases with distance from the solar surface, though it is thought this is the effect of magnetic energy from the Sun's powerful magnetic fields being converted to kinetic energy.

"heliopause" at perhaps 100 AUs, where the pressure from the solar wind is exactly balanced by the pressure of the interstellar medium. It is here that the heliosphere, the Solar System, and the "local bubble" in which our Solar System exists are all said to end.

Once the Sun's nuclear furnace had ignited and the initial wave of energetic photons had begun streaming out into space, the solar wind began pushing the remaining free gas and dust away from the Sun, some of which ended up in the accretion disk. Over the next ten million years or so, the solar wind pushed this gas and dust out past the orbits of what were becoming the rocky inner planets of Mercury, Venus, Earth, and Mars, and into the farther reaches of the Solar System. The gas giant planets that were forming there (Jupiter, Saturn, Neptune, and Uranus) captured some of this frozen dust and gas. Today, the solar wind continues to carry ions from the Sun out into space. Some of them strike our planet, creating the beautiful displays in the upper atmosphere near the poles that we call the auroras, but also at times playing havoc with our power grids.

While the Sun was forming, the gas and dust revolving around it in the accretion disk continued to evolve. As we have seen, gravity flattened the disk until it became a relatively thin and dense plate-like structure orbiting around the Sun's equator. This resulted in more collisions between the dust grains and frozen gas, some of the slower-moving of which began to stick together and form larger grains due to the "electrostatic effect"—the same mechanism that causes dust bunnies to form under your bed. Gases froze onto some of these dust grains as ice. Once these conglomerations of dust and ice grew to about 1 km (0.62 mi), they began to exert a gravitational force on the surrounding matter, drawing it in to become part of its growing mass. If some of these slightly larger bodies collided with sufficient force, they shattered into smaller pieces again; but if they hit with lesser velocities, the heat generated by their collisions, combined with the force of gravity, caused them to stick together, forming a still larger body.

Over time, gravity caused these bodies of greater mass to continue to grow as they traveled in their orbital paths around the Sun. Eventually, they formed ring-like structures within the accretion disk at various distances from the Sun, with spaces between these orbits containing relatively little matter. For millions of years, these larger bodies orbited the Sun, drawing in more matter as they passed through their rings, gradually thinning them, and so forming the ever-larger bodies astronomers call "planetesimals."

When some of these planetesimals grew sufficiently large and collided, gravity would cause the molten body produced by the collision to become roughly spherical in shape. Continuing accretion caused these to grow still larger and become protoplanets, and eight of these grew to such an extent that they became the planets that dominate the Solar System today. Other kinds of planetesimals that formed included dwarf planets (smaller than true planets but which, like planets, revolve around the Sun), moons or satellites (smaller planetesimals captured into orbits around planets), meteors (relatively small rocky bodies that typically formed early in the Solar System's history but were not integrated into other, larger bodies), comets (relatively small bodies made up mostly of ice but containing some rocky material as well), and asteroids (stony bodies occupying the orbital space between Mars and Jupiter).

The main period of planetary accretion in our Solar System evidently lasted only a few tens of millions of years, beginning perhaps 4.57 bya, so that by about fifty to sixty million years after the solar nebula began to collapse, or some 4.5 bya, the planets as we know them today were essentially formed. The inner rocky planets and the outer gas giants evolved in different ways, due to the differences in their orbital locations. The inner planets accreted slowly from the rocky debris in their orbital paths, but the combination of their relatively small size, resultant relatively weak gravity, and the strength of the solar wind so close to the Sun meant that they were unable to hold the light gases, hydrogen and helium, that made up the majority of the material in the solar nebula. They, therefore, became composed almost entirely of metal and silicate rock. The outer planets, however, formed where there was some ten times more gaseous ice, owing in part to the solar wind blowing the gas away from the Sun and inner planets, which combined with rocky debris and dust in the outer Solar System to form the large, gas giant planets. Being larger meant that their gravitational force was greater—enough to hold the frozen hydrogen and helium gas that the inner planets could not hold—so these planets continued to draw in matter, thereby growing ever larger. The outer reaches of the debris cloud that constituted the early Solar System was in this way greatly thinned by these gas giants.

For the next several hundred million years, the planets continued to sweep up planetesimals, the rate of accretion dropping as the numbers of planetesimals in their orbits steadily declined. By about 4.3 billion years ago (bya) the inner planets, including Earth, had largely cleared their orbits of this debris, so the surfaces of these planets stabilized. Then, somewhere around 4.1 bya, the Solar System began to undergo a cataclysmic change.

According to our best understanding today, when this new phase began, the outer gas giants revolved in significantly tighter orbits around the Sun than they do today, though they were still outside the orbits of the rocky inner planets. Jupiter was the closest to the Sun, with Saturn the next closest, then Neptune, and finally Uranus. Then, for some reason, Jupiter's orbit gradually began moving closer to the Sun, while the other gas giants moved slowly outward. At some time between five hundred and six hundred million years after the Solar System formed, the orbits of Jupiter and Saturn became such that, for every two revolutions of Jupiter around the Sun, Saturn orbited exactly once. This "orbital resonance" meant that every two years Jupiter and Saturn were orbitally aligned. As a result, those orbits slowly shifted from their earlier, almost circular paths to more elongated, elliptical paths. In addition, because when the planets were aligned their gravitational forces were aligned as well, their combined strength caused the other two gas giants, Neptune and Uranus, to move out of their former orbits and into the part of the accretion disk that lay still farther from the Sun. Here their new orbits perturbed the planetesimals greatly, scattering them in every direction.

During the next one hundred to two hundred million years, chaos reigned throughout the entire Solar System. Neptune and Uranus eventually changed positions as their orbits restabilized, making Neptune, not Uranus, the outermost planet in the Solar System. Asteroids, meteors, and comets were hurled everywhere, some flying outward past Uranus, while others were swept up by the four gas giants. Some were caught by Jupiter's and Saturn's gravity and became the leading and trailing collection of small moons that continue moving in tandem around those planets today. However, many of these planetesimals continued inward toward the four rocky planets and the Sun, colliding with them in great numbers. Astronomers have called this roughly two-hundred-million-year period of intense planetesimal-planet collisions the "Late Heavy Bombardment Period." By the end of this period, around 3.85 bya, the planets and smaller planetesimals were all firmly locked into the orbital positions around the Sun that they occupy today.

Thus, today our Solar System consists of eight planets, defined as bodies that orbit the Sun, are large enough to have been rounded to a spherical shape by their gravities, and travel in distinct orbital planes around the Sun's equator. Smaller bodies that orbit around the Sun, but whose orbits do not fit this description, such as Pluto and Haumea, have recently been classified (or, in the case of Pluto, reclassified) as dwarf planets. In addition to these two kinds of planets are the many satellites or moons that revolve around most of the planets and dwarf planets in the solar system, and the smaller asteroids, meteors, and comets that orbit the Sun in their own distinctive places and ways and occasionally collide with the planets, dwarf planets, their moons, and one another. Except for comets, which have so little mass that they are easily perturbed by the motions of the larger members of the Solar System, these smaller bodies all move in the same direction and in roughly the same orbital plane—a reflection of their common origin in the protoplanetary disk.

The innermost planet, *Mercury,* is the smallest, with a diameter of 4,878 km (3,031 mi) and a mass only about a twentieth that of Earth. It is unique among the planets in that, unlike the others, which all lie in essentially the same equatorial plane as they orbit the Sun, Mercury's orbit is tilted seven degrees. It is also unique in that its period of rotation—about 56.6 Earth days—is quite close to the length of its year (the time it takes to orbit the Sun once), which is about 88 Earth days. This means that, for every three rotations, it circles the Sun twice. Orbiting only 57.9 million km (34.9 million mi) from the Sun, Mercury experiences huge temperature extremes. Its sunny side reaches a searing 430° C (806° F) at the equator. In addition, the Sun-facing side is bathed in intense solar radiation carried by the solar wind, which long ago stripped away whatever atmosphere and water Mercury might have had in its early history and, along with them, any possible life. Meanwhile, the half of Mercury that faces away from the Sun averages a frigid -173° C (-280° F). With neither an atmosphere to protect it nor any native means to erode its surface features, Mercury has remained pockmarked by its collisions with planetesimals, most of which occurred in its early years. It is one of only two planets in the Solar System to have no moons.

Venus orbits roughly halfway between Mercury and Earth at some 108.2 million km (67.1 million mi) from the Sun and takes almost 225 Earth days to complete a single revolution around it. Its diameter, at 12,102 km (7,520 mi), is only slightly less than that of Earth. Unlike Earth and the other planets in the Solar System, it rotates on its axis in a clockwise direction, indicating that it must have been struck very hard and/or very often in its early history to alter its original counterclockwise rotational direction. Venus is also unique among the inner planets in that it is permanently shrouded in thick clouds made up almost entirely of carbon dioxide (96 percent) plus about 3.5 percent nitrogen and small amounts of water vapor, sulfuric acid, and other gases. Sulfur evidently belches from the planet's surface into the atmosphere, where it combines with water vapor to form the sulfuric acid that rains down on the planet's surface. The thick cloud cover also creates a greenhouse effect that prevents heat from radiating out into space and so keeps the planet's surface temperature at a scorching 480° C (896° F), even hotter than the sunny side of Mercury. Because of these conditions, Venus has an atmospheric density some ninety times that of Earth. Below the clouds, however, the Venusian sky is clear, with some sunlight penetrating the heavy cloud layer, tinting the sky orange.

Radar imaging of Venus' surface has revealed a landscape consisting of hilly plains and two enormous highland areas, named Aphrodite and Ishtar. Radar mapping has indicated that one mountain range in the Ishtar highlands, called Maxwell Montes, rises above the surface to about the same height as Mt. Everest on Earth (nearly 9,000 m/30,000 ft). Volcanoes dominate these mountainous regions and, while they may no longer be active, they have resculpted the surface of Venus almost totally since the Late Heavy Bombardment Period. Thus, as on Earth, almost nothing remains of the planet's original crust. Evidence also suggests that Venus has experienced tectonic activity, like that on Earth; whether it continues is an open question.

There appear to be numerous hot spots around the planet—more than are found on Earth—where plumes of magma rise up from within the planet and flow up through the crust and out onto to the surface. The only photographs of the Venusian surface, taken by an unmanned robotic craft, show a generally flat, barren landscape of such volcanic rock. Scientists believe the interior of Venus is much like that of Earth, with an almost pure iron core and silicate mantle. However, Venus lacks a global magnetic field like the one Earth has, meaning that its interior must differ somehow from our planet—most likely, the lack of a liquid outer core. Venus is also the only Solar System planet besides Mercury to have no moons.

Earth lies between Venus and the last of the inner rocky planets, Mars, revolving around the Sun at an average distance of 149.6 million kilometers (93 million mi). Because, like most planets, Earth's orbit around the Sun is not perfectly circular but slightly elliptical, its orbit ranges from about 91.5 million miles at its closest point to the Sun to about 94.5 million miles at its farthest point. Earth's diameter is 12,756 km (7,926 mi) at the equator and, like all massive spinning objects, a bit less—12,715 km (7,901 mi)—measured from the North to the South Pole. The slight equatorial bulge of roughly 41 km (25.5 mi) is caused by the outward centrifugal force on the planet as it rotates on its axis at some 1,600 km/hr (1,000 mi/hr) at the equator. Earth tilts 23.5 degrees from the vertical relative to its orbital plane around the Sun. We will cover other features of Earth—its history, interior composition, and moon—in chapter 11.

Mars was named for the Roman god of war, whose symbolic color, not surprisingly, was red. The association of the fourth planet with the god Mars was natural given its rusty red color, the result of the oxidized, iron-rich minerals that cover most of the planet's surface. Mars's diameter is 6,794 km (4,222 mi), or a bit more than half that of Earth, with a mass only about a tenth as great as our planet's. It orbits the Sun at an average distance of 227.9 million km (141.6 million mi), more than half again Earth's distance, and takes 687 Earth days, or almost two Earth years, to complete one revolution around the Sun. The planet rotates on its axis an average of 25.2 degrees from the vertical, slightly more than Earth's tilt; but, because it lacks a large moon to stabilize it, its tilt cycles back and forth from 11 degrees to 35 degrees every hundred thousand years or so. Mars has two small moons, but each is only about 20 km (12.4 mi) across, which indicates that they are almost certainly asteroids captured by the planet's gravity at some point early in its history.

Like Venus, Mars is Earthlike in much of its interior structure and surface features. Like Earth, it has a crust, predominantly silicate mantle, and an essentially iron core, part of which is probably molten, though it lacks as strong a magnetic field as our planet. Its smaller size would have caused the planet to cool much faster than Earth after its formation. The apparent lack of surface geological activity over the past three billion years or so seems to support this judgment. The surface of the planet has some scattered volcanoes, all apparently inactive, including the largest volcano in the Solar System, Olympus Mons, which reaches 22 km (13.2 mi) above the surrounding plains—more than twice the height of Mt. Everest—and has a girth that would cover most of the state of Arizona. Mars also has canyons and valleys, including the enormous Valles Marineris, which, at about 5,000 km (3100 mi) in length, would stretch across the United States from coast to coast. Valles Marineris is also about 100 km (62 mi) wide and 10 km (6.2 mi) deep, making it, to the best of our knowledge, the largest valley in the Solar System.

Mars's most tantalizing features, however, are the numerous canyons and valleys that look like dried river beds and seem to have been carved by flowing water at some time in the planet's past. Recent evidence not only confirms this, but also suggests that, in the distant past, Mars had seas and even shallow oceans over significant stretches of its surface. This means that, during its first billion years, Mars was significantly warmer and wetter and, therefore, more hospitable to life than it is today. In fact, because the planet cooled more quickly than Earth, and it once had a warm climate and flowing water, life might have evolved on Mars hundreds of millions of years before life was even possible on our planet.[3]

Some scientists have suggested that life on Earth might actually have originally come from Mars, carried here on a rocky body that a large impact blasted from the Martian surface, which our planet's gravity captured, and which subsequently crashed into its surface, seeding the planet with life.

3. See Drake, "NASA Finds 'Definitive' Liquid Water on Mars" and "Water on Mars."

Even today, about one out of a thousand meteorites (meteors that collide with Earth and survive in fragments) comes from Mars. As we have seen, possible evidence of life in the four-billion-year-old Martian asteroid fragment, found in the Allan Hills of Antarctica in 1984 and named ALH 84001, has spawned a running, if inconclusive, debate within the scientific community for years.

Whereas water clearly flowed over the Martian surface in the early days of the planet's history, the relative coldness of its interior today, its distance from the Sun, and its thin carbon dioxide atmosphere, mean that water can no longer flow across its surface as it once did. The warmest the planet ever gets, during its summer and at the equator, is only slightly above freezing, while most of the time, and over most of the planet, it is much colder—about the same as in Antarctica. Recent evidence indicates that water remains frozen in some areas just beneath the planet's surface as massive glaciers, where it could conceivably support living entities that evolved in Mars's distant past. There is also photographic evidence that a liquid, perhaps water, occasionally emerges from the sides of certain cliffs and flows short distances down their faces before freezing. Liquid water, therefore, might lie beneath the Martian surface, filled with perchlorate salts (a variety of salts produced from perchloric acid, $HClO_4$) that permit it to remain liquid down to as much as -68° C (-90° F). Scientists have also recently discovered that methane gas pockets exist just beneath the surface of certain areas of Mars. Since methane is a gas often created by biological processes, the existence of these pockets, combined with the possibility of liquid water underground, offers a tantalizing hint that microbial life might still exist beneath the frigid surface of the planet. The search for such life is the principle focus of Martian exploration and will remain so in the coming decades.

The atmosphere on Mars today is very thin: only about 1 percent as dense as Earth's. It is composed predominantly of carbon dioxide (about 95 percent) with some nitrogen (about 3 percent) and traces of oxygen and water vapor. Early in its history, when water moved across its surface, Mars had a richer atmosphere. Over time, like the flowing water, it gradually disappeared, lost to space, either gradually or as the result of a cataclysmic impact from a large asteroid or dwarf planet. Despite its thin atmosphere, however, wind blows across the Martian surface, creating gentle breezes, occasional dust devils, and windstorms, the last of which sometimes become so strong that they enshroud the entire planet in reddish dust. Clouds occasionally form in the atmosphere, mostly made of carbon dioxide (CO_2) in the form of dry ice, though apparently water-clouds also sometimes appear.[4] Mars also has two substantial polar ice caps that recent evidence indicates are composed primarily of water-ice, with a relatively thin covering of frozen CO_2. Estimates are that each polar cap contains some 1.6 million cubic kilometers (384,000 cubic miles) of water-ice and is as much as 3 km (1.8 mi) at its deepest. Such huge deposits of water could easily support a significant population of humans should we decide to colonize the planet.

The *asteroid belt* forms the dividing line between the inner rocky planets and the outer gas giants. This region contains a million or so asteroids, of which about 160,000 have been named. They range in size from small, irregular bodies a few meters across to larger, more spherical dwarf planets, such as Ceres, which, at about 1,000 km (620 mi) in diameter, is nearly half the diameter of Earth's Moon and the largest body in the asteroid belt. This region mirrors what the Solar System probably looked like during its earliest stage. Scientists think that nearby Jupiter, with its enormous size and correspondingly large gravitational field, has, from its inception, perturbed the motions of these asteroids and so prevented them from ever combining into the planet they otherwise might have become.

Beyond the asteroid belt lies the domain of the large gas giant planets. *Jupiter* orbits nearest to the Sun and is the largest planet in the Solar System, with a mass some 318 times that of Earth. In fact, Jupiter's mass is greater than the masses of all the other planets in the Solar System combined. Its diameter is 142,984 km (88,846 mi), or more than eleven times that of our planet, and its volume is roughly one thousand times as great. Jupiter rotates on its axis a mere 3.1 degrees from the vertical and takes only ten hours to rotate once, meaning that its speed at the equator is 45,000 km/hr (28,000 mi/hr), about twenty-eight times the speed of Earth at its equator. Jupiter orbits an average of 778.3 million km (483.6 million mi) from the Sun and takes nearly twelve Earth years to revolve once around it.

Jupiter is best known for the colorful bands of gas that parallel its equator and for the famous Great Red Spot that rotates in the planet's southern hemisphere. The banded clouds are composed mostly of ammonia and other frozen gases, which are heated by the planet beneath them, rise, cool, and sink back again in convection cells, like those operating in the Sun's outer layer. The Coriolis effect forms the clouds into bands, with the clouds of different compositions (and so colors) being dragged along by the rapidly

4. Frozen CO_2, or dry ice, when melted, has no liquid intermediate state but converts directly from a solid to a gas. The atmospheric CO_2 comes, in part, from the warmer-weather evaporation of dry ice at the poles. A similar amount is deposited back again during the colder winter months.

spinning planet beneath them.[5] The Great Red Spot is a huge swirling storm, somewhat like a hurricane, but one so large it could contain several Earths within it. It has endured at least since the early 1800s and possibly as far back as 1665, when astronomer Giovanni Cassini described it. It might well have existed even long before the invention of the telescope and humanity's ability to observe it.

Planetary scientists think that, some 10,000 km (6,200 mi) below the surface of this cloud layer, Jupiter's massive gravity causes these atmospheric gases to become liquid, forming a mostly hydrogen-helium ocean. Farther down, this liquid hydrogen and helium mix is compressed into a metallic state. Deeper still lies the core of the planet, a combination of iron and other heavy elements that were drawn deep into Jupiter's center during the planet's formative stage. Due to Jupiter's great gravity, its core is highly compressed, causing it to be some 30,000° C (54,000° F), or about five times as hot as Earth's core. This massive heat source is the ultimate cause of the convection cells roiling the frozen upper atmospheric gases.

Jupiter is also noteworthy for its many moons. Some sixty of them have been identified to date, though most are quite small and are probably captured asteroids. Three moons, however, are among the largest in the Solar System. Galileo was the first to discover them, in 1610, along with the next largest moon, Europa: hence, their collective name, the Galilean moons. All four orbit in the same equatorial plane around Jupiter.

Io is the Galilean moon closest to Jupiter and so experiences the strongest gravitational force. This force causes friction within the rocky moon, creating the heat responsible for the intense volcanic activity observed on its surface, the most extreme in the Solar System today. The liquid sulfur emitted along with lava during these volcanic eruptions colors the moon's surface a rich red and yellow.

Europa is the next closest moon to Jupiter and is somewhat smaller than Earth's Moon. It is covered in ice scarred by long cracks with reddish mineral material scattered around them, which has likely oozed up from below. In fact, several lines of evidence suggest that beneath Europa's ice lies a liquid saltwater ocean heated by the same kind of Jovian (Jupiterian) gravitational forces that warm Io. Some even think that life might have arisen and evolved there, and that it perhaps continues to thrive. Europa has thus become a prime target, along with Mars, for scientists looking for life elsewhere in the Solar System.

5. This same combination of convection cells and Coriolis dragging of the atmosphere also largely determines air currents and so the weather on Earth.

Ganymede, the largest of Jupiter's moons, is even larger than the planet Mercury. Along with *Callisto*, the outermost of the Galilean moons, it is made up of about half rock and half ice, with a thick icy crust that is some four billion years old. Beneath Ganymede's crust there is apparently a saltwater ocean warmed by the same tidal forces that keep Europa's ocean liquid, though it is thought to be less likely than Europa's to contain life. Callisto, on the other hand, is too far away for Jupiter's gravity to disrupt and thus to heat its interior significantly, resulting in its apparent lack of liquid water. As with our own Moon, numerous impact craters pockmark both of these outer moons, which appear to be essentially dead.

Jupiter's gravity has locked these four Galilean moons in their orbits, causing them always to present the same faces to the planet, while their other sides face perpetually outward into space.[6] Besides these four major moons and its other sixty or so small satellites, Jupiter is also, like all the gas giants, orbited by a debris ring at its equator. In Jupiter's case, however, the ring is so inconsequential that it can barely be seen even from nearby.

Saturn is the next gas giant as one moves outward from the Sun. Saturn is the second largest planet in the Solar System, with a diameter of 120,536 km (74,898 mi), or about nine times larger than Earth's, and a mass ninety-five times that of Earth. Saturn rotates on its axis 26.7° from vertical, similar to Earth, but revolves at an average distance of 1.427 billion km (886.7 million mi) from the Sun—about twice that of Jupiter and almost ten times that of the Earth. Saturn takes nearly 29½ Earth years to orbit once around the Sun.

Saturn's internal composition apparently mirrors that of Jupiter, with successive layers of predominantly ammonia ice, liquid hydrogen, and metallic hydrogen covering a core of iron-nickel and silicon rock. Its visible surface also shows a kind of banding similar to that of Jupiter. What sets Saturn apart are the stunning concentric rings that orbit around its equator, the main one of which extends out farther than the planet's radius, some 70,000 km (43,500 mi). Yet the rings are incredibly thin, probably less than 100 meters, or somewhat less than the length of an American football field. Scientists think the rings formed within the last million years, probably from the collision of two small moons orbiting Saturn, or from the collision of an asteroid with one of its small satellites. The rings are made up of bits of rock and ice averaging from a few centimeters (a couple inches or so) to a few meters (yards) across, each travelling in its own orbit around the

6. Most moons in our solar system are "gravitationally locked" to their planets, so that they show the same "faces" to their planets.

planet. Astronomers think that these rings will eventually dissipate, probably within the next million years or so.

Saturn, like Jupiter, has many moons, sixty-one of which have been identified to date. Like Jupiter's moons, they are mostly small bodies that are probably captured asteroids. Saturn's largest moon, *Titan*, however, is larger than Mercury, having a diameter of some 5,000 km (3,100 mi). Titan is one of only two moons in the Solar System with an atmosphere, one that is 50 percent thicker than Earth's and, like ours, made up primarily of nitrogen. Clouds perpetually cover the planet. Some planetary scientists believe that water-ice, with lakes of liquid ethane and methane, make up the surface. These methane lakes have fueled speculation that life might exist in caverns just beneath the surface, making Titan another chief target in the search for life in the Solar System.

The two most distant gas giants in the Solar System, Uranus and Neptune, are roughly the same size and mass, just large enough to retain their gaseous atmospheres. *Uranus* has a diameter of about 51,118 km (31,763 mi) and a mass about seventeen times that of Earth. It orbits about twice as far from the Sun as Saturn, at an average distance of some 2.87 billion km (1.78 billion mi), and takes about eighty-four Earth years to revolve once around the Sun. Unique among the planets of the Solar System, Uranus rotates 97.9° from vertical, meaning the planet essentially rotates lying on its side, with its rotational equator running through what would normally be its north and south poles. Planetary scientists believe that a dwarf planet colliding with Uranus, sometime early in its history, probably caused this radical difference. This theory is strengthened by the fact that Uranus, like Saturn, has a thin, though narrower, ring surrounding it as well as several moons that all orbit the planet around its nearly perpendicular equator. Uranus's atmosphere is thought to be composed predominantly of methane, hydrogen, and helium, the first of which gives the planet its rich blue color. Most astronomers think that beneath the clouds lies a layer mainly composed of methane, ammonia, and water-ices covering an iron-nickel and silicon rock core that is perhaps a bit larger than the size of Earth. Uranus is known to have twenty-seven small moons, most of which have been named after characters that William Shakespeare and Alexander Pope created. Its great distance from the Earth has made it, like Neptune, difficult for astronomers to learn much about it.

Neptune is the outermost planet in the Solar System, orbiting more than half again as far from the Sun as Uranus, at about 4.497 billion km (2.794 billion mi), and taking nearly 165 Earth years to complete one trip around the Sun. At 49,528 km (30,775 mi), Neptune's diameter is slightly less than that of Uranus; its mass is about fifteen times that of Earth. Neptune's rotation is 28.3° from vertical, similar to that of Earth and many of the other Solar System planets. Its atmosphere, like that of Uranus, is probably composed mostly of frozen methane, hydrogen, and helium, giving Neptune a blue color similar to that of Uranus. On the other hand, unlike Uranus, Neptune shows some banding like Jupiter and Saturn. Also like Jupiter, Neptune once had a large swirling storm vortex, called the Great Dark Spot, in its upper atmosphere, which broke up and vanished in 1994.

The planet's deeper interior structure is likely the same as that of Uranus, with a layer probably composed primarily of frozen methane, ammonia, and water overlying a small rock and iron core with perhaps an admixture of water-ice. Also like Uranus, it has a narrow ring structure around its equator, as well as a dozen known small moons. Unlike Uranus, however, Neptune also has a large moon, *Triton*, which is unique in the Solar System, because it orbits in a direction opposite to that of its planet's rotation and the orbits of Neptune's other moons. This anomaly was probably caused by Neptune's gravitational field capturing Triton, formerly a dwarf planet that had originally orbited farther out in the Solar System and that was traveling in the opposite direction from Neptune's rotation. Like Saturn's moon Titan, Triton also has an atmosphere that is likely created by eruptions of gases, such as nitrogen, from beneath its surface. Titan and Triton are the only two moons in the solar system that have atmospheres.[7]

Beyond the eight Solar System planets lies *Pluto*.[8] Formerly classified as a planet, it was recently reassigned to the new category of dwarf planet, due mainly to its relatively small size, but also due to other ways in which it differs from the eight true planets. For example, its orbit lies well outside the orbital plane those planets occupy, tilting more than 17° from the horizontal. In addition, its orbit is highly elliptical, ranging from thirty to forty-nine AU (4.4–7.4 billion km/2.7–4.6 billion mi). Pluto crosses inside the orbit of Neptune at its closest approach to the Sun and stretches out into the Kuiper Belt (see below) at its farthest orbital point. Current theory suggests that Neptune's gravity was responsible for capturing Pluto and setting it into this highly elliptical orbit. Pluto's diameter is roughly 2320 km (1442 mi), or only about two-thirds that of our Moon. Its mass is only about 0.24 percent that of Earth and one-sixth that of the Moon, meaning that it is probably composed of water, ice, and rock, with perhaps polar caps of frozen methane, carbon monoxide, and/or nitrogen. It may have a very thin atmosphere

7. Saturn is the Roman name for the Greek god Cronos, the leader of the *Titans*, offspring of the gods Gaia/Earth and Uranus/sky. *Triton* is the name a Greek god of the sea, the son of Poseidon, whose Roman name is Neptune

8. Editor's note: For the New Horizon spacecraft's 2015 photos of Pluto, see NASA, "NASA Spacecraft Returns New Images of Pluto En Route to Historic Encounter."

made up mostly of nitrogen and carbon monoxide, with perhaps a trace of methane, but it is so small and distant that planetary scientists cannot be sure. Pluto takes nearly 248 Earth years to travel once around the Sun.

Perhaps the most interesting thing about Pluto is that it has a moon with a diameter about half its size, named *Charon*. Their similar sizes mean that, rather than Charon orbiting around Pluto, the two orbit around some point between them, making them more a double-dwarf planetary system than a moon orbiting a dwarf planet. Pluto also has four small moons orbiting it, two named *Nix* and *Hydra*, and two recently-discovered moons that have yet to be named.

Other Pluto-sized dwarf planets have been found somewhat farther out in this remote region of the Solar System. One, *Makemake*, is perhaps the largest planetesimal in the Kuiper Belt, having a diameter estimated to be two-thirds that of Pluto. Its surface is likely frozen methane and ethane, and possibly also nitrogen. Another, *Haumea*, has two apparently solid-ice moons, *Quaoar* and *Sedna*. *Eris* is the final known dwarf planet. It has a moon orbiting it and is about the same size as Pluto, or perhaps somewhat larger. What makes it particularly interesting is that it has an orbit roughly three times farther from the Sun than Pluto's, out beyond the Kuiper Belt and into the Oort Cloud (see below) at some 96.6 AUs from the Sun.

For an image of the Solar System with relative sizes of the Sun and planets to scale (relative distances from the Sun are not to scale), by the International Astronomical Union (IAU), see http://www.iau.org/public/themes/pluto.

Beyond the eight planets lies the large and relatively dense cloud of planetesimals known as the *Kuiper Belt*. The Kuiper Belt is much like the asteroid belt between Mars and Jupiter, except that it is some twenty times as wide and twenty to two hundred times as massive (or about thirty-three times the mass of Earth). It reaches from just beyond Neptune's orbit at about thirty AUs out to about fifty AUs from the Sun. Over a thousand planetesimals occupying this region have been identified, including Pluto and the other dwarf planets noted above, with another 70,000 or so small bodies over 100 km (62 mi) in diameter thought to exist there. Mostly, though, it is believed to contain small bodies composed primarily of frozen gases such as methane, ammonia, and water, all likely to have formed in the early days of the Solar System's history.

Still farther out is another band of planetesimals called the *Oort Cloud*. The Oort Cloud has never been directly imaged but is believed to stretch out to between 50,000 and 100,000 AUs, or 1–2 light-years, from the

Sun—1,000–2,000 times farther than the Kuiper Belt. This means that the Oort Cloud's outer limit is one-quarter to one-half the distance to *Proxima Centauri*, the star nearest to us, some 4.24 light-years away. Current theory understands the Oort Cloud to be made up of two regions: the disk-shaped *inner Oort Cloud*, also named the *Hills Cloud*, and the spherically shaped *outer Oort Cloud*. Both clouds are believed to be composed of trillions of comets made of frozen water, carbon dioxide, ammonia, and methane, mixed together with dust. The Oort Cloud is far enough from the other members of the Solar System that its planetesimals are essentially unaffected by their gravity.

Occasionally something, such as the interaction of several Oort Cloud comets, causes one of these bodies to shoot in toward the inner Solar System where, if it does not collide with the Sun, another planet, dwarf planet, moon, or some smaller planetesimal, it assumes a highly elongated elliptical orbit around the Sun. Some of these, such as *Halley's Comet*, which returns to the inner Solar System every seventy-five to seventy-six years, are termed "short-period comets," while other, "long-period comets" draw near to the Sun only every two hundred years or more. In fact, a number of astronomers think that some long-period comets may even take as much as a million years or more between appearances in the inner Solar System. Comets heat up as they approach the Sun, and the ever-present solar wind causes the frozen gas and dust particles of which they are composed to stream out from them and away from the Sun, forming the long, highly visible tails, which are their most visible features. It is believed that the gas and dust that compose the comets of the Oort Cloud were expelled from the early Solar System by the gravitational forces of the gas giant planets.

10

The Nature of God as Infinite Agapic Spirit

At the end of chapter 3, we began to reflect on the nature of God. It is important that we flesh out this understanding more fully, not only because it is important to correct some common misconceptions about God (including those of many Christians), but also because a proper understanding of God will be crucial to our later discussion of how God might be conceived to act in the world.

Before expanding our discussion about the nature of God, though, we must reiterate what we said in the first chapter concerning thought and language about God: that the Christian tradition has consistently maintained that humans are fundamentally incapable of understanding and, therefore, describing the full nature of God. As suggested by the distinctions already made—that God is Being (and Becoming) itself and we are beings, and that God exists necessarily while we exist contingently—God is not *quantitatively* different from humans, exceeding us somehow in degree (though this is how we often think of God), but *qualitatively* different from us, different in kind. God is *Infinite* Spirit; therefore, the finite human mind cannot adequately comprehend God. From this awareness, we noted, arose apophatic or negative theology, which historically has sought to say what God *is not*, rather than to make positive affirmations about what God *is*.

Yet we also saw that Christians have wanted to say more than simply what God is not—especially, as we have seen, that God is Perfect Agapic Love. However, when we make such a positive assertion about God, as the Christian theological tradition reminds us, we must be aware that we are using analogies. We must understand that, in asserting that God loves us, what we are saying is that God's Love is *somewhat like* human agapic love, in which at certain times we are so concerned with the welfare of the one we love that we forget about our own welfare. Yet, although it is similar in this respect, God's Agapic Love is also *unlike* human agapic love in that, for example, it extends to all creatures of all places and times, is always perfect, and, unlike ours, this Agapic Love is the very essence of God's nature.

We should also note that the term *God* is itself loaded with conceptual baggage that most of us assume is true but which makes it difficult for us to reconsider our concept of God more thoughtfully. One such image is that God is a *person*, analogous to humans; and so God has tended to be imaged in human form, as in Michelangelo's Sistine Chapel ceiling. I suspect this picture of an old, bearded man in a white robe is the first image that comes to mind for most of us, even the more theologically sophisticated of us, when we hear or think the word "God."

However, God is not a person in the usual sense of that term. According to the mainstream Christian tradition, God is an Infinite *Spirit* that nevertheless is possessed of personal characteristics, such as thought and love. To help us get away from our anthropomorphic (human-like) picture of God, I think it would be better if we renamed God something like *Infinite Agapic Spirit*. That would be closer to what Christian theologians have wanted to assert about God and, if widely employed, might gradually alter some of people's incorrect ideas about God. However, given our human, or at least Western, preference for monosyllabic words, I expect that most Christians and others will continue to use the term "God" when they refer to the divine being.[1]

If God is best conceived as Infinite Agapic Spirit, then we must say something about God as Infinite Spirit, so that the rest of what we say about God will be properly grounded in a less inadequate, less anthropomorphic image of God.[2] To say God is Spirit is to say that God is not a physical being like all the things in the universe around us, but rather exists in a different form of being, one that is immaterial. We really do not know what spiritual existence entails in addition to immateriality, as we have no direct experience of it. One might venture an analogy, that spiritual existence in some way resembles the mind's existence, which seems to be immaterial, to transcend the limitations of our physical selves. However, whatever this

1. Jewish scholars have attempted to address this problem by using G-d, rather than God, to refer to the divine being.

2. In this respect, theologians are rather like our scientific counterparts, in that we seek to say things less false about God, just as they seek to say things less false about nature. See p. 13, above.

alternate form of existence entails, the point is that God's form of being as Infinite Spirit means that God is not limited in space or time, as the universe and its creatures are, but instead is unbounded with respect to both.

God, then, is not a *person* who is "up in heaven," does not have a face or body, even though such statements do occasionally appear in the Bible. To take these images as literally true would be to misconceive God and God's relation to the universe and its creatures. God is Spirit and so transcends the limitations of physical being and the universe. That is the meaning of the traditional Judeo-Christian doctrine of divine transcendence. As we have seen, this is one of the essential affirmations of the Genesis 1 creation story. Yet God as Spirit is also present everywhere, permeating the universe, with and even within each creature, as stated in the Judeo-Christian *doctrine of divine immanence,* which we saw illustrated in the Genesis 2 creation story. These two complementary claims are reflected in a third, related doctrine, the *doctrine of divine omnipresence,* which asserts that God is "present everywhere" simultaneously. This doctrine is incompatible with any understanding of God as localized in only one particular place (or time). That God can be present everywhere within the universe, and yet transcend it infinitely, is possible only because God is Infinite Spirit, not a person.

Christians have traditionally thought of God as transcendent—as being "up in heaven"—but have been less likely to think of God as immanent, being all around us. Clearly, this is an inadequate understanding of God in relation to the universe. On the other hand, there have been some philosophers, such as Spinoza, who have thought of God and the universe as being essentially the same thing, a view termed "pantheism" (literally, "all [is] God"). However, there is another way of imaging God's relation to the universe that is more consistent with the traditional claims of divine immanence, transcendence, and omnipresence. If the Infinite Agapic Spirit transcends and yet is present everywhere within all physical reality, then it is logical to conclude that the universe exists *within* God, rather than outside God, as has often been thought. Nothing exists outside of God.

This understanding of the relation of physical reality to God is sometimes called *panentheism*, since "everything" (Greek, *pan*) exists "within" (*en*) "God" (*theos*). This means that we must understand ourselves and all other things in the universe as quite literally existing within God. This recalls the claim in Acts 17:27–28: ". . . indeed He (God) is not far from each one of us. For 'in him we live and move and have our being.'"

From this vantage, we might venture a further suggestion: we might think of the universe analogically as the body of God, so that the way God exists in relation to the universe is analogous to the way our minds relate to our bodies, including to every cell in our bodies.[3] Our mind unifies the activity of all these cells, causing them to work together for the common good of our bodies. Philosopher Charles Hartshorne has reflected insightfully on the nature of this relationship:

> The open secret of the mind-body relation is this: *our cells respond to our feelings (and thoughts) because we respond to their feelings* (and would respond to their thoughts if they had any). Hurt my cells and you hurt me. Give my cells a healthy life, and they give me a feeling of vitality and at least minimal happiness. My sense of welfare tends to sum up theirs, and their misfortunes tend to become negative feelings of mine. I feel what many cells feel, integrating these feelings into a higher unity. I am somewhat as their deity, their fond heavenly companion. They gain their direction and sense of the goodness of life partly from intuiting my sense of that goodness, which *takes theirs intuitively into account*.[4]

Thus, we might think of God as the Perfect Agapic Love in this way, giving unity to the activity of all things in the universe, causing them to behave within certain boundaries by means of natural laws, and influencing them to work together toward the common good of universal harmony. Much more needs to be said about how God relates to the world, and we will do so in chapters 13 and 17.

With this understanding of God as Infinite Spirit in mind, then, we can begin to develop further a proper Christian understanding of God. We will do so by returning to that which we have claimed already to be the most essential and nonnegotiable thing that can be said about God from a Christian perspective: that God is Perfect Agapic Love, that it is God's essential nature to be selflessly loving and giving.[5] The Bible and Christian tradition frequently use words such "grace" (God's giving of free, unmerited love and forgiveness, which leads to salvation/well-being both in this life and the next), "mercy" (God's giving us better than we justly deserve, because of God's great love for us), and "forgiveness" (God's freely forgiving our failures to love, because of God's great love for us) to flesh out the

3. Those who are acquainted with classical Greek philosophy might see in this analogy an echo of Plato's World Soul and its relation to the world.

4. Hartshorne, *Omnipotence*, 80.

5. That God is Perfect Love does not rule out the biblical theme that speaks of the wrath of God. God is angered when any creature is intentionally harmed by another, just as any parent is when his or her child is hurt by another. God's wrath is thus a manifestation of God's Love, and so should never be interpreted to mean that God hates anyone.

dimensions of God's Agapic Love in relation to humans. All these terms intend to assert that God is above all else concerned for the welfare of every creature, not Godself, and therefore attempts to bring about the best possible, most fulfilling existence for each creature in every moment of its life.

For this reason, Jesus used almost exclusively the metaphor of "Father" to refer to God; for, like (though exceeding infinitely) the most perfectly loving human father or parent, God treats each of us like God's own beloved child. Also, I would argue, this is the treatment of every creature in the universe that is capable of having some form of relationship with God.[6] This understanding of God as pure, infinite, selfless, parental love is, again, the *sine qua non* of Christian theology. The problem, as we shall see, is that, although this view is typically asserted by Christians, it has not often been consistently maintained, given other claims about God that many Christians, theologians included, have made.

The ultimate source of these other claims about God is, ironically, the Bible itself—or, more precisely, the misunderstanding many Christians have of the nature and proper interpretation of the Bible. In the Bible, as we have seen, we find many different writings produced by a variety of authors over the span of more than a thousand years. Not surprisingly, each book has some views particular to it; and not all the views of these biblical authors agree. Most reflect understandings of reality—and, to an extent, of God as well—that were commonplace in the times and cultures in which they lived. This should not be surprising, since all people are situated in particular places and times, and their thoughts and writings reflect those historical and cultural situations. Because of this, as previously noted, we also should not be surprised to see some development of thought about God and God's relation to the world over this thousand-or-so-year period, and we do—for example, in the movement from emphasis on God as the God of Israel in a world of multiple gods, to God as the God of the whole world and as the only God there is.[7]

What makes addressing these writers and their sometimes different ideas problematic is that many Christians treat the Bible as though it were written entirely by God, rather than that some parts reflect divine revelation

6. In using "Father," Jesus reflects the patriarchal outlook of the times. Had he referred to God as Mother, he would probably never have had an audience to hear the essential point of his message about God—that God is Perfect Agapic Love.

7. Editor's note: Views about God in the Bible do not always "develop" in one direction. In 1 Cor 8:4–6, the Apostle Paul seems to admit that "many gods" exist. In Phil 2:9–11, Paul declares that the heavenly, earthly, and chthonic beings, which are "gods," will acknowledge that "Jesus Christ is Lord" because "God the Father" gave Jesus a status above all others. See p. 52 n. 16 above.

while others reflect the culturally and historically influenced ideas of the biblical writers themselves—a problem discussed in the third chapter. This has led Christians to attempt to reconcile biblical statements about God and God's relationship to the world that are logically incompatible, ending up with some incoherent and highly questionable understandings of God. For this reason, the clear and most fundamental claim of Jesus and the church, that God is Perfect Agapic Love, has become adulterated and confused in the minds of many Christians.

One example of a problematic doctrine created in this way is the traditional Christian belief that there will be a final judgment for all people, at which God will judge them and send some to everlasting joy and peace in the presence of God in heaven, and send the rest to everlasting torture in hell. There are a number of problems with these ideas about salvation and damnation. For one, they treat salvation as something that occurs only after death, when one gets to "go to heaven." Salvation in this sense is a reward that a person might receive at the *end* of his or her life. However, this is clearly not all, or even primarily, what salvation means in the Bible. There, salvation is something available to people *in this life*, something one can experience here and now each day, and which, at least in the NT, might then continue and become even more perfect after one dies.

In fact, throughout the entire period during which the HB/OT was being written, the Jewish people understood salvation to be something that could be experienced only in this life. They had not yet come to believe that life after death in heaven with God was possible for them. Instead, as seen consistently in the Hebrew Bible/Old Testament,[8] they believed that

8. When referring to books contained in the Hebrew Scriptures, I use the dual term Hebrew Bible/Old Testament (abbreviated HB/OT). The Hebrew Bible is what Christians term the Old Testament, written primarily in the ancient Hebrew language. "Hebrew Bible" is a term used primarily by academics to avoid a Christian theological claim that God established a "new" covenant through Jesus that supersedes the "old" covenant that God established with Israel through Moses. Jews call their scriptures "*Tanakh*," which is an acronym formed from the first Hebrew letters of each of the three components of the Jewish canon: *Torah* [=Pentateuch], *Nevi'im* [=Prophets], and *Ketuvim* [=Writings]. The arrangement of the Jewish canon differs from the order of the books of the Christian "Old Testament." In addition, the Jewish canon contains one book each of Samuel, Kings, Chronicles, and Ezra-Nehemiah, whereas the Christian "Old Testament" divides each into two books; and the Jewish canon counts as one book "the Twelve Prophets" (sometimes called "Minor Prophets"). In other words, both "Hebrew Bible" and "Old Testament" refer to the first thirty-nine books of the Christian Bible, whereas the Jewish canon (Tanakh) consists of twenty-four books. The Christian Bible adds to the Jewish Bible a collection of twenty-seven books written primarily in ancient Koine Greek and authored by early Christians, which is called the New Testament. Thus the Christian Bible contains the Hebrew/Jewish Bible, or Old Testament, plus the New Testament (NT). The Roman Catholic Bible includes the Apocrypha as part of the OT.

when people died they went to *Sheol*, the shadowy abode of the dead, a place neither of punishment nor of reward. One only sees a hint that an afterlife in heaven (or hell) might be possible in the last chapter of the last book written in the Hebrew Bible, Daniel 12, composed around 165 BCE. This belief became commonly accepted within Judaism over the next century or two, though not universally. Even in the time of Jesus, some Jews, including the group known as the Sadducees, refused to believe in this novel doctrine (see Mark 12:18–27 and the parallels in Matt 22:23–33 and Luke 20:27–40). For Jesus and his followers, however, salvation was something one could experience after death—but also, and importantly, in this life as well. Jesus's acts of healing and teaching were signs of salvation happening through his ministry, especially for the neediest people. Christians need to reclaim this teaching of Jesus in order to correct their inappropriately one-dimensional understandings of salvation.

A second problem with the final judgment doctrine is that it seems to make salvation something that people must earn, rather than being the gracious gift of a loving God, as Paul, the important early Christian theologian and missionary, taught. If salvation is something people must earn, then a host of questions appear that the churches and theologians have not historically answered in the same way. For example, what precisely is required for people to attain salvation? With regard to beliefs, exactly which are necessary? That is, which doctrines and which understandings of those doctrines must be believed in order for one to be saved? Christians historically have differed greatly in their answers to this question. How do we know which view is the right one? Moreover, given the ultimate unknowability of God, how can we even be sure we have gotten much right in our beliefs about God at all? Jesus really did not say much about God, other than to emphasize God's father-like love, mercy, and forgiveness of "his" children.

Similarly, with regard to good deeds, how many and which ones must one do to get to go to heaven? Is there a certain percentage of people's lives that must be moral or loving for them to be saved? Which is more important to be saved, one's beliefs or one's good deeds, or are both of them equally necessary? The church has not provided a clear answer, either historically or today, for any of these questions, and so the individual Christian must live in uncertainty regarding what to do to be saved and therefore whether he or she has managed to qualify for everlasting life in heaven or not.

This notion of salvation as something one must earn by one's beliefs and/or good deeds has obscured the core Christian understandings of God as Perfect Agapic Love and salvation as God's free, undeserved gift.

Standard abbreviations will be used for the books of the Bible.

In addition, they have had the destructive effect of driving people to focus their lives on attaining this salvation (again, conceived as something after death) for themselves. This is not surprising, given the fear generated by the thought that one must earn one's salvation or spend forever in hell. Yet, it is precisely this self-centered preoccupation that the Christian tradition has consistently defined as the essence of all sin. Sin is living one's life centered on oneself and what one wants. This self-centered preoccupation destroys the possibility of having a proper relationship with God and so experiencing salvation in this life.

To center one's life on God and what God wants is what is meant by the term "faith." It is this faith, along with God's gift of grace in offering to relate to us in this way, that Paul claims brings about salvation. Faith is not the same thing as belief, even though many people, Christians included, incorrectly think it is. Rather, faith is a fundamental life-orientation in which one seeks to have a proper *relationship* with God. This involves not only what one believes but also, and more importantly, the love, trust, and commitment one gives to God—and, because of this, the love with which one treats nature, other creatures, and other people. Faith is living in a *proper* relationship with God, as creature to creator, with God and God's will at the center of our lives, or at least striving to do so to the best of our ability in each moment of our lives. Sin, as we have seen, is living in an *improper* relationship with God, in which we put what we want, and ourselves rather than God, at the center of our lives. According to the Christian tradition, it is from this sinful, improper life-orientation that all the various sins we commit arise, most importantly, the failure to love others.

Understood this way, to live in a sinful life-orientation is, according to both the Bible and Christian teaching, stupid, in that it is ultimately self-defeating. In so living, we—who do not even really understand ourselves and have to spend a lifetime trying to do so—act as though we know better what is good for us and how to achieve it than God does. In this way, we cause our lives to have more suffering and less personal fulfillment than they would have had we listened to and trusted the all-knowing and all-loving God. The possibility of living in the agapic way God wants means that, to the extent we do so, we are already, in a very real sense, living in heaven.

From this it should be clear that Jesus, core biblical teaching, and the best strains of Christian thought past and present, all agree that salvation is not a matter of us getting what we deserve after death according to some uncertain divine standard. It is not earning a ticket after death to some holy Disneyland. Rather, salvation has to do with our living in a loving relationship with God and others here and now and, because of this, experiencing the most rewarding and fulfilling—i.e., salvific—life possible, both now and after death. This is why certain biblical authors claim that living in obedience to God is wisdom, for it is the only way to live if we are to attain the happiness and fulfillment we truly desire. This is why, for the biblical writers, any other way of living was ignorant and foolish.

Still another problem with the notion of God sending people to hell involves the question of where hell could be. We have seen that, according to the doctrine of divine transcendence, there is no place that exists or could exist apart from God. If this were the case, then hell would have to be a "place" within God. However, wherever God is, heaven exists, as the tradition rightly intuited. This heaven consists in being in an intimate, loving relationship with God. Therefore, if God exists quite literally everywhere, and God is Perfect Agapic Love, then there is no place where hell could possibly exist. We will consider this in greater detail in chapter 20 when we discuss the notion of life after death.

Another thing must be emphasized: this salvific relationship with God is something that God offers to all people of all places and times, not just Christians, as the church has typically claimed throughout its history. God did not start offering salvation to people only after Jesus appeared, as though it was not available to those who lived before Jesus or who lived after he did but without ever having heard of him. God *must* have offered the opportunity for a saving relationship with Godself to *everyone* who has ever lived—not just on this planet but on every planet that has produced creatures capable of entering into such a relationship, in this universe and in any others that exist or have existed—precisely because of who God is. God and the reach of God's infinite love are far greater, both in scope and in kind, than our parochial minds have ever imagined, simply because it is God's very nature to be *Infinite* Agapic Love.

Still another major problem that the doctrine of the final judgment has created is the idea that those who do not attain whatever is required to go to heaven are sent to hell, where they will be tortured and suffer forever. Clearly, the understanding of God here is *not* that of a Perfectly Loving Father, for not even a halfway-decent human parent would do such a thing, even to a child he or she did not like, let alone a perfectly loving Parent-God. Rather, we are dealing here with a very different image of God, that of God the Judge,[9] the ultimate purveyor of justice. Here, based on certain objective criteria (whatever they are) and God's absolute knowledge of every person's life, God supposedly metes out perfect justice, giving people precisely what they deserve. The most common result of this view has been some form of

9. In the ancient Near East, the final judge of a person's case was the king; therefore the two images are conjoined in much of the literature from that time and place, including the Bible.

legalism, in which Christians focus on biblical laws, believing them to be God's laws, and seek to obey them in order to earn salvation after death. This legalism is a far cry from the teaching of Jesus, who said that all God's laws are summed up in only two fundamental commandments: love God and love all people, even your enemies (e.g., Mark 12:28–31; Matt 5:43–48; 22:34–40; Luke 10:25–28; also see Rom 13:8–10 and Gal 5:14). In love, the focus is on others; in obeying laws to gain heaven, the focus inevitably is on oneself—and so we are back to sinful self-centeredness.

Certain important theologians in the history of the church have emphasized this view of God as judge, while trying to combine it with God's merciful love, most famously by having Jesus bear the burden of other people's sins in his suffering and death on the cross. But what kind of justice is punishing an innocent person instead of the guilty ones? Not even our admittedly flawed human legal systems permit such a thing, even if an innocent person volunteers to take the guilty person's punishment. Moreover, where is the love and forgiveness? Does God really require a bloody corpse before being willing or able to forgive people their sins? I would not, and I hardly think I am more loving than God is!

This view becomes even more unsatisfactory when one considers the orthodox Christian understanding of Jesus as the incarnate Second Person of the Trinity. In this light, what we have is God in fact punishing Godself—or God the Father torturing and killing God the Son—so that God can or will forgive.[10] Why does not God just forgive people without requiring torture and murder? Because, adherents of this view typically say, God's justice requires that *someone* be punished for sin. It appears, then, that for such Christians it is the image of God as judge, not God as Loving Father, that describes God's essential nature. However, both cannot be true, for one cannot give people what they deserve (God the Judge), while also giving them better than they deserve, which is forgiveness and mercy (God the Loving Father). Quite simply, if God is the Judge, God is not the Loving Father.

These problems with the ideas of salvation and a final judgment become even more acute given the Western doctrine of original sin (see chapter 3 above) and, especially within certain strains of Christianity, the doctrine of divine predestination.[11] Developed by Augustine in the late fourth and early

10. Christians who hold this view would probably object to putting the matter in this way, preferring, perhaps, to say God allowed the Romans to torture and kill Jesus. However, if Jesus's death was necessary for God to forgive and was God's idea, then the way I have phrased this is certainly, if uncomfortably, the case. Feminist theologians in particular have emphasized that this understanding of salvation leads to an intolerable picture of child abuse on the part of the Father.

11. Eastern Orthodox Christianity rejects the doctrine of original sin—namely, that

fifth centuries from earlier Western thought, these doctrines are logically related. According to the doctrine of original sin, after Adam and Eve's first sin their natures were changed such that they were "not able not to sin," to use Augustine's phrase. Therefore, those descended from them, meaning everyone (except Jesus, due to his virgin birth, it was said), are also born with this same sinful nature and so are unable to do the good. For this, Augustine claims, God, the righteous judge, justly condemns them to hell. But how can it be just to punish people for their sins when they can do nothing but sin because of the sinful nature they inherited, which was not their fault but that of their ancestors? Can people justly be blamed for sinning when they lack the freedom to do otherwise?[12]

The situation becomes even worse when you add to the mix Augustine's doctrine of divine predestination, according to which God had already predestined which people would go to hell even before the creation of the world! Still, God is also loving and merciful, Augustine says, and so God predestined a minority of these future sinners to go to heaven, even though they too were born with original sin and so would be every bit as sinful as those going to hell. Augustine admits that he can give no reason why God chose to predestine these particular people, and only these people, to go to heaven; it is a divine mystery, he says, beyond our ability to understand.

Both doctrines have significant and, to my mind, irresolvable problems. Regarding the original sin doctrine, the first problem is its lack of biblical support. Nothing is said in the Genesis 3 story or anywhere else in the Bible that Adam and Eve's sin caused a change in human nature such that all subsequent people were "not able not to sin." If this doctrine were true, it would make the prophets and Jesus some of the greatest idiots that

---

all humans are guilty of the original sin of the first humans. Adam and Eve harmed their relationship with God but did not change human nature. However, the Orthodox belief in divine predestination logically negated the belief that they continued to have free will, though they asserted only God's predestination of the elect to heaven, not the damned to hell. Logically, of course, the result is the same: only certain people go to heaven, and the rest do not.

12. Augustine used Rom 5:12 as a way to justify God's punishing all future descendants of Adam and Eve, but he was using the Old Latin translation of the Bible (before the Vulgate), and that particular verse had been mistranslated. Augustine understood Paul to be speaking of Adam "in whom all sinned," and so claimed that all subsequent humanity was somehow present in Adam and chose with him to disobey God. The Greek original speaks of all people being like Adam "because all have sinned." This is not the only time that a Christian doctrine has been grounded in a mistranslation of a Bible verse! Editor's note: Jarvis seems to be referring to Augustine's "Treatise on the Merits and Forgiveness of Sins, and on the Baptism of Infants" (*De Peccatorum Meritis et Remissione, et de Baptismo Parvulorum*), Books 1–3. See Reasoner, *Romans in Full Circle*, 46–47.

ever lived for having told their fellow Jews to repent and sin no more when it was, in fact, impossible for them to do so! Second, there is the problem just noted of the injustice of blaming people for others' sins and punishing them for the sins they do commit when they lack the freedom to do otherwise. Third, in light of contemporary science, which has demonstrated convincingly that all life on this planet, including humans, has evolved from earlier creatures, taking the Adam and Eve story as historical fact is simply wrong. Finally, as we have already seen, taking the biblical creation stories as literal history or science, rather than as the "theology in story form" that they are, is to misconstrue radically both the nature and meaning of these texts.

The doctrine of divine predestination has its own fatal problems.[13] It too suffers from a lack of biblical justification. In only two places in the entire Bible is any form of the word "predestine" used, and both of them appear in a single passage in Paul's letter to the Roman church (8:29–30). The proper interpretation of these verses is notoriously unclear, but one thing is clear: in them, Paul ties God's predestination to God's foreknowledge. Augustine, however, specifically denied that God's predestination of people to heaven or hell was based on God's foreknowledge of their future lives, thereby depriving his view of any genuine biblical support.

There are crucial theological problems with the doctrine of predestination, too. For one, how could a just God send some sinful people to everlasting damnation and other, equally sinful people to everlasting salvation? What is just about that? If God were going to be merciful to some, why not to others? Why not to everyone?[14] Is not this what an infinitely loving God would do? What is pivotal to understand is that God can have only one essential nature: either a justice-oriented nature or a loving and merciful one. In God's very essence, God cannot be both just and merciful because, as we have seen, to be merciful is by definition not being just; it is giving people not what they deserve but better than they deserve. To claim otherwise is to make God seem capricious, which is what the predestination doctrine does—seeing God as just on this occasion or to these people and merciful on other occasions or to other people.

13. The doctrine of predestination is often misunderstood to mean that God controls what happens to people in this life. That is a different doctrine, the doctrine of divine predetermination, a view associated with the doctrine of divine omnipotence, but which has not been widely accepted by theologians or churches, past or present. The doctrine of predestination concerns only a person's fate after death.

14. At least two theologians in the history of the church have either asserted or suggested strongly that God, being a perfectly loving God, predestines all people to heaven: Origen in the third century and Karl Barth in the twentieth. Their ideas have garnered little support from other theologians or the church.

As we have seen, Augustine and those who have followed him have claimed this to be a divine mystery beyond our ability to comprehend. The Christian tradition, however, has consistently used logic as a means for understanding things about God. In discussing God's power, for example, the tradition has consistently stated that God cannot do what is logically impossible (nor what is evil, since that would go against God's essential nature). Apparently, only with those doctrines that go against logic have the church and its theologians suddenly given up the use of logic and cried "mystery!"

Even more important than the evident injustice and lack of biblical justification for the doctrine of predestination is that it makes no sense whatsoever in light of Christianity's most fundamental claim that the essential nature of God is Perfect Agape. How, then, can any rational Christian claim God to be essentially Love itself, the absolutely Perfect Parent who offers mercy and forgiveness to all even before they ask for it, and in the next breath say that God predestines some of them to everlasting suffering? How is such a disgusting thought even possible? That some theologians and churches can believe and teach such a thing is frankly unbelievable. No wonder many thoughtful people have left the church! If I thought being a Christian meant that I had to believe such ugliness about God, I would leave the church too.

Finally, it is important to point out that both of these doctrines serve to illustrate an important point about biblical interpretation: that it is a grievous error to create doctrines based on only one or a couple of biblical passages. Traditional Christian theological wisdom has taught, as we noted, that what the Bible says about anything is what it states repeatedly and in various books spanning both Testaments. Using only (a misinterpretation of) Genesis 3 and (a mistranslation of) Rom 5:12 to justify the doctrine of original sin, and only (a misinterpretation of) Rom 8:29–30 to establish the doctrine of divine predestination was quite simply a wrongful use of Scripture by Augustine and those who have continued to uphold these doctrines.

All these problematic beliefs are quickly seen for the false doctrines they are when one holds fast to the fundamental and incontrovertible Christian teaching that God is Perfect Agapic Love, and when one recognizes that not everything one reads in the Bible is divine revelation. I find it highly significant that, on the one hand, scholars of Jesus and the Gospels today hotly debate whether Jesus himself spoke of a last judgment or whether it was a creation of some first-century Christians in addressing situations in their communities (especially persecution, which was the historical genesis of such views).[15] On the other hand, scholars do not doubt that Jesus, according to the Gospels, spoke of God as a loving and merciful father.

Also, the apostle Paul points out, especially in his letters to the churches in Rome and Galatia, that God gives salvation, not as the result of human works, but as a free, undeserved gift of grace offered to all people who would enter into a loving relationship with God through faith—and this solely because of God's great love for all people. This makes sense, therefore, of the oft-repeated Pauline claim that people are saved "by grace through faith" (Eph 2:8). Salvation is possible only because God comes to people with the gracious offer of a loving relationship that will be their salvation if they enter into that relationship. Since faith is a right relationship with God, people are saved "by grace through faith," through having responded to God's offer and having chosen to engage in that relationship with God.

In 1 Timothy (a letter attributed to Paul but probably written one to three generations after his death), the author speaks of ". . . God our Savior, who desires *everyone* to be saved . . ." (2:4; emphasis added). That the loving God wills that all people be saved should be obvious, given that God loves all people and views them as beloved children. If these understandings of these important biblical texts are correct, as I believe they must be if one holds to the core teachings of Jesus, the NT, and the church (at its best), and God loves all creatures infinitely, wills that they be saved, and offers salvation freely to all, then it is inconceivable to me that God would send anyone, no matter how much they had rejected God and done evil in their lives, to everlasting torture, as the traditional doctrine has it. This is not something the God who *is* Perfect Love could do. Rather, the God who is Love acts creatively and salvifically in every moment and for every creature in the universe. (What happens to the people who reject God's offer of a loving relationship is discussed in more detail in chapter 20.)

If salvation in this life and the next is fundamentally about living in a loving relationship with God and others, and gaining the benefits that one experiences in such relationships, rather than some prize to be attained at the end of one's life by one's own efforts, then, as we have said, God must surely offer to be in such a relationship with *all* people—all who have ever lived on Earth and, indeed, all creatures everywhere in the universe capable of entering into such a relationship—in every moment of their lives. That is

15. Matthew repeatedly mentions the severe punishment that God and God's angels mete out to wrongdoers (see, e.g., 8:12; 13:42, 50; 22:13; 24:51; 25:30, 31–46). In John 12:47–50, Jesus speaks of the final judgment at which the word he has spoken, which God gave him to speak, will be what judges people. I will address this issue of the last judgment more in chapter 20.

what a perfectly loving God, seeking to create the best possible life for those God loves, would most certainly do.

If Jesus teaches that God is Love and always acts lovingly toward us, seeking to enable our salvation, and that God wills that we return that love by loving all people, even those who are our enemies, in the same, self-giving ways, then that is perhaps all that any Christian really needs to know. Instead of spending so much time reading the Bible and trying to make consistent sense of everything that is said in every verse—things that are not consistent—maybe it would be better for Christians to ponder the Bible less and simply begin with Jesus's teaching that God is Perfect Agapic Love, believe only what flows logically from and is consistent with that truth, and seek to put it into practice in their lives. This would probably result in an immediate improvement in Christian theology—and hopefully the behavior of Christians as well—and a more positive perception of Christianity by those who are not Christians.

11

The Early History of the Earth

LIKE ALL THE OTHER planets in our solar system, Earth was formed from the gradual accretion of planetesimals, due to the attractive force of gravity. However, some scientists claim that the majority of the planet's mass, perhaps as much as 75 percent, came from the impacts of several large bodies with the nascent Earth, including especially the Mars-sized body that struck the Earth and whose detritus formed the Moon. Using radiometric dating, scientists estimate that the majority of the process of planet formation took perhaps one hundred million years. Because rock from that early period has disappeared, due to Earth's continuous recycling of its crust, most scientists date the age of the Earth according to the age of the meteors that occasionally fall to Earth. These meteorites are of three basic types: stony meteorites, which are largely composed of (the dense silicate[1]) olivine; iron meteorites, composed essentially of iron; and stony-iron meteorites, which are a combination of the two. Because Earth formed from the accretion of and collisions with such meteors, it is not surprising that it consists mostly of these two substances: the core of the planet is predominantly iron, and the mantle (the region between the core and the outer crust) is largely olivine.

Dating of these meteorites by radiometric analysis has revealed a very consistent picture. Whether using the rubidium-strontium, potassium-argon, or uranium-thorium method, the dates of these rocks all cluster around a date of 4.55 bya +/- 1 percent. In addition, American astronauts brought rocks back from the Moon that date to around 4.55 bya. Thus scientists are

1. Silicates are compounds of silicon and other atoms.

confident that the essential creation of the solar system, including Earth, was completed by about 4.55 bya.

The history of the early Earth's subsequent evolution seems generally well established, again with only some slight variations in the dating of when the individual events occurred.[2] The earliest history of the planet, from around 4.55 to about 3.85 bya, is termed the Hadean Eon, from the term Hades, referring to the hellishly hot initial stage in the planet's history.[3] The constant bombardment of the earliest Earth by large and small planetesimals generated enormous heat as the kinetic energy of these moving bodies converted to heat energy upon impact. This heat, supplemented by the decay of radioactive isotopes inside nascent Earth that had been deposited by such colliding planetesimals, as well as the gradual compression of materials within the planet, probably kept the entire planet molten, including its surface, for tens of millions of years. This molten state permitted the denser elements, especially iron and nickel, to sink slowly toward the center of the planet, while the lighter materials—especially silicon, magnesium, aluminum, and oxygen-bonded compounds—gradually moved upward.

This process added yet more heat to the planet's interior from the friction caused by these moving materials. The result of this process, after some hundred to several hundred million years, was a *core* made predominantly of iron (with traces of other heavy metals, especially nickel), a *mantle* composed largely of dense silicate minerals, and a *surface* of less dense silicates, predominantly *basalt*, an igneous rock with high magnesium and iron content. Earth's surface, a molten ocean, gradually cooled through radiation into space. As the planetesimal bombardment steadily declined over time, possibly within a few tens of millions of years, it began to become Earth's first solid crust. All the while, gases trapped within the planet gradually rose through the thick viscous material, finally reaching the surface and being released, or "outgassed," to form Earth's earliest atmosphere.

At a very early stage, perhaps forty to one hundred million years after it had formed, and while it was still molten (though possibly just beginning to form a crust), Earth experienced the largest cataclysm in its history. According to recent evidence, an enormous object roughly the size of the planet Mars collided with Earth. This massive body evidently struck Earth with a glancing blow, imparting enough energy to melt any of Earth's crust that might have already formed and causing both bodies to eject enormous

2. See the Geologic Time Scale in the appendix.

3. This terminology wrongly equates hell, the place of everlasting punishment in traditional Christian thought, with Hades, the abode of the dead in ancient Greek thought. The latter was originally considered the place where all the dead went, not only the evildoers.

quantities of molten material into space. The heavy iron core of the colliding body was drawn into Earth, sinking through its mantle to merge with and substantially increase the size of its core. The glancing collision added momentum to Earth's rotational spin, causing it to rotate on its axis faster than it previously had, resulting in shorter Earth days than was previously the case. In addition, this blow, perhaps aided by other smaller yet still significant collisions early in Earth's history, caused a shift in the planet's axis of rotation, tilting it from the perpendicular to the 23.5 degree angle it has today.

The remnants of the colliding planetoid's mantle, and whatever crust it had created, cooled rapidly through radiation into space, forming a debris disk surrounding Earth, which gradually accreted to form the Moon. Like Earth's, the Moon's molten surface eventually solidified, leaving a less dense, feldspar-dominated crust similar to Earth's crust. The Moon's interior is made of heavier materials, like those of Earth's mantle, though of a different chemical composition. This implies that the Moon's mantle is composed almost exclusively of the planetoid's former mantle. Finally, the Moon lacks any significant iron core, which causes the Moon's overall density, at 3.3 grams per cubic centimeter (gm/cm^3), to be considerably less than Earth's 5.5 gm/cm^3. Once formed, the Moon's relatively small size meant that its gravity was weak, allowing most of the gases that percolated up through its molten interior to the surface, which might otherwise have formed an atmosphere, to escape into space. Similarly, nearly all the water that had existed in its rocks also evaporated into space. Today, only a trace of water-ice has been detected on the Moon's surface in places where the Sun's light never shines, such as the shaded walls of deep craters.

Throughout the Moon's early existence, it was pummeled, like Earth, by space debris, creating the crater-packed lighter regions that astronomers term the lunar highlands. Rocks from these areas, dating to about 4.55 bya, have remained largely as they were at the end of the early bombardment period. However, as the crust continued to cool, several large asteroids struck the Moon, probably during the late bombardment period (around 4.0–3.85 bya), perhaps the result of the reshuffling of the orbits of the outer gas giant planets (see chapter 9). Some of these asteroids likely reached 100 km (62 mi) across, creating huge blast holes in the Moon's crust, as much as 2,500 km (1,550 mi) in diameter. Over perhaps the next billion years, these craters were gradually filled with basalt that oozed to the surface, flooding them and forming the flat, dark areas that Galileo termed *maria*, plural of *mare* (Latin for "sea"), because of their smooth appearance. With neither water nor an atmosphere to drive erosion, and with the Moon's interior too cool for further volcanism or other geological activity to alter it, the

Moon's surface has remained largely as it was at the end of the late heavy bombardment period (around 3.85 bya). Since this time, only an occasional meteor or asteroid collision has altered the Moon's appearance, as can be seen by the relatively few, small impact craters within the maria. Earth today would look essentially the same as the Moon, were it not for the geological, hydrological, and atmospheric processes that have effectively erased that primordial landscape.

The massive collision that resulted in the creation of the Moon was evidently unique within the inner Solar System, for Earth is the only inner planet with a large satellite. The Moon has a radius of 1737.4 km (1,077 mi) and a mass of some 7.18×10^{22} kilograms (7.9×10^{19} tons), or slightly more than one-hundredth that of Earth. It orbits Earth at a distance of 384,400 km (238,300 mi) and takes 27.32 days to make one trip around the planet. The Moon's surface layer is pulverized rock that extends about 10 m (almost 40 feet) below the surface. The crust lies beneath this layer, extending down about 60–70 km (37.2–43.5 mi) on average. It is made predominantly of silicate rocks with little iron. Below this, the Moon's mantle, made of solid rock, stretches down more than 1,000 km (620 mi). As we noted, the Moon apparently lacks any significant core.

The Moon's presence has been of great importance for the subsequent evolution of Earth and its life. Its presence was critical in stabilizing Earth's tilt to within 1° more or less than its 23.5° average, which ensured the long-term climatic regularity essential to the evolution of life. Its gravity is also responsible for the tidal raising and lowering of Earth's oceans, which creates a frictional force that has gradually, if almost imperceptibly, slowed the rotational speed of the planet throughout its history. This has in turn caused Earth and the Moon to move slowly apart from each other at a rate currently of about 4 cm (1.5 in) per year.

Earth has exerted a similar, if much greater, influence on the Moon's rotation, because of its far stronger gravitational force. Over millions of years, it gradually slowed the Moon's rotation until finally it became locked in a synchronous orbit, such that it takes the same amount of time to rotate once on its axis—27.32 days—as it does to revolve once around Earth. This is the reason we always see the same face of the Moon and its other side remains perpetually hidden from us on Earth. It also makes the Moon's orbital and rotational relation to Earth analogous to that of the planet Mercury and the Sun.

After Earth had reformed following the cataclysm that formed the Moon, perhaps some 4.4 bya, it began to cool and generate a crust. In addition, it began forming a thin atmosphere, due to the outgassing of interior water vapor and gasses through such things as volcanic eruptions. Within a few hundred million years, this atmosphere had become significant, encasing the planet in a thick cloud composed mainly of water vapor, hydrogen, hydrogen chloride, nitrogen, carbon monoxide, and carbon dioxide. From the outside, Earth would have looked much like Venus today, with its surface completely hidden and lightning probably occasionally lighting up the cloud layer from below.

As the simple gases mixed in the atmosphere, they combined to form copious amounts of ammonia (NH_3) and methane (CH_4), along with smaller amounts of oxygen. This oxygen was created when water vapor was chemically split by the Sun's ultraviolet radiation streaming through the atmosphere, severing the oxygen from its hydrogen partners. Any free hydrogen in the atmosphere that did not react with other gasses to form heavier compounds gradually escaped Earth's gravity and floated out into space. The relatively lightweight oxygen atoms liberated in this process rose above the heavier gasses and combined to form Earth's first *ozone* (O_3) layer. Over time, that created a progressively thicker barrier around the planet that shielded it from the Sun's ultraviolet radiation. Since this radiation is harmful to living organisms, the creation of the ozone layer was critical in permitting life's eventual origination and evolution.

Earth and its atmosphere continued to cool over the next several hundred million years as heat slowly rose up from inside the planet, cooling its interior, while the declining bombardment of planetesimals melted the crust wherever they hurtled into it. At the same time, the shorter-lived radioactive elements within Earth decayed, decreasing another source of internal planetary heat. Slowly and fitfully, then, the planet cooled. Eventually the molten surface of Earth solidified sufficiently to produce what was perhaps the second beginning of a crust in the planet's brief history. This process of crust liquefaction and reformation, due to planetesimal collisions, likely was repeated many times but on smaller, more localized scales before the bombardment effectively came to an end.

As this process of planetary cooling proceeded, at some point the water vapor in the atmosphere became cool enough to form droplets, which then rained down upon Earth's surface. Because the ground was still too hot for the water to remain liquid, it evaporated as quickly as it hit the ground. However, over the succeeding tens of millions of years, as Earth continued to cool, the rain eventually could remain liquid longer, dissolving minerals in the rock and beginning the weathering of the planet's surface. As it evaporated, the water cooled the surface further, gradually lowering its temperature, until finally the water was able to remain liquid on the surface. There it gathered to form the first streams, rivers, lakes, and oceans in the low places. These new watery features dissolved salts of various minerals, eventually depositing them in the oceans, where the salt levels continued to rise gradually throughout the Hadean Eon. Scientists estimate that the oceans began to form around 4.2 bya and had filled to roughly their present volume by about 4 bya. Some evidence suggests that these early oceans covered the entire surface of the earth and continued to do so for about 200 million years, until the first continental crust appeared.

Meanwhile, outgassing from the planet's interior continued, adding somewhat to the water supply, as did comets that occasionally crashed into the surface.[4] Asteroids, meteorites, and comets kept bombarding Earth as it swept up the last planetesimals, as well as copious amounts of cosmic dust, in its orbital path. The first small plates—the enormous slabs of rock that roam slowly but constantly across Earth's surface—probably formed during this period, which might be when the most primitive forms of life arose in the oceans.

Around 4 bya, however, the era known as the Late Heavy Bombardment Period began, due to the alteration of the orbits of Uranus and Neptune. This change in the outer solar system sent a massive new wave of debris hurtling toward the inner planets, and collisions with some of this debris remelted significant portions of Earth's crust. This bombardment of Earth, as well as the Moon and other inner planets, would have been far greater had it not been for the presence of Jupiter farther out in the solar system, with its huge gravity gathering up much of the space debris that would otherwise have pummeled Earth. In fact, astronomers have estimated that Earth's bombardment during the Hadean Eon would have been a thousand times greater without Jupiter's presence. Nonetheless, the steady rain of asteroids and meteors onto Earth—estimated to be a thousand times more often then than today—combined with the resultant violent volcanism, kept the crust in a state of constant turmoil. If life had arisen previously in the oceans, all but the hardiest, best protected of these single-celled creatures living deep in the oceans would have been annihilated during this period.

Once this late bombardment had essentially ended, around 3.85 bya, the interior structure of Earth stabilized, assuming the form it has today, aside from changes on its surface due to plate tectonics. The planet is composed mostly of just four elements: by weight, about 35% iron, 30% oxygen, 15% silicon, and 13% magnesium, 2.4% nickel, 1.9% sulfur, 1.1% calcium, 1.1% aluminum, and 0.5% all other elements combined. Its interior is

4. This outgassing of water vapor from Earth's interior continues today, largely in volcanic eruptions, and adds about 0.1 cubic kilometer of water to Earth's oceans each year. Today our oceans have a volume of about 1.37 billion cubic kilometers (329 million cubic miles), an average depth of 3,796 meters (12,451 feet), and an average temperature of 3.9° C (39° F).

composed of a series of concentric shells of material that grow hotter and denser as they approach the planet's center, with their chemical composition changing abruptly at the boundaries of each nested shell.

For an illustration of the internal structure of the Earth, by Kevin Song, see http://commons.wikimedia.org/wiki/ File:Earth_poster.svg.

We have no direct evidence of the planet's structure, because it has been impossible to drill very deeply even through Earth's crust. We must infer this structure from several different types of evidence. Most important are the motions of seismic waves, caused by earthquakes, traveling through the earth, which tell us much about the size and composition of its interior structure. We also make inferences from such things as the mineral composition of the meteorites that accreted to form the planet. This means that, while geoscientists agree on the general picture of the planet's interior structure, debate continues about certain details of that picture.

The outer layer, or *lithosphere*, is quite thin and not very dense, compared to the other shells, ranging from about 70 km (43.4 mi) in thickness beneath the oceans to about 125, and perhaps as much as 250, km (77.5–155 mi) beneath the continents. The lithosphere is broken up into the dozen or so large plates and several smaller ones that move around Earth's surface, giving shape to its features.

Although it is the coolest and least dense layer, the lithosphere quickly becomes hotter and denser with increasing depth, rising some 20° F for every kilometer beneath the surface. The deepest drilling to date has reached more than 12 km (or about 7.5 mi), at which point the temperature was 245° C (473° F) and the density was so great that it squeezed the drill-hole shut. The idea that humans might one day travel deep enough to find habitable caverns or other such structures, as in certain science fiction movies, is indeed purely fiction.

The crust[5] is composed predominantly of three light elements: by weight, oxygen (46.6%), silicon (27.7%), and aluminum (8.1%). There are lesser quantities of iron (5%), calcium (3.6%), sodium (2.8%), potassium (2.6%), magnesium (2.1%), and all other elements (1.5%). These elements combine in various ways to make the nearly four thousand naturally occurring minerals found on Earth, and these minerals combine to form the manifold varieties of rock found in our planet's crust. Altogether, the crust forms only about 0.1% of Earth's volume and 0.4% of its mass. It has two

5. The crust is part of the lithosphere, the other part being the uppermost mantle, called the lithospheric mantle.

distinct forms: a thinner *oceanic crust*, made up largely of magnesium- and iron-rich silicate minerals that have risen up from the mantle as magma to ooze out onto the ocean floor as basalt (a rock composed of silica, SiO_2, bonded to magnesium and iron); and a thicker *continental crust*, made up mostly of feldspar, a mineral composed of one atom of potassium, one atom of aluminum, three atoms of silicon, and eight atoms of oxygen ($KAlSi_3O_8$), and other silicon compounds. Because the atomic symbols for magnesium and iron are Mg and Fe, the rock forming the oceanic crust is called "mafic"; and because the continental crust is made up predominantly of feldspar and silicon in the form of granite (an igneous rock made up of feldspar and quartz, a form of silica) and related minerals, its rocks are called "felsic."

The oceanic crust is relatively thin, ranging from about 5 to 8 km (3–5 mi) in thickness with an average of about 7 km (4.3 mi). As noted above, it is predominantly basalt, a fine-grained igneous rock, though evidently with a coarse-grained version of basalt (called "gabbro") beneath it, which together have an overall density of about 2.9 or 3.0 gm/cm^3. This relatively young basaltic crust forms at ridges on the ocean floor, and then spreads outward as the lithospheric plates it adjoins move away from one another. Over time, this ocean crust becomes overlaid with sediments that rain down upon the seafloor from above, including wind-, rain-, and river-born debris from the continents; the remains of sea creatures that die and sink; and the majority of the cosmic dust that Earth continues to sweep up as it orbits the Sun, estimated to total about three tons a day. The plates carrying this oceanic crust converge with plates carrying continental crust, where they slip under those plates, sink down into the mantle, and eventually become molten rock once again.

The continental crust is significantly older than the oceanic crust, with some rock dating back to the Hadean Eon some four bya. It is also thicker than the oceanic crust, ranging from about 20 to 70 km (12.4–43.4 mi) in thickness and averaging between 30 and 50 km (18.6–31 mi). It is also more diverse and complex than oceanic crust, consisting not only of igneous rock but also of sedimentary and metamorphic rock. Typically, it contains a base layer of granite covered by various types of sedimentary rock at the top and, below this layer, metamorphic and other igneous rock.

Only about 20 percent of the continental crust's thickness lies above sea level, with some 80 percent of it lying below—rather like icebergs in the ocean. Continental crust is thicker where it has higher topographical features. As tectonic plate interactions cause mountains to rise up in the crust, their enormous weight causes the crust to sink down into the molten layer below, so the continental crust is thickest directly beneath Earth's tallest mountains. Because the continental crust, with a density of about

2.7 gm/cm^3, is less dense than the oceanic crust, when plate movements cause the two of them to converge, the oceanic lithosphere slides beneath (or "subducts" under) the continental lithosphere.

Beneath Earth's crust lies the *mantle.* Because oceanic crust is thinner than continental crust, the mantle lies closer to Earth's surface under this crust, perhaps only 5 km (3.1 mi) below the ocean floor. Overall, the mantle stretches from about 5 to 70 km (3.1 to 43.4 mi) below the surface down about 2900 km (1800 mi) and constitutes some 83 percent of Earth's volume and 68 percent of its mass. Its upper layers are composed chiefly of the igneous minerals olivine ($MgSiO_4$ or $FeSiO_4$) and pyroxene ($MgSiO_3$ or $FeSiO_3$). Deeper in the mantle, the greater pressure there crushes these to become the calcium-titanium mineral perovskite ($CaTiO_3$).

The *Mohorovicic discontinuity,* or *Moho,* forms a boundary between the lithosphere and the *asthenosphere* (from Greek *asthenēs,* "weak"). The asthenosphere is a weak layer of solid rock in the top part of the upper mantle which, probably due to the greater pressure there, is heated close to its melting point and so flows fairly easily. The thickness and depth of this layer are debated today. There is some general agreement that its depth under the oceanic crust is from about 70 km (43.4 mi) to as much as 200 km (124 mi), but there is significant disagreement on its extent under the continents, where some think it begins at around 250 km (155 mi) and extends much deeper. The asthenosphere, like the rest of the upper mantle, is composed primarily of olivine and pyroxene.

Although the viscoelastic asthenosphere is essentially solid (if plastic) rock, pockets of magma form within it. This magma is a combination of solid rock particles with relatively high melting temperatures floating within a viscous liquid made of water, dissolved gases, and rock that melts at relatively lower temperature at about 50 km (30 mi) beneath the surface. Due to its lesser density, this magma percolates up through cracks in the lithosphere, melting more rock on the way, until it finally erupts through the crust as a volcano. Volcanoes spew out large quantities of water vapor and other gases, lava, and *tephra.* Tephra is magma shot into the air, where it cools to form solid particles of ash and rock of various densities. These range from airy pumice that can float on water to the glassy obsidian used in traditional Native American arrowheads and the dense basalt that covers the ocean floor. Lava solidifies quickly under water but can flow significant distances above ground. The ash emitted from particularly large volcanic eruptions can form long-lasting clouds of fine particles that blanket the world, reflecting sunlight back into space and so cooling the planet's surface for as much as several years. One example is the eruption of Mt. Tambora in

Indonesia in 1815, which sent 40 cubic kilometers (9.6 cubic miles) of rock and ash into the air. The following year came to be known as "the year without summer," as snow and frost persisted into the summer in Europe and North America, ruining crops, killing livestock, and so creating the worst famine of the nineteenth century in the northern hemisphere.

The majority of volcanic eruptions occur underwater at the mid-ocean ridges that form the boundaries between two plates spreading apart. The rest form above plate subduction zones and over hot spots in the interiors of plates. Magma's viscosity varies with its location, composition, and temperature. At subduction zones more water and gas are added to the magma, thinning its viscosity and so making these volcanoes particularly explosive. Sometimes even whole sections of the volcano are blown away, as happened to the Mt. St. Helens volcano in Washington State in 1980. Significantly, most of Earth's minerals are crystallized forms of magma, with different minerals crystallizing at different temperatures. These then combine in a wide variety of ways to form different types of rock with different grain sizes and appearances.

The relatively thin asthenosphere at the top of the mantle forms only 6 percent or less of the total thickness of the mantle. However, it is extremely important, because the lithospheric plates float on it and have been driven in their various directions throughout Earth's history by the convection currents within this molten rock. These currents move rather like what we saw in the upper, radiative layer of the Sun, where material—in this case, molten rock—is heated from beneath (the mantle), causing it to rise, and then, after being cooled by its contact with the lithosphere, to sink, only to be reheated again. Because this movement in the asthenosphere drives the movements of the lithospheric plates, this layer is ultimately responsible for the appearance and constant changing of Earth's surface.

Below the lithosphere and asthenosphere lies the rest of the *upper mantle*, which may range from about 200 to about 670 km (124–415 mi) beneath Earth's surface. Below this lies the *lower mantle*, which stretches from perhaps 670 km down to around 2,900 km (1800 mi) below the surface and so constitutes the majority of Earth. Material in the mantle grows ever hotter and denser the closer it is to the core. The combination of these effects on the melting points of the various minerals spread throughout the layers of the mantle may be the reason, geologists think, for its layered structure. At a depth of 670 km (415 mi), the pressure becomes so great that olivine and peroxine (major components of the mantle, as we have seen) are compressed to form the denser mineral perovskite, the predominant mineral of Earth as a whole. This means that the chemical composition of the entire mantle is essentially the same but that its mineralogical composition differs between the upper and lower layers. The increasing density within both the upper and lower mantle causes these layers to be increasingly more viscous with greater depth, making the convective movement of the molten rock within each successive layer downward slower than in the one above. In reality, the mantle is solid rock, yet it can flow, even if at an incredibly slow rate, because of the enormous heat and pressure within the mantle. This almost imperceptible motion means that the heat from the core convects extremely slowly upward through the lower mantle, increasing in speed somewhat in the upper mantle, until it is finally transferred to the asthenosphere, where the magma can move at a humanly perceptible speed. The overall rate of molten rock movement through the mantle as a whole is about equal to the rate at which hair grows. The temperature at the mantle-core boundary is estimated to be some 3,800° C, and its pressure to be roughly three million times the pressure at the surface.

Earth's *core* extends from about 2900 km (1800 mi) below the surface to some 6370 km (3949 mi) at the planet's center. It comprises only 16 percent of the planet's volume, but, because of its great density, it makes up about 31.5 percent of its mass. The core probably consists of about 90 percent iron, with trace amounts of other materials, foremost of which is likely nickel, in addition to some oxygen, silicon, and sulfur. The core consists of two layers, termed, logically enough, the outer and inner cores. The *outer core*, which reaches downward some 2250 km (1395 mi) below the core's boundary with the mantle, is molten, heated from beneath by the inner core. Because it is molten, it contains convection currents, in addition to the slower conduction that moves heat away from the inner core upward to the mantle-core boundary. It is this quite significant movement in the outer core that scientists believe causes the enormous magnetic field that surrounds Earth, protecting it from the solar wind and other life-damaging radiation from space. Interestingly, geo-scientists have recently discovered that, over long and irregular time intervals, the magnetic field, reversing Earth's positive and negative magnetic poles.

Within the outer core, at the very center of the planet, lies the *inner core*, which reaches upward about 1220 km (756 mi) from the planet's center. This inner core is an incredibly dense and hot solid composed almost entirely of iron, which may exist there as a single enormous crystal. The pressure there is some 3.64 million times greater than at sea level, causing the core's enormous density of somewhere between 12 and 16 grams per cubic centimeter, or about five or six times that of granite. At the boundary between the outer and inner core, the temperature has recently been estimated to be 6,300° C +/- 800° C (11,350° F +/- 1,400° F), and at the center perhaps 6,400° C +/- 600° C (11,550° F +/- 1,110° F), which is even hotter than the surface of the Sun. Because the core is solid, however, its heat cannot be convected upward, as in the layers above it; it can only be conducted. Since convection is a more efficient means of heat transfer, this means that heat leaves the core relatively slowly. Nevertheless, this gradual rising of heat from the inner core through the outer core, and ultimately to the surface, means that the planet is gradually cooling. In fact, Earth constantly radiates an enormous amount of energy into space—some 44.2 trillion watts, or enough to keep 700 billion 60-watt incandescent light bulbs constantly shining—primarily as infrared radiation.[6] Recent evidence suggests that this solid inner core rotates slowly in a counterclockwise direction (when viewed from above) and at a rate slightly faster than the rest of the planet.

Despite this neatly layered depiction of Earth's structure, the reality is that the planet is, and always has been, in constant motion both within and without, and has other processes that disrupt somewhat this neat structure. For example, molten *plumes* rise up through the mantle, probably originating at various levels within the mantle, some from all the way down at the core boundary (see the image Earth's internal structure at the QR code above). Current theory suggests that, as the molten rock of the plume rises up through the mantle, it forms a broad head, looking rather like the mushroom cloud of an atomic bomb. When these plume heads reach the lithosphere, they typically lift and stretch it, causing basalt to seep up through cracks in the seafloor and volcanoes to rise from hot spots below the oceans and continents.

For an image of the Hawai'ian island chain (including the string of under water mountains) stretching west and north to Korea, by the National Geophysical Data Center/USGS and U.S. National Oceanic and Atmospheric Administration, see http://commons.wikimedia.org/wiki/File:Hawaii_hotspot.jpg.

Such seafloor hotspots are responsible for the creation of many island chains. For example, the Hawaiian Islands were formed over the past few million years as the Pacific plate gradually moved northwestward over a hot spot in the mantle (see the QR code above). Although the last island created, Hawaii, continues to experience volcanic activity caused by this hot spot, another island has already begun to grow up from the ocean floor to the southeast of Hawaii. It has even been given a name, Loihi, and will at some point in the future grow tall enough, by its gradual accumulation of lava, to break through the surface, where it will continue to build up until

6. Interestingly, we also radiate away energy, at infrared wavelengths (as heat), into the atmosphere around us.

the plate's motion decouples it from the hot spot. At that point, yet another Hawaiian island will be rising up from the ocean's floor. Meanwhile, the northwestern most islands in the chain will continue to diminish by erosion and eventually disappear beneath the ocean's surface, as so many have done before. These earlier, submerged Hawaiian islands reach all the way to Korea. Plumes rising under continental crust also cause volcanoes. Such plumes might also play a role in the splitting up of continents by causing the plate on which a single continent had ridden to split into two plates. These then begin moving apart, creating a nascent ocean in the void between them. Geologists think the breakup of the ancient supercontinent Pangaea might have been caused or at least aided by such hot spots, and the Afar hot spot beneath the East Africa-Arabia boundary is probably aiding the growing split between those two continents, as currently evidenced by the formation and widening of the Red Sea.

The neat picture of the Earth's structure is also retouched, as we have seen, by the convection that occurs in the mantle, which apparently causes the lithospheric plates that form the surface of the planet to be in constant, if rather slow, motion, moving only a few centimeters (a couple of inches) or so per year. Why the plates move in the particular directions they do is actually much more complex than the simple picture of mantle convection I have presented. For example, it seems that just as the motion of convection cells in the asthenosphere moves the Earth's plates, so the motion of the plates influences the convection cells in the asthenosphere. Thus, Earth scientists continue to study and debate the nature and causes of this plate flow. However, what is certain is that these plates do move, that they interact in the three very specific ways to be described below, and that these movements and interactions are the major reasons for the large-scale appearance of the planet's surface.

One way these floating plates interact is by moving apart from one another. Places where this occurs are called *spreading zones* and can be found both beneath the oceans and within the continents. They are major places of tectonic activity in the form of earthquakes and volcanic eruptions. One major example of an oceanic spreading zone is found in the middle of the Atlantic Ocean, where the North American Plate and South American Plate are moving westward away from the Eurasian Plate and African Plate, widening the Atlantic Ocean in the process. Molten rock from the asthenosphere pushes upward along the spreading zone, creating a ridge.

The plates then are moved apart by two main forces. In *ridge push*, gravity causes the plate to slide down off the ridge caused by the hot, elevated mantle. In *slab pull*, the far side of the plate sinks beneath another plate and is then pulled down into the mantle by gravity, dragging the rest of the plate along with it. As the two plates move apart, new lithosphere forms in the cracks created between the spreading plates as mafic material rises from the asthenosphere to fill the void. This occurs principally through volcanic eruptions. These undersea spreading zones form an almost unbroken chain of ridges and volcanoes all around the globe, in all totaling about 70,000 km (44,400 mi) in length.

Because of the constant spread and subduction of Earth's seafloors, the oceanic crust is at most only about 180,000 years old, whereas, as we have seen, some continental crust is dated all the way back to the early days of the planet some 4 bya. *Intracontinental* spreading zones are also important in shaping Earth's surface, functioning principally to tear continents apart. A current example is the aforementioned Red Sea, which continues to widen due to the mantle pushing up under it and so gradually moving the African plate and continent apart from the Arabian plate and its continental landmass, the Arabian Peninsula. Eventually the small land bridge remaining at the north end of the Red Sea will disappear, at which point the Red Sea will become joined to the Mediterranean Sea, eventually to become a new ocean.

For the ways in which tectonic plates move and interact with other plates, see http://pubs.usgs.gov/gip/earthq1/plate.html/, José F. Vigil's illustration, a wall map produced jointly by the U.S. Geological Survey, the Smithsonian Institution, and the U.S. Naval Research Laboratory.

When plates pull apart in this way, they either crash into other plates, typically forming *subduction zones*, or slide past other plates, creating *transform faults*. At subduction zones, as noted above, the denser oceanic plate slides under the lighter continental plate. This movement has major geological repercussions. First, sediment sitting atop the oceanic crust piles up against the continental crust, forming an *accretionary wedge*, which can join to the continent, creating a new coastline, or remain offshore and possibly form islands. Second, the subducting plate creates a deep oceanic trench as it slides beneath the continental plate, such as the famous Mariana Trench in the western Pacific, almost 11 km (6.8 mi) deep. The subducting plate also causes earthquakes as it scrapes the underside of the subducted plate. At some places, parts of the ocean sediments and crust will melt from the friction, forming magma containing large amounts of water and carbon dioxide from the crust and subduction process. This magma then rises up through the continental crust, forming Earth's most explosive volcanoes. Finally, the oceanic plate, being relatively cold and rigid, pushes down into the mantle when it subducts, retaining its slab-like form as it sinks, disturbing the flow within the mantle and, in some cases, reaching all the way down to the mantle-core boundary. There it affects the movement within the molten iron outer core and, therefore, Earth's magnetic field. The slab then ceases its downward motion, gradually melts, and is recycled back into the mantle of the planet from which it originally formed.

In less frequent cases, two plates of continental crust of essentially the same density converge. In such a case, neither plate subducts under the other; instead, both plates compress and fold or break (*fault*) the rock layers within themselves, usually uplifting the convergent boundary of the plates to form mountain ranges. The most spectacular current example of such convergence is the Himalayan mountain range, formed when the India-Australian plate carrying India collided with the Eurasian Plate. This convergence continues today, slowly pushing the Himalayas ever higher.

Transform faults result when two plates slide past one another, with one plate's edge grinding against the other's edge, creating earthquakes. While transform faults constitute the least frequent form of plate interaction generally and rarely appear on continental crust, they do form frequently at spreading zones on the ocean floor between segments of the spreading ridges. Probably the best-known transform fault is the 1300 km- (800 mi-) long San Andreas Fault in California. It is the result of the Pacific Plate moving northward relative to the North American Plate, causing Los Angeles (on the Pacific Plate) to move gradually toward San Francisco (on the North American Plate). At the present rate of movement, about three centimeters (a bit over one inch) per year, the two cities will be neighbors in some 13 million years, after which Los Angeles (or whatever it is called by then, if it still exists) will begin moving out to sea.

Before the discovery of plate tectonics in the 1960s, it had already been noted that earthquakes and volcanoes tended to cluster in certain areas, such as the so-called Rim of Fire surrounding the Pacific Ocean. Since then it has become clear why these patterns exist. Most earthquakes and volcanoes occur at plate boundaries and owe their existence to the three types of plate interaction—spreading, collision, and transform sliding.

However volatile the Earth may seem to us today, it is sedate compared with the early Earth, at least until the end of the late heavy bombardment period some 3.85 bya. Even after that violent period ended, heat generated by the core and mantle continued to be significant, due to several processes, including the decay of radioactive materials within the planet and the friction caused by the settling out of Earth's interior as the lighter minerals rose upward and the heavier ones sank. However, as the shorter-lived radioactive material decayed away, and the rate of decay of longer-lived radioactive isotopes—particularly uranium-235, uranium-238, thorium 232, and potassium-40—slowed over time, the amount of energy, and therefore the heat produced from this source, declined. Geoscientists estimate that the heat from radioactive decay within Earth today is only about half of what it was when the planet was young. In addition, the settling out of the heavier from the lighter minerals within Earth gradually reduced the friction heat caused by those movements. The diminution of the heat created by these two processes, in addition to the declining rate of asteroid and meteor collisions and the continuous release of heat from the planet's surface, caused the lithosphere to begin to solidify. This prevented the crust from sinking back into the mantle as it cooled and provided a solid base for the formation of the lithospheric plates. Still, hot spots rising up through the mantle caused numerous fractures in the lithosphere, keeping these plates small.

The subduction of these small plates under one another generated widespread volcanic activity at their boundaries. The magma from these subduction-caused volcanoes was a combination of mafic and felsic rock. Volcanoes that were located over hot spots spewed material that was more felsic in nature. These volcanoes were predominantly responsible for the formation of the first patches of felsic continental crust. These first protocontinents were perhaps about the size of Iceland today.

Over time, plate collisions caused some of these protocontinents to combine and form larger landmasses. Evidence suggests, though, that these landmasses remained relatively small compared with today's continents, due to the continued plate rifting caused by heat rising up from within the planet. This heat also prevented the continental crust from becoming as thick as it typically is today. Geologists have located remnants of the earliest protocontinents today in a number of places around the world and on almost every continents. The exception is Antarctica, which is almost entirely shrouded in ice and snow, which prevents most geological investigation.

Geologists term these ancient continental crustal remnants *cratons* or *shields*, the oldest of which, the northwestern part of the Canadian Shield, dates from 3.8 to 4.0 bya. Some zircon crystals from Western Australia date from perhaps as early as 4.4 to 4.1 bya and are the oldest remnants ever found of what was once continental crust. These archaic cratons constitute about 7 percent of Earth's continental crust today. The amount of continental crust around these first cratons increased relatively rapidly, so that by the end of the Archean Eon 2.5 bya, the total amount of continental crust was perhaps roughly equal to what it is today. Since then the new crust created, for example, by volcanic action, has been balanced by the amount lost to the sea by erosion. The rest of Earth's surface at the end of the Archaean Eon was, as now, covered with oceans, which grew steadily saltier until about this time. Ever since the end of the Archaean Eon, the amount of salt entering the oceans has been roughly the same as the amount of salts precipitating out and sinking to the ocean floor.

From the end of the Hadean Eon on, Earth's surface would be sculpted fundamentally by plate tectonics, but also significantly by what geologists term the rock and water cycles. The *rock cycle* is about the creation of new crust, both oceanic and continental. In its simplest form, the rock cycle is when magma rises up from the asthenosphere through cracks in the lithosphere and crust and is deposited either on the seafloor or on land, where it cools to form igneous rock. If igneous rock cools quickly at the surface, it is called *extrusive igneous rock*, which is generally a finer-grained type of rock. If it cools more slowly beneath the surface, it forms the more coarsely grained *intrusive igneous rock*, which may then rise later, because of tectonic action, and become exposed by the erosion of rock layers above it.

This intrusive igneous rock is then exposed to the action of wind and water and so gradually erodes to form sediment, which in the ocean sinks to the bottom. On the land, wind, streams, and rivers may carry this sediment into the oceans, where it also sinks to the floor. There more sediment will usually cover it, solidifying to form *sedimentary rock*, the second fundamental type of rock. This sedimentary rock will typically be buried under still more layers of sediment and become cemented together. If the rock is eventually buried under several kilometers/miles of sediment, the pressure and heat can become sufficient to transform (morph) it mineralogically, structurally, and chemically into a new kind of rock, *metamorphic rock*, which is the third and final fundamental type of rock. If this rock is pushed still further down by the deposition of additional layers of material or is subducted beneath a continent, it will eventually reach its melting point, in which case the rock has returned to its original state, magma, and so is available to begin the rock cycle again. Tectonic forces can and do alter this simple picture, so that not all rock follows precisely this straightforward path. Indeed, there are many possible pathways through the rock cycle; but eventually, virtually all of Earth's crust is recycled in this way.

Weathering is the slow process by which the rock created in the rock cycle is gradually torn down. It occurs in two forms: mechanical weathering and chemical weathering. *Mechanical weathering* is the physical disintegration of rock, and it happens in several ways. It can be caused by the wind, in which cases small airborne particles crash into rocks, sandblasting away their rough outer edges before finally reducing them to small rocks or sand grains and soil. Extreme changes in air temperature, such as are caused by forest fires, can cause rock to expand and crack, contributing to its decomposition.

Water, however, is the primary agent of mechanical weathering, operating within what geologists term the *hydrologic* or *water cycle*. The force of raindrops driving into rock can contribute to breaking it down over time. Water in streams and rivers can carry along rock particles, driving them into other rocks, which slowly abrades them. Ocean currents and surface waves crashing into shores function similarly. Water can also seep into cracks in rock and then freeze, prying open these cracks further, in a process that repeats over and over, slowly breaking the rock into small pieces. The rapid growth of salt crystals where briny water evaporates quickly can similarly crack rock apart. Glaciers also move across the land, crushing rock as they creep downward through the mountains. Eventually, the water cycle concludes as water is evaporated back into the air, from either the land or lakes and oceans, where it will eventually cool and fall as rain once again, continuing its weathering of the land.

Tectonic action at plate boundaries also breaks down rock mechanically through the friction caused by two plates rubbing against each other. Even gravity can even play a role in mechanical weathering, as when rocks roll down from higher to lower places, striking other rocks along the way.

Biological organisms also contribute to mechanical weathering. Plants' roots grow into cracks in rock, prying them apart as they grow and so gradually breaking them down. Burrowing animals open up passages underground, enabling water and air to enter and cause both mechanical and chemical weathering. Humans also contribute in major ways, both intentional and unintentional, to cause the breakdown and erosion of rock.

Chemical weathering is the breakdown and alteration of the chemical composition of rock due to chemical reactions that occur in the air and water. Here again air and rain play important roles. Some minerals, when exposed to the air, react chemically with oxygen molecules to form different minerals, as with the rusting of iron-containing minerals to form *iron oxide* (Fe_2O_3), as in the mineral *hematite*, or, if combined with water, *limonite*. Water can dissolve certain minerals, such as salts and limestone, from rock, causing them to crack apart. It can also break down certain minerals by reacting chemically with them, such as feldspar, which water turns into clay.

In addition, naturally occurring acids cause chemical weathering. Some are strongly corrosive, such as *sulfuric acid* (H_2SO_4) and *hydrofluoric acid* (HF), which volcanic eruptions can spew out. However, the acid most responsible for global chemical weathering is actually the weakest naturally occurring acid, *carbonic acid* (H_2CO_3). This acid forms when carbon dioxide (CO_2) is dissolved in water. Since the world has an abundance of both, carbonic acid is formed readily almost everywhere. For example, soil naturally contains as much as 10 percent carbonic acid, due to biological processes such as acidic plant root secretions, and this mixes with water to weather rock chemically beneath the surface. More significantly, Earth's atmosphere contains about 0.03 percent CO_2 which dissolves in the water droplets of rain, making most rain slightly acidic. Human pumping of CO_2 and other gases into the atmosphere has increased the acidity of rain, which has in turn caused an increase in chemical weathering, as well as significant damage to the biosphere.

In these varied ways, the majority of the rock that is created from magma in the asthenosphere and makes its way to the surface is gradually altered and broken down, finally to return to the oceans, where it eventually subducts back into the mantle to melt and begin the rock cycle once again. In this way, almost since Earth's beginning, the planet has recycled its surface features. This recycling, along with the movements of the planet's massive plates, has made our home the endlessly dynamic place it is. Since it is the Sun and rotation of Earth that drives the wind- and ocean-current patterns, it becomes clear that not only are all of Earth's natural systems, physical and biological, interrelated, but that that interrelationship extends to the Sun, the Moon, and, to a lesser extent, the rest of the solar system as well.

12

The Origin of Life on Earth

We are fortunate that Earth is where it is in relation to the Sun and other planets and is the size it is. Were it not, we would not be here! In fact, without a whole series of fortunate circumstances, life could never have arisen on our planet. For example, were Earth much closer to the Sun, it would be too hot for water to exist in its liquid state, and so life as we know it could not exist. If it were much farther from the Sun, it would be too cold, and water would always be frozen, again prohibiting our kind of life. In addition, if Earth did not rotate on its axis, one side would be too hot and other too cold for liquid water and so, again, life could not have developed. Similarly, if Earth were much larger, its increased gravity would have resulted in a denser atmosphere, which would have blotted out too much of the Sun's energy for life to arise. If Earth had been much smaller, its weaker gravity would not have held onto the oxygen that is necessary for larger and more complex forms of life, like us, to have evolved. Furthermore, a smaller Earth would have cooled off sooner, as Mars did, making it much more difficult for life to arise. In addition, as we have seen, without Jupiter and its massive gravity to attract asteroids and other space debris, bombardment by these solar-system wanderers would have been far greater throughout Earth's history. This too might have made the development of life, or at least any more complex form of life, impossible. For all these and still other reasons, Earth has indeed been fortunate in having just the right set of circumstances for life to originate and evolve, making it a special kind of planet in the universe.

The earliest pieces of continental crust were barren landscapes, totally devoid of living things. Nevertheless, even before these first landmasses had formed, life had originated and begun its long, slow evolution in the oceans. Teasing out precisely when, where, and how life originated has been, for much of the past century and even today, the focus of much scientific research. At present, however, these questions remain unanswered; indeed, we might never know the precise circumstances under which life first began.

Part of the problem in determining when and how life began has to do with the very definition of what constitutes a living entity. Many definitions have been proposed since the dawn of the modern scientific era some four hundred years ago. Noam Lahav has compiled forty-eight of these from the past one hundred fifty years alone, including many recent definitions. None has gained anything approaching universal acceptance within the scientific community. However, several general features do appear in some form or other in many of these definitions, such as that, to be considered alive, a thing must have an organized chemical structure, get energy from metabolism, reproduce and pass along information to the next generation, and produce variations upon which natural selection can act.

The problem may well lie in the either-or assumption behind such a question—that a thing is either a living being or a nonliving being. Many scientists now believe there is no such clear line between life and non-life. Rather, they suggest, life originated through a gradual series of biochemical steps, each of which resulted in a new level of emergent complexity. With each stage, the biochemical process became more like what we know as life today, yet without a clear point of delineation between life and non-life.

The approach scientists take to discerning *when* the simplest single-celled organisms arose—those creatures that all would recognize as living entities—involves at least two steps. The first is to determine when the conditions on Earth would probably have been such that life might have arisen. The second is to follow the fossil record backward as far as possible to find traces of that life embedded in the most ancient rocks. This means that the answer cannot be precise but must be within a range of dates. Scientists generally think that life could not have existed prior to 4.2 bya, due to the massive bombardment of Earth by giant asteroids before that time. In addition, the high temperature of the early Earth would not have permitted liquid water to remain on its surface.

Had life originated during the next 200 million years, it would then have had to endure the second round of asteroid impacts during the late bombardment period. These impacts would have generated so much heat that the planet's surface would have been sterilized. Asteroids as small as 500 km (300 mi) in diameter colliding with Earth would have temporarily evaporated all the planet's oceans, so, almost certainly would have destroyed whatever early life-forms that might have lived there. Therefore, the end of the late bombardment period 3.85 bya seems to be a reasonable earliest limit on when simple life, which eventually evolved to form all the living beings on Earth today, could have begun to exist.

The latest date for the appearance of life also defies precise identification. We have clear fossil evidence of single-celled organisms existing by about 3 bya, though these are already relatively sophisticated forms of life. At this point, life had already evolved a myriad of forms from small spherical and rod-like single cells to more complex chain-like and corkscrew-shaped single cells. This suggests that more primitive forms of life had preceded them, though by how much remains unknown. Such direct fossil remains of early life, then, take us back only so far.

Another means that geologists use in searching for the origins of life on Earth is to seek out certain chemical compounds in ancient rocks that must have been used in, or created by, the metabolic processes within living cells. All life on Earth is carbon-based, and living organisms use only a small percentage, or several hundred, of all the possible carbon-based molecules that exist. Some of these organic molecules have very distinctive characteristics. For example, *hydrocarbons* (molecules made from carbon and hydrogen) are plentiful on Earth, but life-forms use only those with even numbers of carbon atoms strung together in chains. Similarly, *polycyclic molecules* (molecules composed of two or more closed rings of atoms) are found almost everywhere, including on meteorites and floating in space. Living cells, however, create only very specific ones, such as the *sterol cholesterol*.

The *sugars* used by life are also idiosyncratic. Sugars are carbohydrates (molecules containing carbon, hydrogen, and oxygen) that typically have the formula $C_nH_{2n}O_n$ (or $C_mH_{2n}O_n$). Thus, *glucose* (the simple sugar that cells use for food/energy) has the chemical formula $C_6H_{12}O_6$. Sugars typically exist in both right-handed and left-handed configurations, meaning that they exist in mirror-imaged forms, like human hands. Life, however, uses only right-handed sugars. Similarly, life uses *amino acids* (the building blocks of proteins) only in the left-handed configuration.

Only living organisms produce *hopane* (another kind of polycyclic molecule), which is a particularly good biochemical marker for life. In addition, living cells preferentially use certain isotopes of some elements, such as carbon-12, as distinct from carbon-13. Some of these molecules, by their distinctive properties, can even tell us something about the lifestyle of the organism that made it. All these different *chemical bio-signatures* offer the potential for finding traces of ancient life in rock even where the fossils of the single-celled organisms that made them have long since disappeared. They therefore enable biologists to find even more traces of ancient life.

Using such biomarkers to determine the presence of ancient life is not without its difficulties. The main problem is that most such molecules, with the exception of hopanoids, are relatively fragile, so that geological processes that have occurred over the intervening billions of years have easily destroyed them. Further complicating this picture is the possibility that such molecules are actually more recent in origin but have slowly seeped into the ancient rock, contaminating it and thus giving the erroneous impression that the bio-signatures are a lot older than they actually are. Nevertheless, geologists have had some notable successes in teasing out the chemical clues from certain ancient rocks, thereby pushing our knowledge of the antiquity of living beings still farther back into the past.

These chemical signatures seem to indicate that life existed on Earth by at least 3.5 bya. Ancient *stromatolites* (see the QR code below) also provide important evidence about that date. Stromatolites are rock-like structures built up of layers of ancient *cyanobacteria* with sediment between the layers. The oldest such stromatolite that has been confirmed to be of ancient cyanobacterial origin has been dated to more than 2.7 bya. A more recent discovery has been dated to nearly 3.5 bya, although some questions about its evidentiary value remain. Nevertheless, what is particularly interesting about this find is that stromatolites are already quite sophisticated entities, which suggests that the simplest form of life must have arisen significantly earlier. For this reason most geologists today seem to date the origin of life on Earth to between 3.8 and 3.85 bya, or at about the end of the Hadean Eon or the beginning of the Archaean Eon.[1]

For a photo of ancient stromatolites by Prof. Mark A. Wilson (Department of Geology, The College of Wooster, Wooster, OH), see http://commons.wikimedia.org/wiki/File:SoegininaStromatolitesEstonia.jpg. Note the layers in the rocks.

The question of *where* life started has become a lot more controversial in recent years. While researchers are still virtually unanimous that life must have arisen in the oceans, several new theories have recently emerged, each with its own logic, suggesting that life might not have arisen first on Earth at all. As we have seen, some scientists have suggested that simple life-forms could have arisen on Mars early in its history, when the planet still had an atmosphere and oceans of water on its surface. At times, large bodies

1. See the Geologic Time Scale in the appendix.

that collided with Mars gouged out chunks of rock and the microbial life that lived on them, sending them into space. Some of these, in this theory, seeded our planet with its first life. The fact that we have some Martian meteorites that were gouged out of the Martian surface, such as that found in the Allan Hills of Antarctica, might support this theory.

On the other hand, the chemist Leslie Orgel and Francis Crick, co-discoverer of the double-helical structure of DNA, proposed the *directed panspermia theory*. According to this theory, aliens may have intentionally seeded Earth with life in the distant past as they sought to spread life throughout the galaxy. Both theories are interesting, but finding empirical evidence to support either will be extremely difficult—in the case of the latter hypothesis, perhaps even impossible.

For this reason and others, most origin-of-life scientists believe that life arose on its own in Earth's early oceans. For one thing, all the earliest fossils on Earth come from organisms that lived in the oceans. In addition, water appears necessary for life to occur in any form, because it is the necessary medium for organic chemicals and complex macromolecules to mix and react with each other. This is fundamental to all life as we know it. Further, all living cells today are encased in what is essentially a salt-water membrane, within which biochemical reactions and processes take place. All these factors point to the oceans as where life must have arisen. The next question is where in the oceans this might have happened.

Early in origin-of-life research, the consensus among scientists was that life probably arose in warm tidal pools along shorelines. There organic molecules could have become concentrated as the water evaporated, and lightning, or perhaps ultraviolet light from the Sun, could have provided the energy to enable their combination into the ever more complex organic molecules that finally resulted in life. The foundational support for this theory came from the famous Miller-Urey experiment, done in late 1952 by graduate student Stanley Miller for his doctoral research under the supervision of his advisor Harold Urey and published in 1953. Miller attempted to duplicate the conditions of the early Earth by filling a flask with water vapor, hydrogen gas, ammonia (NH_3), and methane (CH_4), which Urey had previously argued formed the early atmosphere, and then shot electric sparks across two electrodes to simulate lightning, which might have provided energy for the reaction of the gases. A two-day run of the experiment resulted in the creation of *glycine*, the simplest of the amino acids used by life; and a weeklong run produced at least five more amino acids essential to life as well as other useful biomolecules. Subsequent similar experiments over the years yielded similar results, producing a variety of organic molecules necessary to life. Thus, the tidal pool theory provided an interesting possibility regarding how life might have arisen on Earth.

However, there were also some problems with this approach. One was that scientists seriously questioned the kind of early Earth atmosphere rich in ammonia and methane that Urey proposed and Miller used in his experiment. Based on geological evidence, the atmosphere around 4 bya seems to have been mostly nitrogen and carbon dioxide gas, with only a trace of methane and ammonia. Another problem was how all these different amino acids and other hydrocarbons, having been spread throughout and highly diluted within the early oceans (that also contained huge quantities of other compounds not used by life), could have been brought together and sufficiently concentrated to combine and form larger macromolecules, such as proteins and RNA, on which life depends. Moreover, even if somehow they could have formed, the lightning and the ultraviolet light that occur at Earth's surface tend to break down such long-chained macromolecules, making it unlikely they would have had time to form the stable, more complex emergent molecular systems that were necessary to life's creation.

More recently, a rival theory has arisen following the 1977 discovery of rich ecosystems at *hydrothermal vents* (often called *black smokers*) at the East Pacific Rise plate-rifting zone eight thousand feet beneath the sea. The conditions there would seem to preclude the possibility of life—at least life as we have come to know it in the environments we normally experience. The pressure at hydrothermal vents is twenty-five times greater than the atmospheric pressure we experience at sea level; the temperature is as high as 400° C/752° F (water is kept from boiling only by the enormous pressure at the seafloor); and the water that emerges from such vents is highly acidic, rather like vinegar.

Although such conditions are lethal to most life-forms on Earth, the vents were teeming with life, including unusually large, red tubeworms up to six feet in length, spindly shrimp, and enormous clams. Such life survives, not by using energy from the sunlight, which is essentially nonexistent at such depths, but by using chemosynthetic processes to convert the rich minerals that spew from the vents into energy. Bacteria use the minerals in their environment to create energy by means of simple chemical reactions. Then the more complex beings around them consume these bacteria. Finding life in such a seemingly impossible location revolutionized the scientific understanding of the possible conditions under which life might flourish. Since then, it also has led to the discovery of other such *extremophilic* (extreme loving) *life* in ice a mile beneath Antarctica's surface, deep within dry deserts, and within rock formations of every type miles underground.

The hydrothermal vent theory proposes, for a variety of reasons, that such places would have been logical sites for the formation of life. Spreading zones are enormous in size, as we have noted, reaching around the world, and so providing numerous places where life might have taken hold. Vents would have provided a plentiful and consistent source of chemical components necessary for life, such as phosphorus, metals like zinc and nickel, and other elements that could be synthesized to create the energy needed for the cell's life-processes. Such sites deep beneath the ocean surface would have been protected from the lethal effects of the Sun's ultraviolet radiation, which was particularly intense before the formation of Earth's protective ozone (O_3) layer hundreds of millions of years later, not to mention the periodic asteroid and meteor collisions that would have thrown the planet's surface into disarray. Finally, these sites are rich sources of the clays that can function as useful substrates for assembling complex biomolecules and for providing protection for these fragile compounds, within their small pores and cavities, from the damaging effects of their extreme environment. Support for this theory has come from the discovery of fossil bacteria in ancient hydrothermal rock deposits dating to 3.5 bya. Additional support comes from the numerous other ecosystems that have been found around hydrothermal vents along the Pacific and Atlantic rifts.

Nevertheless, this theory has problems as well. The high temperatures at hydrothermal vents would tend to break down amino acids and other important biomolecules, and the generally volatile conditions at hydrothermal vents might have made the stable environment necessary for life to evolve impossible. In addition, the ecosystems at such vents depend today on oxygen created by plants to fuel their reactions, but the early oceans would have been largely, if not entirely, devoid of oxygen. This would have retarded, if not totally prevented, some of the chemical reactions necessary for life. The debate between the proponents of these two theories has at times been acrimonious. Without some major new discovery it is unlikely that a consensus will arise among origin-of-life researchers anytime soon, if ever. However logical our theories, it may be impossible to discern with any real certainty where the first life actually arose on Earth.

Precisely *how* life arose—the exact order in which the biochemicals in the primordial oceans synthesized and eventually led to life—remains a mystery, though the probable general sequence of steps that must have taken place seems fairly clear. The main debates among scientists center on the details of this evolution.

The first step must have been the formation of the simpler organic molecules—the sugars, amino acids, and hydrocarbons—necessary for life from the primordial molecules present in the early Earth. This is the best-understood phase in the process of life's origins. Atoms of carbon, oxygen, and hydrogen that make up the fundamental constituents of these organic molecules are plentiful in molecular gas clouds in space, where simple organic compounds form on grains of dust when energized by radiation from a star or some other source. In fact, not only are relatively simple molecules—such as water, carbon monoxide, and carbon dioxide—formed, but also some 140 other molecules, some composed of a dozen atoms or more, including such important biomolecules as ammonia and methane. As we have noted, such space dust continues to rain down unseen onto Earth at a rate of several tons per year, continuing to add to the planet's store of these organic molecules. Comets and meteorites also bring them to Earth, somehow surviving the extreme heat encountered when they fall through the atmosphere. Experiments recreating the conditions of meteorite impacts have shown that, while the enormous heat and pressure created in such collisions drastically reduce the number of amino acids, some amino acid molecules actually react with each other to form pairs of amino acids. Also, peptide bonds were formed, which caused more complex organic molecules to join together. The *Murchison meteorite*, which crashed to Earth in 1969 and was found to be filled with various types of organic molecules, attests to the survival of such biomolecules after collisions with Earth.

Thus, a variety of simple organic molecules would have arrived on Earth during its formation and others added in subsequent meteorite collisions. Evidence also suggests that geological processes added to the store of simple organic molecules in the early Earth, including their creation at hydrothermal vents on the ocean floor. A form of natural selection, called *molecular selection*, was already underway among these organic molecules. Some were better adapted to their surroundings than others. Some were more stable, others more unstable. Some dissolved more easily than others did, while some would have attached themselves firmly to the surfaces of minerals or clumped together into masses, making them no longer available for reactions with other molecules. Eventually some kind of molecular equilibrium was reached in various oceanic environments, with some molecules favored and others not.

The presence of these simple biomolecules in various ocean environments would have enabled the second step on the path to life: getting just the right organic molecules together and concentrating them to such a degree that they could combine to form the larger, more complex organic *macromolecules* that form the basis for all life on Earth. The simple sugar glucose ($C_6H_{12}O_6$) would have combined with other simple sugars to form long-chained molecules, or *polymers*, which we call *carbohydrates* ($C_mH_{2n}O_n$), as well as *starches* (for energy storage) and *cellulose* (necessary for plant life).

Amino acids would have been joined together by peptide bonds to form the larger *proteins* that govern cellular activity and exist in thousands of forms, all made up from combinations of twenty amino acids.

Nucleic acids, such as *ribonucleic acid* (*RNA*) and *deoxyribonucleic acid* (*DNA*), would have formed through the combination of large organic molecules called *nucleotides*. These feature a combination of a 5-carbon sugar, either *ribose* or *deoxyribose*, a *phosphate group* (a phosphorus atom surrounded by four oxygen atoms), and one of five *bases*. Bases are small, flat, cyclical molecules that come in five varieties: adenine, cytosine, guanine, thymine, and uracil. Other hydrocarbons would have combined to form the long-chained *fats* and *oils* that constitute cell membranes, provide energy for cell processes, and participate in other important cell functions.

As we have seen, the early Earth would have been full of the simple sugars, amino acids, hydrocarbons, and nucleic acids, which could combine into more complex biomacromolecules. However, the early Earth was enormous, and its oceans would have contained only an extremely diluted mix of these biomolecules. The key question is how just the right biomolecules would have come into proximity to one another and concentrated to such an extent that the chemical reactions necessary to form the macromolecules essential to life could occur. This problem is made more acute by the fact that, as we have seen, only certain organic molecules out of the many hundreds that exist are used by life. For example, of the hundreds of sugars that exist naturally, living cells use only a few. Also, recall that many organic molecules exist in right- and left-handed forms, yet life uses only one or the other of these. The important sugar ribose is an example here: living cells use only the right-handed form.

As already noted, one suggestion has been that the chemical needed for life might have been trapped in tidal pools, where gradual evaporation of the water would have concentrated the biomolecules so the needed reactions could occur. Water, however, is also a solvent and naturally breaks down carbohydrates into their constituent sugars, and proteins into their constituent amino acids. The presence of rock surfaces, though, could have mitigated these tendencies. Both above and beneath the ocean surface, organic molecules could have affixed themselves in particular ways to the mineral surfaces, both on the outside of the rocks and in the tiny pores within them, where they might have combined into organic macromolecules.

This could have occurred with a variety of minerals, with different minerals tending to attract different kinds of molecules and molecules with different handedness (or *chirality*). For example, the sheet-like structure of clays allows water and organic molecules to concentrate between the sheets. The sheets are sometimes electrically charged, which causes some organic molecules to be attracted to and concentrated on their surfaces. Amino acids have been shown to combine on clay particle surfaces to form longer, protein-like molecules, and nucleotides have been seen to assemble into RNA chains on at least one kind of clay. Clay particles also provide the catalyst for *phospholipids* to self-assemble into cell-like spheres, as we will see below. These important functions of clay minerals, when added to the fact that clays have also been shown to be capable of storing information by the way their various sheets are composed and layered, has led one researcher to assert that fine-grained clays were the original form of life on Earth, a geochemical kind of life that predated the emergence of the biochemical type we recognize as life today.

At some point these macromolecules would have become encased in membranes like the cell walls that exist today in order to have a protective, chemically concentrated environment within which complex chemical processes could occur. Here we seem to have a rather good understanding of how this might have arisen. Cell walls are made of phospholipid molecules, in which two long hydrocarbon chains (the lipid) on one side are attached to a phosphate group on the other. The lipid end naturally seeks to avoid water (remember the saying, oil and water do not mix)—that is, it is *hydrophobic*—while the phosphate end is attracted to water, or is *hydrophilic*. When large numbers of such phospholipid molecules are put in water, they spontaneously join together and arrange themselves into double-layered spheres, called *vesicles*, in which the hydrophobic lipid ends of each layer join together in the center of the two layers and the hydrophilic phosphate ends face inwards and outwards, touching the water inside and outside of the sphere. These vesicles are very similar to cell membranes today. In addition, these vesicles naturally grow in size by adding more phospholipids from the water around them and can even divide spontaneously into two smaller vesicles. Such lipid building blocks are known to have formed in space and have been found in carbon-rich meteorites, including the Murchison meteorite, meaning they would have been plentiful in the early Earth. Since the kind of self-assembly, growth, and reproduction seen in phospholipids are key features of processes within living cells today, origin-of-life researchers believe they now understand this crucial aspect of the processes that finally resulted in living cells.

The next step would have been the creation of self-replicating molecules capable of metabolism. This is currently an area of significant debate. Some scientists believe metabolism came first, then self-replication; others believe self-replication preceded metabolism, and some suggest both may have arisen at essentially the same time. *Metabolism* is the process by which energy is created and used to power processes in the cell or make new molecules from other, simpler molecules. *Self-replication* is, in its simplest form, the ability of a molecule to make copies of itself. In living things, self-replication involves systems of molecules that together form a self-replicating cycle, as in the cycle of steps that cause the DNA ladder to be split into two single strands, which are then both duplicated and recombined to form two identical DNA molecules.

Those researchers who favor the metabolism-first model argue that metabolism must have come first in order to synthesize the simpler organic molecules into complex self-replicating biomolecules. The general scenario suggested for life's evolution in this model is as follows. First there were simple molecules in the oceans, and with minerals providing the chemical energy for reactions to occur, these simple molecules combined into more complex molecules. These became organized, perhaps on rock surfaces, to form metabolic chemical cycles that could self-replicate by drawing more molecules from their environment. Variations on these metabolic cycles competed with each other for resources, with the more efficient winning out. Over time such variation led to the creation of cell walls and genetic molecules, with the more effective forms out-competing the others and surviving. In this scenario, then, life evolved first on the surfaces of rocks where this chemical evolution occurred.

Theorists who believe self-replication came first claim that, without the prior ability of biomacromolecules to self-replicate, the complexity of structures and processes that exist within cells could not have continued in subsequent generations of cells. In this scenario, polymer chains were created first, aided perhaps by mineral templates, and gradually evolved to make copies of themselves. The first genetic molecule must have been simpler than RNA, yet functioned similarly in that it was a polymer carrying the same bases as RNA and creating proteins from amino acids by patterning these bases in various ways. Variations developed between such molecules, and the ones that were most efficient survived, gradually evolving into the more complex self-replicating RNA and DNA molecules that form the basis of all life today.

The third general proposal is that both metabolism and genetic molecules developed together. This scenario argues that neither process could have become established on its own in the prebiotic oceans, because metabolic processes tend to be too random without the direction provided by some genetic molecule, and self-replicating molecules are too unstable to exist on their own, and their nucleotide molecules do not seem to form under primitive ocean conditions. The proposal is that a primitive genetic molecule might have attached to a primitive metabolic molecule, with each supplying the other what it lacked alone. Competing symbiotic pairs then would have formed, and evolution would have gradually produced more efficient and complex molecules and processes like those that exist today.

Origin-of-life researchers have put forth numerous specific proposals as to how each type of system might have evolved chemically in the prebiotic soup. However, so far the experimental evidence has failed to compel acceptance of any one theory. Important holes remain in each argument, and so this crucial intermediate step on the way to life remains the most active and hotly contested area of origins research today.

Once self-replicating molecules had formed, however, things changed in an important way in the prebiotic environments. Self-replicating organic molecules take simpler molecules from their environment to reproduce themselves; but the supply of such simpler organic molecules was finite, in which case those self-replicating molecules that were more efficient at capturing these molecules would have had an advantage over other, competing self-replicators. Such molecular selection would have established the more competitive self-replicators as the reigning biomolecules in the seas. However, variations on these molecules would have continued to form, so that, as in biological evolution, the struggle for existence would have continued, with ever more efficient molecules evolving over time. Many researchers believe that, once this kind of molecular selection and evolution began to occur, it is appropriate to say that life had originated on Earth.

The final step in the emergence of life as it exists today would have been the evolution of these simpler biochemical systems into more complex, self-replicating ones. In modern cells, DNA serves the functions of storing and copying information and assisting in the process of forming the complex biochemical molecules necessary to life. Segments of DNA called *genes* carry the information and provide the template for the creation of proteins, which in turn are composed of shorter stretches of DNA called amino acids, each of which is three nucleotides long. RNA comes in three forms: messenger (mRNA), transfer (tRNA), and ribosomal (rRNA). After an enzyme splits the DNA ladder down the middle, a strand of mRNA binds to and copies the bases constituting a gene, with each mRNA strand copying a single gene. The tRNA molecules consist of a particular amino acid on one end and the three bases that complete the molecule on the other end. The rRNA then links the tRNA's amino acids together in the pattern provided by the mRNA to form the particular protein encoded by that mRNA molecule. In this way, each gene along the DNA strand is copied and the proteins are made that govern the functioning of the cell. Eventually two complementary RNA strands became joined together to form the ladder-like, spiraling DNA that carries genetic information for the formation of subsequent generations of proteins. Each DNA molecule is composed of millions to billions of bases. RNA, among other things, provides the template for creating and assembling the three-base amino acids on these nucleotides in various ways to form the twenty amino acids and, from them, all the various proteins that cells need to maintain metabolism and reproduction.

The earliest living cells were probably rather like today's cells but with a simpler structure and biochemistry. Many, perhaps most, researchers agree that, prior to our DNA world, a simpler RNA world must have existed. In this world, RNA need only have been able to self-replicate, not perform the other functions that the three types of RNA do today. Experiments have shown that lipid shells containing RNA not only protect that molecule from destructive outside chemical influences, but that these shells also grow to accommodate the new, self-replicated copies of the original RNA by taking lipid molecules from other lipid shells with no RNA inside. In this way, lipid vesicles containing RNA out-compete other lipid vesicles for survival, eventually growing so large that they divide, and a new generation of RNA-containing vesicles is formed. Something like this probably happened in the gradual movement toward life in the early oceans.

There is also evidence suggesting that there may have been an intermediate step between the RNA and DNA worlds, in which only two bases existed, guanine and cytocine. The four possible combinations of these would have formed four simple amino acids: glycine, alanine, proline, and arginine, which could have combined in various ways to form a host of proteins. This kind of gradual progression from the RNA world to this intermediate phase to the DNA world is appealing, because it suggests the kind of series of gradual steps of emergent complexity that most origin-of-life researchers expect formed the movement from simpler to more complex biomolecules in life's formative stage.

What these steps highlight is the gradual emergence of increased complexity of the molecules and systems that occurred in the progression from simple organic molecules to the first living cell, with new properties emerging at each step. If this process occurred easily, it would probably happen elsewhere in the universe. If it occurred only by an improbable sequence of events, we might be alone in the universe. We do not yet know which of these is true. My own intuition is that that life starts readily wherever the conditions for it exist. Further exploration of the places in our own solar system that might have evolved life, especially Mars and Europa, may help us answer this most consequential of human questions.

13

God as Everlasting, Omniscient, Omnipotent

Over the course of the past two theological chapters I have been sketching an understanding of God that I believe to be not only truer to the fundamental teachings of Jesus and the Bible taken as a whole but also more logically consistent and religiously satisfying than some understandings of God that have existed in the Christian tradition. I have imaged God as the Infinite Agapic Spirit that permeates the universe yet is not contained by it, One who loves *all* creatures, seeks to have an active and loving relationship with them, and, out of selfless love, does whatever can be done to "save" them—that is, to minimize their suffering and maximize their enjoyment and fulfillment. Those teachings within the Bible and Christian Tradition that in any way weaken this understanding I have rejected as being inauthentically Christian and, as such, examples of errant human ideas about God rather than teaching based on divine self-revelation as mediated by Jesus and other biblical witnesses.

We now need to focus more closely on that loving relationship, particularly, what must be the case for God to engage in such relationships and how God seeks to bring about creaturely salvation. To do this we must first focus on God's relationship to time, the nature of God's knowledge, and the nature of God's power.

The Christian understanding of God's relation to time has had a complex history. The major cause of this complexity arose during the first century after Jesus's life and death as early Christians sought to spread the *gospel* ("good news") about Jesus throughout the Mediterranean region. Like most missionaries throughout history, they found it helpful to express their message in ways that would be readily understood by those with whom they spoke. In this case, the essentially Jewish message of God's Messiah having come needed to be expressed in such a way as to be meaningful to *gentiles* (non-Jews) throughout the Mediterranean region, whose worldview was shaped by the dominant Greco-Roman culture.

An example of this attempt appears in an account of the apostle Paul's missionary preaching in Athens as described in the NT book of the Acts of the Apostles (or simply Acts) 17:22–31. Here Paul notes how he saw an altar that was inscribed "to an unknown god." Paul tells his Athenian hearers that he has come to tell them who this god is, and that it is in fact the only God who exists, the one whom Jesus made known.

Later Christian writers confronted several other problems—namely, the accusations that Christians were atheistic, immoral, unpatriotic, and irrational. The charge of atheism stemmed from the fact that Christians refused to participate in the common worship of the traditional Roman gods as well as the fact that they had no images or statues of their God, which was totally counter to traditional pagan practice. The charge of immorality was based on such things as their practice of sharing the "kiss" between the Christian "brothers and sisters," which was taken as evidence of promiscuity and incest. They were also accused of cannibalism, based on their practice of the Eucharist/Holy Communion/Lord's Supper, in which they ate "the body and blood of Jesus." The charge that early Christians were unpatriotic stemmed from the refusal of many, if not most of them, to participate in the honoring of the emperors as possessed of divine virtues and/or acting as agents of the gods in their rule of the empire. They were also charged with generally "acting contrary to the decrees of the emperor" (Acts 17:7).

Important too was the charge that Christians believed and practiced an irrational superstition, that the crucified insurrectionist Jesus was God's chosen prophet and, for some it seems, the incarnation of God. This offended not only those who participated in the traditional worship of the Roman gods but also the Roman intellectual elite who subscribed to the religious ideas stemming from Greco-Roman philosophy. Finally, when the slow collapse of the Roman Empire in the face of barbarian invasions had become clear to most people by the middle of the third century, some Roman writers blamed this disintegration on the Christians. They claimed that the Christians had led people to desert the old Roman gods who had long protected them. This led to significant persecution of Christians under the emperors Decius (250–251) and Diocletian (303–311).

In the face of such accusations, some educated Christians began to write responses, beginning in the early second century, in which they refuted the charges, both moral and intellectual, brought against them. Together these early defenders of Christianity are called *apologists* (from Greek *apologia*, a "defense"). From the beginning, one of their chief strategies in defending the rationality of Christian faith was to incorporate the metaphysical perspectives of certain strands of the Greek philosophical tradition into Christian theology, particularly its view of the divine.

The Greek understanding of the divine as developed in the Platonic and Aristotelian traditions was based on their understanding of what we might term the logic of perfection. If the divine is perfect, they reasoned, it could not possibly change, since any change from perfection must be in the direction of imperfection. If the divine never changes, then it must be *eternal*—existing *outside of time*—apart from the world and everything in it, all of which exist *in time* and are constantly changing. Thus, according to Aristotle, God exists eternally, in a changeless present moment, contemplating God's own perfection. Because God cannot change, God must be *impassive*, unaffected by changeable emotions as we are and uninfluenced by the world. Thus Aristotle's God is absolutely uninvolved in the temporal universe and totally unconcerned with it.

What should be clear is that this understanding of God and God's relation to the world is vastly different from the Judeo-Christian understanding. The biblical God is deeply involved with the world, seeking to have relationships with its creatures and to make the world a better place for all. As such God *cannot* be eternal, as in the Greek understanding. Rather, God must be conceived as *everlasting*, existing *in* time—in some sort of meta-time which transcends the space-time continuum of our universe—yet without having a beginning or an end. Thus, God must be understood to be infinite in relation to time, just as God also is in relation to space.[1]

That God is everlasting is precisely what is taught in both Testaments of the Christian Bible, though this is at times obscured by mistranslations of the Hebrew and Greek terms used in the two Testaments. The term used throughout the HB/OT is *'wlm*, the root of which means "long duration, everlasting, forever." The term used in the Greek NT is *aiōnios*, which means "very long time, without beginning or end." The point is that both words refer to a process in time, one that has long if not infinite duration, and therefore neither of these terms should be translated as "eternity," when that term properly means the absence of time. The biblical God does change,

1. How time is experienced by God is unknown, given that in our universe, and perhaps all others (should they exist), space and time are inextricably tied together. For our purposes we needn't understand precisely *how* God experiences time, only *that* God does experience it in some way analogous to the way we experience it—meaning that "before" and "after" exist in God's experience.

based on what transpires in the world, though this change occurs only in God's feelings and specific behavior in relating to the world.[2] In God's essential nature—God as Perfect Agapic Love—God is unchangeable, and it is only in this sense that God can be spoken of as eternal.

Because God is *essentially* Love, God *cannot* be unaffected by what happens to creatures in the universe. God must feel something analogous to happiness and sorrow, and likely other emotions as well, in response to events in the lives of God's beloved creatures, just as we do when things go well or badly for the ones we love. Because God's experience encompasses all the creatures that have ever lived in however many universes there are and/or have been, the extent of God's experiences of sorrow, like those of God's joy, must be, unimaginably great.

Although the Greek philosophical and biblical understandings of God in relation to time, the world, and God's Love are radically different, the apologists nonetheless attempted to combine them into a single perspective, one that they thought would demonstrate the rationality of Christian thought to the Roman emperors and intellectuals and hopefully bring an end to the condemnation of the Christian ideas and practices that followed from it. For example, Justin the Martyr, an early-second-century apologist, taught that God was eternal and apart from the world, per the Greek understanding, and that Jesus was his divine-human-composite son—divine because of his virginal conception and birth, in which God's Spirit functioned as his father, and human because Mary was his mother. Jesus therefore was rather like Hercules, who, because he had a god, Zeus, for a father and a human mother, was part divine and part human. As such, Justin claimed, Jesus had a foot, as it were, in both worlds, the divine and human, and so functioned as mediator between them; how Jesus could be simultaneously outside time and in time was left unexplained by Justin.

Throughout the history of the church such attempts to combine these two logically contradictory understandings of the divine have failed to resolve their inherent incompatibility. How could a God who exists outside of time, in a changeless single moment, possibly have relationships with creatures? Relationships require interaction over time, the back and forth exchanges of experiences; and, as we have seen in physics, temporality and relatedness are part of what it means for something to be real. A God outside time could not be temporal or related to other things and could not even be said to exist in any meaningful sense. In fact, could such a God even be thought of as personal, a Thou rather than an It? If one follows out the

2. Biblical examples of God "changing God's mind" include when Lot persuaded God to alter "his" plan to destroy Sodom and Gomorrah (Genesis 19). See also Gen 6:6–7; Exod 32:7–14; and Jonah 3:10.

logic of the eternal divine idea, as Platonism did after Plato in its Middle Platonic and later Neoplatonic forms in the early centuries CE, one finds at the end of that tradition the idea of an impersonal being not only unaware of and unconcerned for the world, but also one who does not even think, since thought implies a subject (thinker) and object (the thought) duality as well as change over time. The logical outcome of an eternal God, then, is finally an It. Friedrich Nietzsche evidently recognized this ill-fated attempt to merge Greek philosophical ideas of the divine with the biblical perspective in his rejection of Christianity as "Platonism for 'the people.'"[3]

Throughout Christian history, when theologians have sought to combine these Greek philosophical ideas about God with the biblical understanding of God, what they have produced have been hybrid views that take elements from both understandings yet never succeed in creating a logically seamless view of God, let alone one that is authentically biblical. The problems with these attempts did not go unnoticed by some theologians, such as Tertullian in the third century and the Protestant Reformers in the sixteenth. In fact, one of the fundamental intentions of the Protestant leaders was to remove such Platonic- and Aristotelian-based views of God from Christian theology and to replace them with a purer biblical understanding as expressed definitively in the teachings of Jesus. With their attempt I concur, even if not always with some of the doctrinal teachings that have resulted from this attempt; for by removing these essentially foreign teachings from Christian theology, Jesus's simple yet profound understanding that God is Perfect Agapic Love becomes far clearer and as a result might resume the central place in Christian thought about God that it must hold if such thought is to be authentically Christian.

This issue of God's relation to time is also of significance for interpreting properly the traditional Christian *doctrine of divine omniscience*, according to which God "knows everything."[4] Essentially all Christians have believed that God knows everything that has ever happened in the past and everything that is happening in the present. What has been debated is whether God knows the future, as well.

3. Nietzsche, *Beyond Good and Evil*, xii; cited in Ogden, "Love Unbounded: The Doctrine of God," 10.

4. To say that God knows everything is to say that God is not restricted in the knowledge of what can be known but that, unlike us, God is not limited and so experiences all of reality directly. Thus, God experiences and knows the universe and its creatures at all size and time scales and experiences the entire electromagnetic spectrum simultaneously. It also means that God's memory is perfect: nothing that ever occurs is forgotten by the Divine Mind.

Whether God knows the future depends on the nature of the future—specifically, on whether it can be said to exist already or not. If the future is dependent upon what creatures freely decide to do in the successive moments of their existence, then the future is open and does not yet exist. This means that God *cannot* know the future because God does not know what the creatures will do. God can only know what the *options* are for any given creature in any particular moment and the *probabilities*, based on God's complete knowledge of who that creature is and what it is has done in the past, of what it will do. This recalls the Copenhagen interpretation of quantum mechanics—that only probabilities can be known regarding the behavior of subatomic particles because they have the freedom to behave in several possible ways. Given an open future based on the freedom of creatures, then, God should be understood as knowing the past and present as actual but the future only as possible.

This understanding of God's knowledge has important ramifications for our understanding of the kind of divine-creature relationship with which we are most personally familiar, the divine-human relationship. If God were moving through time with us and did know precisely what each one of us was going to think, say, and do in every moment of the future, and also what God was going to do in response to our actions, the result could hardly be anything like what we mean by a real relationship. God's experience would be like participating in a play whose every part had already been written and whose outcome was already known. The boredom God would have to endure from an infinity of such relationships would be unimaginable and surely intolerable. The value to God in participating in relationships is created precisely by *not* knowing what will happen in the future, just as it is with us. These relationships are, again, risky for God as they are for us in our relationships with others, and this is precisely the tale that is told throughout the Bible in the history of divine-human relationships. There is something at stake for God and for us—the possibility of rich relationships where the acts of both partners will together create something good and true and beautiful in the world, and fulfilling for all involved. But it is also the case that we will fail at times—maybe most of the time—to cooperate with God's calling for us to love, to the detriment of the experiences of all concerned. This back-and-forth exchange between God and us over time seems to me to be the only kind of relationship worthy of the name, and this is what is consistently illustrated throughout the Bible.

The issue of divine knowledge also has relevance for our deep intuition that we are free. If God knows what we are going do in the future, then does that mean we *have to* do it? And if we have to do it, then are we in fact free? If God is eternal, the answer is yes, we are free. God only knows our futures because for God, seeing all history as a completed whole outside of time, they have actually already occurred, and no lack of human freedom is implied. But if God is everlasting, then were God to know our futures, the answer must logically be no, we are not free, because we have to do what God already knows we will do. Some theologians have suggested that God knows our future thoughts and acts and yet we still do them freely, but this does not strike me (or most theologians past or present) as a rationally coherent perspective. However one answers this question, though, it raises yet another related theological issue: that of the nature of God's power.

As expressed in the traditional *doctrine of divine omnipotence*, Christians have typically believed that God is "all powerful," though they have not all agreed on precisely what that term means. Those who hold to the most radical interpretation of this doctrine claim that God can do absolutely anything God wants to do and in fact does so. In this God is like the ancient kings that lived during biblical times, and the image that goes with this view is God the King. If God's power is unlimited, then God is "in control" in the universe. Things happen because God wants them to, and therefore everything "happens for a reason," as the current Christian platitude has it. It is out of this basic view that the doctrines of *predetermination* (that God causes everything that happens in the world to occur) and predestination (that God determines unilaterally who goes to heaven and who goes to hell) logically arise.

There are several serious and, I would claim, insuperable problems with this view. One is that the Christian tradition has, as we have noted, typically claimed that there are some things that God *cannot* do—particularly, logically impossible and evil things. Some adherents of the radical understanding of God's power, however, take a stronger view. Over the years a number of students in my theology classes have insisted that God can do what is logically contradictory—that God could, for example, make it both pitch black and brilliantly light at the exact same time and in the exact same place, though they have always been at a loss to explain how that could be. For most Christian theologians over the past two thousand years, however, this logically contradictory possibility has been rightly denied.

Second, this kind of radical omnipotence does not seem to be the biblical understanding of God at all. There God is portrayed as very powerful, far more than any creature, but not necessarily omnipotent in the strong sense of that word—that God *controls* everything—nor is the term "omnipotent" ever used anywhere in the Bible. The word "Almighty" does appear in reference to God—in English translations, at least—but here again we encounter a translation problem. The Hebrew text does not say "God almighty" but rather "God of the mountain" or "God of the high places," reflecting the common ancient view that the gods lived atop the highest mountains. The translation of this Hebrew phrase as "Almighty" is therefore incorrect. Similarly, the NT employs the Greek term *pantocratōr*, meaning "ruler of all," in which the analogy is again drawn between God and human kings. Yet this does not imply that God has all the power and creatures have none, any more than human kings have absolute control over the thoughts and deeds of their subjects, yet this is the image created when *pantocratōr* is mistranslated as "almighty" in English versions of the NT.

Third, there are many things in the Bible that contradict the notion that God is in control of what happens on Earth. For example, Jesus taught his disciples to pray, "(may) your kingdom (royal reign) come, (may) your will be done on earth, as it (already) is in heaven" (Matt 6:9–13; cf. Luke 11:2–4). This clearly indicates Jesus's belief that God's will is *not* being done on Earth, and that is the main reason for his ministry: to clarify for his fellow Jews what the essence of the divine nature and will were, as opposed to what some of the Jewish religious leaders (and Roman emperors) were teaching. Jesus's call for people to focus their lives on the love that is God's fundamental commandment would have been absurd if he had understood God to be in control of everything and people therefore not to have been free to heed his call. Indeed, God's initiation of, and desire for, relationships with Israel and all people throughout the HB/OT, and God's disappointment and occasional punishment of them when they did not so respond, make no sense if the Hebrews and other people were not free to enter into such a relationship or not. Clearly God was not understood to be in control of what people thought and did; they were responsible for their decisions precisely because they had made them freely.

For me, however, the strongest possible proof that God is *not* in control of what goes on in the world comes simply from looking at the world around us. One only need look at the horrors in the world—the hatred, war, rape, torture, disease, murder, and more—to know without a doubt that the God who is Love Itself is *not* in control in the world. If God were in charge, none of these things would exist on Earth. In a world filled with such evils, if God is the ultimate cause behind all of these evils, then God is surely the one Christians have historically called the Devil and not the loving God Jesus proclaimed.

If God is not in control in the universe, which is the only rational conclusion a Christian can draw given the above, then what is the nature of God's power? Is God omnipotent? As we have seen, the kind of omnipotence that many Christians have claimed for God is one way of understanding that term, but it is not the only one. As in everything having to do with God, we must begin with God as Perfect Agapic Love. The question then becomes, what kind of power is it that *necessarily* accompanies such Perfect, Self-giving Love?

The key distinction that must be made here is between coercive and persuasive power. *Coercive power* is the power to control things, to make them do what you want them to do. It is the kind of power that we exercise, for example, over the tools we use in the many tasks of life (cooking, cleaning, writing, teaching, and the like). Some people attempt to use this coercive type of control over the people with whom they interact, seeking to force them to think as they think or act the way they want them to act. *Persuasive power* is the other kind of power, the one people typically use in their relationships with others when they seek to change the way those others think or behave while still respecting their freedom finally to choose such things for themselves.

The essential question here is which form of power God uses in dealing with God's creation. Does God control the universe and all that is in it or does God interact with it persuasively, or both? Here we must distinguish between how God interacts with individual creatures possessed of freedom—creatures like ourselves, or amoeba, or even quarks—and how God deals with inanimate conglomerates of individual creatures, such as rocks or galaxies. In the remainder of this chapter we will deal with the first type of interaction, the way in which God interacts with creatures possessed of some degree of freedom, followed in the next theological chapter by a consideration of the way in which God might be thought to deal with conglomerate objects in the universe.

From what we have already said about the nature of God as Perfect Agapic Love and, as a result of this nature, God's desire to enter into loving relationships with all creatures, it should be clear that the form of power God employs in such relationships must be persuasive. *God's power is the power of love*, the power of love's attraction; a number of arguments support this conclusion.

One thing we learn early in life is that we cannot finally control other people. We cannot control what our friends and others think and do, and the reason for this is clear: they, like we, are free to think and do what they will. Parents learn this lesson quickly with their children. You simply *cannot force* them to do what you want no matter what you do, because they "have minds of their own." All you can do, even with children, is to try to *persuade* them to think or act in the way you think would be best for them. With friends you will likely attempt to reason with them, and with children, where reason is so often ineffective, you will likely try offering rewards for desired behaviors and punishments for undesirable behaviors. But regardless of who the person is, whether child or friend or spouse, the fundamental reality remains: he or she is free and therefore cannot be controlled. Therefore the only power we can have in relating to them is persuasion. The most effective way to persuade people to change is to get them to *want* to change, and this is accomplished when they come to see that the thoughts and behaviors you are encouraging them to have and do are the ways that will be the most beneficial to them.

This kind of persuasion is certainly the way the Bible presents God's dealings with humankind—from Adam and Eve to Israel to the world's other peoples. It is the way Jesus interacts with his disciples and all others he encounters in seeking to change their ways. But this is only what one would expect from God given God's Perfect, Self-giving Love, which entails respect for the freedom and integrity of each person, a respect far more perfect than the type we typically show one another because of our less-than-perfect love.

So how have Christians (and others) come to think of God as being "in control" in the world in this coercive kind of way despite the general biblical witness against it and the logic of love? Because this is the kind of God *we want* God to be. We want God to have the power to fix things, to protect us, to make everything turn out right in the end—which was exactly Sigmund Freud's understanding of why we invent God. I would say, rather, not that we have invented God but that we have made God fit our own image of what God ought to be.

For the ancient Hebrews God was envisioned as the perfect king. Monarchs in that time had the greatest possible power a person could have, a coercive power to make their subjects do what they wanted or have them punished if they did not. This is how they pictured God, albeit without the flaws of human monarchs.

What Jesus revealed in a decisive way, and what Christianity has taught at its best—that is, when it has actually taken what Jesus taught to heart—is that God is *not* like some worldly king who uses his power in a coercive manner. Rather, he taught, God is Perfect Agapic Love, a perfectly loving Father, and one who suffers because of the unloving free decisions of humans. This has been the standard Christian understanding of Jesus's death on the cross—an image of God and God's power that is very different from the coercive power that people typically assign to the divine being. Thus, Paul speaks in 1 Cor 1:18–31 of the great reversal of worldly notions of power and strength and weakness that the Christian knows through Christ, and particularly his death on the cross: ". . . a stumbling block to Jews and foolishness to Gentiles" (1:23). A God of absolute coercive power, the kind humans would like God to be, is not the God that Jesus proclaimed. Such a

God of coercive power is not worthy of worship; only a God of Perfect Love, and hence persuasive power, is.

This confusion about God's power persists in much Christian thought past and present. Many still think they can have it both ways, that God is "in control" and that "everything happens for a reason" (and that reason is God's controlling power, not the natural results of prior events and actions). But if God controls the events of the world and individuals' lives, then God must control what people do, and we are right back to the problems with this notion already noted. If God is in control, if everything happens because of God's will, then God, evidently wills each act of horror and evil ever committed in the universe. These must serve God's greater design, some Christians suggest. Perhaps without them the world would lack some greater good. But if God is all-powerful, then God would not need to cause these evils to produce the good God desires. Others say that God does not directly cause these evils, God only allows them to happen (meaning that God does not stop humans from doing the evils that they do). This is supposed to get God off the hook for directly causing evil. But if God *allows* evil to occur, if God could prevent evil from happening and does not, then God is fully responsible for the existence of that evil. If God is powerful enough to control people, then God *must*, if God is indeed who Jesus said he was, make them do what is good in the first place, not evil in order to attain good. The one who seeks to make people do evil has always been defined in the Christian tradition as the devil. God is *not* the devil, but God would be if God's power as coercive omnipotence were true.

All of this is simply to reaffirm what we have maintained from the beginning, that the one nonnegotiable in Christian theology is that God is Perfect Agapic Love, and that whatever contradicts this claim in any way must be rejected as contrary to Christian teaching. As the Christian tradition has consistently taught, God *cannot* do (and so *cannot* will) evil, period. The view of some Christians that God willed the horrors that we find in this world is simply wrong-headed and must be rejected. Again, is it any wonder that increasing numbers of people are now rejecting a religion that proclaims such a callous God? Perhaps they have more truly "Christian" sensibilities than many so-called Christians do.

The "*problem of evil*" has long been understood to be the greatest problem for theism, that which has caused more people to become atheists than any other thing. In its most general form (i.e., ignoring the finer distinctions which need not here concern us here), the problem consists of three assertions people have typically wanted to make about God but which cannot all be simultaneously true: that God is perfectly loving (and so would never will the existence of any hurt or evil), that God is omnipotent (in the coercive sense that God can accomplish anything God wants), and that genuine evil exists. The logically possible solutions to the problem are four: either God is not perfectly good, or God is not in control of everything, or evil does not really exist, or God does not exist (some combination of these is also possible, e.g., that God is neither all-good nor all-powerful).

Historically, the solutions offered by Christian thinkers have almost universally rejected the first option and have focused instead on the second and third options. Those who have opted to deny the second proposition, that God is in control in the world, have typically adopted what is called the "free-will" defense, claiming that because humans have free will, God is not fully in control. What has typically weakened this argument, however, has been the additional claim that God could overrule that human freedom, as it were, because God is still omnipotent in a coercive way. But if this were so, then God is culpable for allowing evil to exist when God could prevent it, and therefore God cannot be thought to be perfectly good. The traditional free-will defense simply cannot give up a God who is coercively "in control."

The second traditional Christian response to the problem of evil has been to deny the third proposition: namely, that evil really exists. Here a distinction is made between two types of evil, *instrumental evil* and *genuine* (or "surd" or "excess") *evil*. Instrumental evil is evil that is necessary for greater evil to be prevented or greater good to occur. Getting vaccinations would be an example of an instrumental evil: having a needle stuck into one's arm is painful, yet it is understood to be necessary to prevent possibly getting a disease that would inflict far greater suffering. Genuine evil is evil that is not instrumental to attaining some greater good or lesser evil but is simply evil. According to this defense, all evil in the world is instrumental, necessary to some greater good, even though we may not be able to see it. One classic example concerns World War II. Tens of millions of people died or were permanently injured in that war, yet some have suggested that all that suffering was necessary for some divine goal, however unclear to us, to be accomplished.

The problem with this second defense is simply that it is incredible once one looks more carefully at it, and this from several angles. First, as already suggested, it neglects the suffering of individuals as it seeks some greater good. The twenty-million-plus dead in World War II, for example, not to mention the tens of millions more whose minds and/or lives were shattered by the horrors of that war, are quickly passed over in search of some justifying cause to get God off the hook. Clearly, if God is Perfect Love, every bit of suffering of each person matters absolutely; and if God could stop it from happening, God certainly would—especially if all it took were simply preventing, for example, the fertilization of the Adolf Hitler egg in his mother's womb. How much more loving would that act have been rather than allowing tens of millions to suffer and die?

Second, and perhaps even more devastating to the traditional argument, is the fact that for all the evil to be instrumental, every single death, every single moment of suffering, would have had to be *necessary* to accomplishing God's desired goal. In our World War II example, that would mean that one single death less, one single act of suffering less, would have meant that God's unknown goal, whatever it was, would not have been accomplished—otherwise, that death or moment of suffering would have been excess, more than necessary instrumentally, and so genuinely evil. To believe that this was the case regarding World War II seems to me to be well beyond the bounds of human reason. God works to bring whatever good is possible out of evil, but God does not create or permit the evil in the first place.

The only reasonable and Christianly appropriate solution to the problem of evil seems obvious, given what we have already said: humanly caused evil exists in the world because humans have and misuse their freedom. Because freedom is inherent in human nature, God could not override that freedom without destroying that nature—i.e., killing them—something a Perfectly Loving God would not do. In the case of those who unleash evil on the world—to a greater or lesser extent and with varying degrees of intention—God must seek to have them healed of whatever sickness or hurt it was that led them to do so, and in extreme cases, such as a Charles Manson or a Hitler, to have them removed from society until such healing can be accomplished, if ever it can. That is what Love does. That is what the Infinite Agapic Spirit does.

But what of that which has traditionally been called "*natural evil,* those evils in the world which cannot be attributed to human causes yet which we define as evil because of their negative impacts on our lives, such as diseases, hurricanes, tsunamis, and the like? Must not God be held accountable for these evils? How these evils and the God of Perfect Love fit together will be addressed in chapter 17.

14

The Evolution of Earth and Its Life through the Proterozoic Eon

At this point in Earth's history, after the first cratons had formed and Earth's constant bombardment by planetesimals had largely ceased—and perhaps at about the time that the first living cells arose in the oceans—the Hadean Eon ended, and our planet entered what geologists call the Archaean Eon.[1] This eon began around 3.85 bya and lasted until about 2.5 bya.[2]

During the early Archaean Eon, Earth continued to cool as the amount of radiation produced deep within it gradually declined, and the degree of volcanism at the surface probably subsided also. After several hundred million years, as we have seen, Earth had cooled sufficiently that rifting of the ocean crust was not so severe, permitting the rise of small continental plates upon which small *protocontinents* formed. Over time some of these protocontinents, driven by the motion of the plates on which they rode, collided and combined to form larger continents. These first continents were probably relatively small by today's standards, separated by ocean basins with hydrothermal vents shooting up gases and debris from the seafloor. Volcanic islands would also likely have been created during this period, as they continue to be today. Sometime around 3.3 bya the first large supercontinent, which geologists have named *Vaalbara*, probably formed. It survived

1. The word "Archaean" is derived from a Greek word that means "ancient," "early," "beginning." "Archaeozoic," the former term for this eon, is derived from two Greek words that, together, mean "ancient, early, or beginning life."

2. See the Geologic Time Scale in the appendix.

until about 2.7 bya, at which point the same kinds of tectonic forces that had formed it caused it to break up into smaller continental fragments again. Vaalbara evidently included the most ancient *cratons*, which, as we have seen, are early continental masses that have continued to exist relatively unchanged since the early Earth. Geologists estimate that by about 3 bya all of the large, modern cratons—found today in certain regions of Canada, Western Australia, and South Africa—had formed.

Today, scientists have solid evidence that, again, even before this first supercontinent began to form, the first living entities had already arisen in the oceans—certainly by 3.5 bya and more probably by 3.85–4.0 bya, or at the dawn of the *Archean Eon*. As we have seen, origin-of-life researchers believe the first living entity was a very simple biochemical system capable of metabolism and reproduction, but beyond that little more can be said at present. Paleobiologists therefore focus more on what they call the *Last Universal Ancestor* or *Last Universal Common Ancestor*, by which they mean the most recent entity that became the parent of all other subsequent living beings. Not all of researchers, however, agree that there ever was a single common ancestor. This seemingly small minority of scientists think of early life as the product of multiple simple entities exchanging genes, with different such entities giving rise to the two earliest domains of life, the Archaea and Bacteria. As we have already seen, when talking of the ultimate origins of things—whether the universe itself or life on our planet—the paucity of evidence allows for multiple theories, which we should entertain only tentatively. As before, we will present what appears to be the majority view at this particular time, with the assumption that some parts of this reconstruction will be altered, and some perhaps dramatically, given further research and discovery.

For an image of a phylogenetic tree of life, by the NASA Astrobiology Institute (modified by Eric Gaba), see http://en.wikipedia.org/wiki/File:Phylogenetic_tree.svg.

Most paleobiologists not only hold to the single-common-ancestor theory of life on Earth, they also generally agree on the fundamental nature of that entity. They think it would have been a very small, single-celled organism, probably only several microns (10^{-6} m) in diameter. Its cell membrane would have been the kind of simple, two-layered lipid structure we encountered in our origin-of-life discussion. This membrane would have had small pores called *ion pumps* in it to allow vital chemicals such as sodium and potassium to pass through and to keep their concentrations within the cell at the proper levels. This entity would also have been a *prokaryote* (a cell without a nucleus), its genetic material spread throughout the cell because it had no nuclear membrane to contain it. Its genes would have been located on strands of DNA, which, as in all life today, were read and copied by RNA. *Messenger RNA* (mRNA) would have brought this genetic information to *ribosomes*, small structures where this information was read in order to stitch together amino acids—the same 20 amino acids we have seen used by all subsequent life—to form the various proteins used by the cell for all its basic functions. These would have included several hundred protein *enzymes*, which functioned to *catalyze* (speed up) vital chemical reactions within the cell. Of great importance were the enzymes that extracted energy from fats, sugars such as glucose, and amino acids to power the cell's metabolism and reproduction. This latter function would have been accomplished by a simple splitting of the cell in two. Apart from its ribosomes, this last universal ancestral cell would have lacked other complex internal structures like the creatures that soon evolved from them. Some think this first living cell was probably an extremophile—a being adapted to live under the kind of extreme conditions of temperature, pressure, salinity, acidity, lack of oxygen, or some combination thereof that existed in various places in the oceans of the early Earth.

Some of the simple, single-celled beings that succeeded this primitive ancestor constitute the members of the domain *Archaea* (formerly called the kingdom *Archaebacteria*), one of the three domains of life on Earth.[3] The second domain is that of the *Bacteria* (formerly the kingdom Eubacteria) which, like Archaea, is also constituted solely by small, single-celled prokaryotes. The members of both domains are further divided into kingdoms, which reflect fundamental differences among the various archaea and bacteria. What these kingdoms are is also controversial today. The third of the three domains is *Eukaryota*. *Eukaryotes* (cells with nuclei) are subdivided into four kingdoms: *Protista*, *Plantae, Fungi* and *Animalia*. Protists are mostly single-celled entities, some of which produce their own food and others of which consume nutrients created by food-producing protists. Protists evolved to produce the final three kingdoms of life: plants, which are multicellular beings that produce their own food by photosynthesis; fungi (mostly multicellular) which feed on living or decaying organic material; and animals (all multicellular) which mostly consume plants or other animals, though a few, like we humans, also consume some fungi (e.g., mushrooms).

3. There is much debate today about the best *taxonomy*, or dividing into groups, for these earliest living entities, and whether the recent three-domain model is the best way of grouping all life on Earth. Here I'm ignoring this complex debate and using the taxonomy that seems favored by most paleobiologists today.

In addition to the domains and kingdoms to which they belong, all living things are further classified into subordinate groups, called *taxonomic groups* or *taxa* (singular *taxon*) according to their similarities and differences from each other. These subordinate groups within a kingdom, from the more general to the specific, are phylum, class, order, family, genus, and species. Humans are members of the kingdom *Anamalia*, the phylum *Chordata* (animals with spinal cords), the class *Mammalia* (chordates whose females provide milk for their young), order *Primates* (mammals with larger brains, shortened snouts and hands and feet adapted for tree-climbing), family *Hominidae* (the great apes and humans), genus *Homo* (humans and their humanoid ancestors), and species *Sapiens* (modern humans). Sometimes other taxa are added when further distinctions are required, such as *suborders* and *superfamilies*. Thus, humans belong to the suborder *Anthropoidea,* which distinguishes them along with monkeys, gibbons, gorillas, orangutans and chimpanzees from lemurs and related primates, and the superfamily *Hominoidea,* which distinguishes humans, gibbons, gorillas, orangutans and chimpanzees from Old and New World monkeys.

THE ARCHAEAN EON: THE RISE OF THE ARCHAEA AND BACTERIA

Ancient archaea, probably like their forebears, were probably extremophiles. Most paleobiologists seem to think they evolved around superheated vents created by rifting on the ancient ocean floor. This made them precisely the kind of organisms that could have evolved and survived the harsh conditions of the early Earth during and after the late bombardment period. Archaea also have the most primitive genetic makeup and are the smallest of any living creatures extant today, with diameters typically on the scale of a few microns. These factors, plus their lack of well-articulated internal structures, provide support for the claim that they were among the earliest forms of life on Earth.

Archaea are sometimes divided into several broad types, most typically three (though this grouping too is by no means universal). The first type are the *Methanogens,* which produce methane as a byproduct of their biological processes. Second are the *halophiles,* which exist in high-salt environments, such as the Dead Sea. Third are the *thermophiles,* which only live in high-temperature water, such as in hot springs or around hydrothermal vents on the ocean floor. These last types survive in the absence of sunlight, getting the energy they need to grow and reproduce chemically by breaking down (or *reducing*) sulfur compounds and so releasing energy needed for the cell's basic life-functions. Archaea are also almost entirely *anaerobic*, surviving only in oxygen-free environments. They exist in a variety of shapes, including spheres, rods, spirals, and plates, and secrete a mucus coating which makes them sticky. They therefore clump together readily, forming mats of various colors, depending on the species of the organism. Green archaea mats, for example, can be seen surrounding hot springs in Yellowstone National Park.

While most archaea today appear to be extremophiles, they have also recently been discovered in less harsh environments, reflecting most probably a later evolutionary adaptation from their earlier extreme habitats. The fact that archaea continue to exist today is testament to the effectiveness of their biological design. Should mass destruction visit Earth, perhaps in the form of a giant asteroid impact, it is the archaea on the ocean floors that would likely survive when all other life-forms had been extinguished. From these simple survivors there would eventually evolve an entirely new variety of life-forms that would repopulate the planet.

At about the same time as (or perhaps somewhat earlier or later than) the appearance of the Archaea was the rise of another type of single-celled beings, those of the domain Bacteria. Like the archaea and the progenitor from which they developed, bacteria had no nuclei to contain their genetic material nor any other large internal organelles like the later eukaryotes would have. Also like archaea, these earliest bacteria were anaerobic. They differed in some important ways from the archaea, particularly in certain of their internal and structural features, such as their ribosomes, cell membranes, lipids, and RNA polymerase. Today bacteria exist in a huge variety of forms. Indeed, biologists can only estimate very roughly how many species of bacteria there are on Earth, with the numbers ranging from ten million to as many as a billion. These species of bacteria are distinguished from one another largely by the structure of their cell walls and whether they are capable of locomotion via *flagella* (tiny whip-like tails) or not. Some, such as *E. coli* and *Salmonella*, cause diseases in humans, while others serve more constructive ends (from a human point of view), such as turning grapes into wine and milk into cheese, cleaning wastewater, and decomposing dead tissue back into its constituent parts, where they can be used again by other life-forms. Helpful bacteria also live inside our bodies, aiding our digestion of food, among other things. Given their extremely small size—most on the order of about five microns, it is stunning to realize that bacteria constitute the majority of Earth's biomass—the total mass of all living beings combined. Some evidence even suggests that certain prokaryotes had already begun to colonize the land by 2.6 bya.

Cyanobacteria (formerly and mistakenly called "blue-green algae") are a particularly important form of bacteria from a human perspective. They were the first creatures able to manufacture their own food, doing so through the process of photosynthesis, in which water and carbon dioxide combine to form sugars such as *glucose* (their food/energy source) and oxygen ($6H_2O + 6CO_2 > C6H_{12}O_6 + 6O_2$). Chlorophyll has the ability to capture photons, which provide the energy to drive the photosynthetic reaction. It is chlorophyll that gives cyanobacteria their characteristic blue-green color. Like archaea, they too secreted sticky outer coatings that caused them to clump together in mats. This outer coating also protected them from the damaging ultraviolet solar radiation that rained down upon Earth before it developed its protective ozone layer. Because of their need for sunlight, they lived close to the surface of the ocean—mostly, it seems, in shallow coastal waters. The more light these mats had, the larger they grew, forming colonies made up of millions of cells.

The most ancient fossils we have found, dating to about 3.5 bya, are the remains of these cyanobacteria. Some of the early cyanobacteria seafloor mats gradually became covered with layers of sediment, which were then colonized and covered over by another layer of the bacteria. Over time, this process was repeated multiple times, resulting in the layered *stromatolites* that we encountered in chapter 12. These multilayered remnants from the Archaean Eon are found today in several places around the world, some of them dated to nearly 3.5 bya (see the QR code below). Stromatolites that form today have been found to have other types of bacteria lying beneath their cyanobacterial mats, some of which break down waste from the cyanobacteria above. This suggests that ancient stromatolites would also likely have constituted similar simple ecosystems, with symbiosis occurring between the different bacteria involved. These bacteria and their ecosystems continued to evolve over the next several hundred million years, producing a wide variety of stromatolite forms by about 2.7 bya. These ranged from small, finger-like structures to large towers and flattened, pillow-like structures. Despite this diversity of forms, however, later examples of these cyanobacteria evidenced no significant structural evolution from their earliest forms, showing only some internal biochemical development. Stromatolites persisted as one of the major features of the early Earth until they began to decline around 1.25 bya, nearly two billion years after they first appeared.

For modern stromatolites in Australia, looking much as they would have during the Archaean Eon, see the photo by Paul Harrison (Reading, UK) at http://commons.wikimedia.org/wiki/File:Stromatolites_in_Sharkbay.jpg.

Cyanobacteria did one thing that was crucial for the future evolution of the more complex forms of multicellular life: the creation of an oxygen-rich atmosphere. In producing their own food by means of photosynthesis they released oxygen into the upper layer of the oceans and atmosphere. It was therefore cyanobacteria that played the major role in altering Earth's atmosphere from its early, carbon-dioxide-rich state to one increasingly rich in free oxygen. A small amount of oxygen was also created as ultraviolet light from the Sun split carbon dioxide molecules, freeing their oxygen, but this paled in comparison to the oxygen created by the cyanobacteria.

This *oxygenation* process, begun by around 3.5 bya with the appearance of the first cyanobacteria, lasted about a billion years before any significant amount of oxygen remained in the air and upper ocean. Most of the oxygen liberated by the cyanobacteria quickly left the water and atmosphere by bonding with such things as limestone, iron, and sulfur. Oxygen combined with limestone to produce *calcium carbonate* ($CaCO_3$), was captured by iron to form iron ores or *rust* (Fe_2O_3), and combined with sulfur to form *sulfur dioxide* (SO_2). These reactions kept the oxygenation of the atmosphere at a very low level until, after about a billion years, these oxygen-binding *sinks* of calcium, iron, and sulfur were essentially used up.

Today, cyanobacteria inhabit virtually every known habitat, from the extremely hot to the extremely cold, freshwater to saltwater, highly acidic to strongly alkaline, and even in the driest deserts, where they desiccate as the ground dries up only to come to life again in the presence of water—even more than a century later. Cyanobacteria can also survive substantial radiation, which was necessary in the ancient world with its ozone-deficient and so weakly protective atmosphere. Given their 3.5 billion years of survival on this planet through all its changes and mass extinctions, bacteria, like the archaea, have proven the successfulness of their bodily design. It is rather amazing that these two domains of simple, single-celled prokaryotes were the only forms of life to inhabit Earth for perhaps a billion years or more following the appearance of the first life on Earth.

While the archaea and bacteria were evolving, the surface of Earth was continuing its own slow alteration. As we have noted, the minute but inexorable movement of Earth's plates that had caused the first great supercontinent, Vaalbara, to form around 3.3 bya eventually caused it to break apart around 2.7 bya. Sediments deposited along the margins of the resulting continents grew more substantial beginning around 2.6 bya, giving rise to broad, shallow seas—new niches where life could thrive and evolve.

At about the same time or perhaps just before this great sedimentation began, the next huge supercontinent, called *Kenorland*, formed.[4] This supercontinent existed for a relatively short time compared to Vaalbara, remaining intact for at most 200 million years, geologists estimate, before breaking apart around 2.5 bya or shortly thereafter. Like Vaalbara, Kenorland would likely have been barren, utterly devoid of life, since apparently all of the protokaryotic archaea and bacteria continued to live only in the oceans. Besides natural weathering caused by wind and rain, the only thing that happened on these first land masses was the rising up of mountains, or *orogenies* ("mountain birth/origin"), created by colliding tectonic plates causing the terrestrial crust to buckle. The first of these, to the best of our knowledge, was the *Kenoran orogeny*. Weathering of these new mountains over millions of years would have carried fresh minerals to the ocean margins, perhaps contributing to the evolution of life in the shallows along the continental margins. From about 2.2 bya on, stromatolites became more abundant, perhaps a reflection of these enhanced, shallow, continental-margin ecosystems. The break-up of Kenorland—if the continent did exist—coincided with what geologists define as the end of the Archaean Eon and the beginning of its successor, the *Proterozoic* ("first animals") *Eon* some 2.5 bya.

By the beginning of the Proterozoic Eon, roughly 2.5 bya,[5] geologists estimate, the oceans would have attained about the same degree of salinity they have today. After this, salts dissolved from the continental surfaces and washed into the oceans by rain and rivers would be balanced by salts precipitating out and settling to the bottom of the ocean.

Also around the beginning of the Proterozoic Eon, free oxygen created by cyanobacteria finally began to build up in the upper layers of the oceans and in the atmosphere as the major oxygen sinks were filled. This unleashing of significant amounts of oxygen into the ocean and air created what scientists have variously termed the *Great Oxygenation Event (GOE)*, the *Oxygen Catastrophe*, the *Oxygen Crisis*, or the *Oxygen Holocaust*. The last three names underscore the fact that oxygen was deadly for the anaerobic organisms that had been the sole living entities on Earth since life began.

4. There is in fact much debate among specialists concerning how many early continents formed and then disappeared again during the Precambrian Era prior to the formation of Rodinia, upon which essentially all paleogeologists agree. Here we will include discussion of the four continents of the maximal view.

5. See the Geologic Time Scale in the appendix.

Indeed, this oxygenation of the oceans and atmosphere was the cause of what was probably the greatest mass extinction of life in the history of our planet. Anaerobic creatures henceforth would be confined largely to the deep oceans.

From this time on, oxygen would gradually accumulate in the oceans and atmosphere, leading to the appearance of *aerobic* bacteria, ones that used oxygen in their life processes. By 2.2 bya oxygen levels had risen to 1–2 percent of the atmosphere, or nearly a tenth of what they are today. Around 1.85 bya the oxygen evened out and stayed roughly the same for the next billion years. Then, roughly 850 mya (million years ago), it began to rise rapidly, continuing to increase markedly for the next 300 million years, until the end of the Proterozoic Eon, when oxygen levels in the atmosphere had risen to perhaps as much as 10 percent, or about half what they are today. After 540 mya oxygen slowed its rate of increase but continued building up, finally reaching its present-day level of about 21 percent.

Shortly after the beginning of this oxygen build-up in the atmosphere, or around 2.3 bya, a major worldwide cooling occurred that geologists call the *Makganyene Ice Age*. The extent of the freezing is debated. At a minimum the freeze caused the continents to be buried under enormous glaciers. Some believe that even the oceans were frozen, hence the name "*Snowball Earth*." What caused it is a matter of debate, but many geologists think it was the result of the oxygenation of the atmosphere. Some of the new oxygen would have reacted with methane to form carbon dioxide and water. Methane is a powerful greenhouse gas, whereas the carbon dioxide that replaced it is weak by comparison. Thus, the replacement of methane with CO_2 would have caused a net cooling of the atmosphere, leading to a chilling effect on the planet as a whole. It is unclear how long the Makganyene Ice Age lasted, but it was likely on the order of one hundred million years.

At some time during this great oxygenation event, and perhaps quite early in it, a new type of *aerobic* creature came into being, the *eukaryote* (cell with a nucleus). This new life-form gave rise to the third and final domain of life on Earth, the *Eukaryota*, which would eventually include all plants, fungi, and animals. The evolution of the eukaryotic cell was therefore one of the pivotal events in the history of life on this planet.

THE PROTEROZOIC EON: THE RISE OF THE EUKARYOTES

The early forms of eukaryotes are termed *protists* and, according to most taxonomic schemas today, constitute their own kingdom. Protists probably evolved from bacteria rather than archaea and did so in the bacterial mats that lived on the bottoms of shallow seas, where they can still be found today. They are almost all *single-celled* eukaryotes, which is the main feature that distinguishes them from the multicellular eukaryotic plants, animals, and fungi that would later evolve from them. Several multicellular eukaryotes are also included among the protists, particularly some algal *seaweed*, though these too are simple forms, lacking the internal tissue differentiation into things like roots, stems, and leaves that would characterize later plants. Seaweeds evolved to survive in the harsh conditions near the shorelines of the continents. They are flexible and strongly attached to the seafloor so that they can survive the jostling caused by the waves above. Also, they do not dry out easily, which enables them to survive the rise and fall of the tides and resultant exposure to the air.

The first eukaryotes and their prokaryote forebears shared a number of features. Both had very similar *metabolisms*, the chemical reactions that maintain the cell's structures and cause it to grow, reproduce, and stay alive. Both used RNA and DNA for their genetic instructions and reproduction. Both used the same twenty amino acids to create the proteins that controlled their cellular processes and ribosomes to convert those amino acids to proteins. Both also had a two-layered lipid (or "*bilipid*") membrane that enclosed their cells.

The main innovation of these eukaryotic protists was that they were the first cells to have more complex, membrane-enclosed internal structures, especially a nucleus, in which their genetic material was housed. Their DNA was also much more complex than that of the protokaryotes. Most eukaryotes also had other internal organelles, such as chloroplasts, *mitochondria* (which generate chemical energy for use within cells), *endoplasmic reticula*, (which form membranes that create different compartments within the cell, transport chemicals around the cell, and synthesize proteins and fats), and the *Golgi bodies/Golgi apparatus* (which work together with endoplasmic reticula to sort, synthesize, package, and send vital macromolecules, such as proteins and carbohydrates, to their proper destinations within the cell).

A crucial and related eukaryotic protist innovation was the development of a more complex *cytoskeleton*, the protein-filament scaffolding within the cell. This new structure not only permitted eukaryotic cells to grow larger without collapsing, it also permitted a new flexibility in the cell wall. Thus, where prokaryotes were enclosed in a rigid cell wall, these new eukaryotes could change their cell shapes easily. As we will see in the case of amoeba, this flexibility permitted the creation of pseudopods, by which the creature was able to move through the water and ingest food in an entirely new way.

In addition, the cytoskeleton also aided the movement of structures within the cell, assisted the process of cell division, and permitted the evolution of yet other new structures by which the cell could move through its environment. These were the *cilia*, small hair-like structures that surround a cell and beat rhythmically to move the cell through its watery environment, and *flagella*, large tail-like structures located singly or in multiples at one end of the cell whose whipping motions powered it through the water. This innovation permitted eukaryotes to move toward food sources and away from potential threats far more effectively than their prokaryotic forebears, which generally relied on floating through the oceans. Given this evolutionary advantage, most subsequent protists came to have cilia or flagella in at least some phase of their life-cycles.

Another significant innovation of these early eukaryotes was the way they reproduced. Unlike prokaryotes, with their free-floating, circular chromosomes, eukaryotes had long strands of chromosomes encased within their nuclei. In both prokaryotes and eukaryotes, these chromosomes were duplicated during reproduction and then separated by being moved along *microtubule filaments*. Prokaryotes and most protists reproduce asexually, by a process known as *mitosis*, in which the single cell with its duplicated chromosomes simply divides into two identical cells, one the parent cell and the other the daughter cell. The parent and daughter cells will therefore be genetically identical. Novel types of prokaryotes and asexually reproducing protists generally arise by gene swapping with others of their kind, a relatively slow and inefficient process for producing the biological diversity upon which natural selection works.

But some protists arose with a new way of reproducing: sexually, meaning by means of male and female reproductive organs. This new kind of reproduction proceeded by *meiosis*, in which each kind of reproductive organ produced its own genetically identical cell within its nucleus. Then the two cells combined into a single *diploid cell*. Within this cell the chromosomes from the two parents underwent *recombination*, in which the chromosomes from each parent were broken down into their constituent DNA sequences, which then combined in new and apparently random ways. This created an offspring that was similar to, yet genetically different from, both parents—a unique individual.

The evolutionary advantage of sexual reproduction was that it produced variation within a species rapidly and efficiently in each generation, thereby enhancing that species' chances of survival in the face of environmental changes as well as the possibility of evolving increasingly complex forms of life. So important to the creation of more complex forms of life was the development of sexual reproduction that famed evolutionary theorist

John Maynard Smith (1920–2004) referred to it as possibly *the* crucial development in the history of life on Earth. Precisely when sexual reproduction evolved is unclear, but a time later in the Proterozoic is a reasonable hypothesis at present.

How the first protist evolved from prokaryotic bacteria is unclear, though two main hypotheses have been advanced. The first is the *autogenous model*, which claims that eukaryotes arose by enclosing certain cellular chemical functions, such as reproduction and energy creation, within membranes to form intracellular structures distinct from the cytoplasm that made up the rest of the interior of the cell. The other model, *endosymbiosis*, asserts that the internal structures of the cells were actually prokaryotic bacteria that were engulfed by other, larger prokaryotes or primitive eukaryotes, the two eventually forming a symbiotic relationship. For example, mitochondria are thought to be aerobic protokaryotic bacteria engulfed by an anaerobic prokaryote or simple eukaryotic cell. The mitochondrian bacterium gained protection from the world outside the host bacterium, while the host gained the mitochondrian's ability to remove poisonous oxygen from its interior and convert it to energy to help run the host cell's biochemical processes. The presence of DNA within mitochondria that differs from the host cell's nuclear DNA is one line of evidence supporting this theory of mitochondrian engulfment. Similarly, chloroplasts have their own DNA and were likely originally cyanobacteria that were engulfed by a prokaryote or primitive eukaryote, with each gaining a survival advantage from the symbiotic relationship. *Spirochetes* are also asserted by some to have joined to primitive eukaryotes, evolving into the flagella we see in many eukaryotes today.

The issue between these two hypotheses is far from settled. It appears that at present most paleobiologists prefer the endosymbiosis model for the creation of the mitochondria and chloroplasts and the autogenous model for the creation of the nucleus and other structures within the cell. It may have been that the first eukaryotic protists were quite primitive, essentially bacteria that had developed a nucleus but otherwise retained the simple bacterial form. In this they would be like modern parasites, which are anaerobic and lack mitochondria and chloroplasts and obtain their energy directly from their host.

Later, one of these early eukaryotic bacteria engulfed a smaller bacterium, which developed into what we call a mitochondrion, thus becoming the first animal-like protist. Probably somewhat later, another eukaryotic bacterium engulfed a cyanobacterium that would become the cell's chloroplast, thus giving rise to the first plant-like protist. Still other internal organelles could also have evolved at this same time, giving rise to yet more complex though still single-celled protists. This process of protist evolution might have happened more than once in the ancient oceans during this several-hundred-million-year period, or it may be that only a single protist became the ancestor of all subsequent life on Earth. We may never know.

Regarding *when* these first eukayrotes evolved, there is much debate but little conclusive evidence due to the paucity of fossil discoveries of these small, unicellular organisms. Suggested dates for the origin of eukaryotes range from as early as 2.7 bya to as late as 850 mya. The majority of paleobiologists today seems to place the rise of eukaryotes between 2.5 and 1.8 bya, or at about the time the increasing oxygenation of the atmosphere and oceans would have provided the environment necessary for the evolution of these new aerobic creatures.

The oldest fossils found to date that are believed to be protists date to about 2.1 bya. One is *Grypania*, a tubular-shaped alga that measured an enormous (for this era) 1 cm (0.39 in) in length. Very recently, what seems to be an animal-like protist has also been found in rocks dating to 2.1 bya in Gabon, West Africa. This as-yet-unnamed species had a flat but somewhat rounded soft body with slits around its edges and complex folding patterns inside. The fossils range from 1 to an enormous 12 cm (5 in), were probably aerobic, and evidently lived in the oceans some 30 to 40 meters beneath the surface.

In addition to these two fossil finds, we also have evidence for some multicellular protists that date to 2.0 or 2.1 bya (see below). If these finds have been correctly interpreted, and given that single-celled protists would likely have evolved before multi-cellular ones, this would give a date of at least 2.1 bya for the appearance of the first, simplest eukaryotes. After 2 bya, organic-walled microfossils called *acritarchs* become significant in the fossil record, perhaps in part because increased oxygen in the atmosphere meant the creation of more nitrates in the oceans, which were important for these eukaryotes. Some evidence suggests that fresh-water protists evolved by about 1.7 bya. If so, this would make them the first living beings on Earth to colonize the land.

The subsequent evolutionary history of the earliest protists is controversial. Indeed, the fossil record is at present so spotty and the biochemical evidence so uncertain that many paleobiologists wonder whether we will ever be able to reconstruct the evolutionary history of the protists with any confidence. Certainly there must have been a number of evolutionary dead-ends among the many varieties of these organisms, with only some types surviving to give rise to the wealth of protists today. Most of the evolution of protists likely occurred within the various mats they created on the shallow seafloors. Paleontologists typically that the period from the middle of the

Proterozoic Eon to nearly its end as a time of stasis, where little happened over a very long time.

The earliest protists evidently existed in two main types: the animal-like *Protozoa* and the plant-like *Protophyta*. Today there is also third type of protist, the fungus-like *slime molds* and *water molds*, but it seems that these were a later development from the earlier fungal protists. Nature does not always cooperate with our attempts at a neat categorization of things, and this is certainly the case even with this rough grouping of protists. Some plant-like algae are apparently more closely related to animal-like protozoa and proto-fungi than they are to other algae, while some protozoa and proto-fungi appear more closely related to algae than to other protozoa or proto-fungi. Therefore these broad categories can serve only as approximations for the kinds of protists that have existed in the past and continue to populate the planet today.

The term "protozoa" literally means "first animals." Technically speaking, this is a misnomer, given that these protists are all single-celled, as distinct from animals, which are defined as multi-cellular creatures with distinctive *tissues*—that is, collections of typically similar cells united to perform a particular task or tasks. On the other hand, the name does properly indicate that protozoans, unlike protophyta, do not manufacture their own food. Rather, they ingest other single-celled creatures, whether protophyta or other protozoa. Today some 30,000 varieties of protozoa are known to exist.

Protozoans are often grouped according to their mode of movement into *amoebas* (or *sarcodina*), *zooflagellates*, and *ciliates*. *Sporozoans* are classed as a fourth group for reasons unrelated to their motion, as we will see.

Amoebas are irregularly-shaped, single-celled creatures. Some recent taxonomic groupings put amoebas in their own kingdom, *Amoebozoa*, rather than the kingdom Protista, though this, like much of the taxonomy of protists at present, is controversial. An amoeba moves by extending its cell wall outward, its cytoplasm flowing along behind it to form what biologists term *pseudopods* ("false feet"). The rest of the cell then follows after it, creating an active if rather slow mode of movement. Amoebas also use their pseudopods in order to eat, moving these "false feet" around a bacterium, alga, or other organic material and enclosing it to create a *food vacuole*. This small vesicle is then brought into the amoeba where the food inside it is digested. Amoebas typically live on the bottom of shallow ponds and reproduce asexually. Interestingly, they, like certain other protists, sometimes have more than one nucleus. When present within humans in great quantities, amoebas can also cause disease, as in amoebic dysentery.

Several other amoeba-related creatures eventually evolved, each characterized by its own particular type of exoskeleton. Fossils indicate they definitely existed in the Cambrian Period of the Phanerozoic Eon, or between 541 and 485 mya,[6] though their origins, according to at least some paleobiologists, are likely to be significantly earlier, perhaps somewhat before the end of Ordovician carbonate around an inner chamber that houses the amoeba. As they grow, forams add new chambers in various geometrical configurations, some growing to several centimeters (or inches) in diameter. The foram's pseudopods reach out through openings in the exoskeleton where they join together, forming nets in which to trap their food. Another type, the *radiolarian*, is similar to forams except that its exoskeletons are made of silicon dioxide and exhibit a wide variety of geometric shapes, many of which are stunningly beautiful. Like forams, radiolarians also have many pseudopods that extend out through their shells to aid in food gathering.

Zooflagellates are protozoans that move by means of a whip-like flagellum. Most take in food by absorbing it through their cell membrane or engulfing it, like the amoeba, in food vacuoles by means of pseudopods. Several forms have even been found that are capable of photosynthesis. Like amoebas, zooflagellates also come in a wide variety of forms.

Ciliates, such as the *paramecium*, are distinguished by having small, hair-like cilia surrounding their external membranes. Not only do these cilia aid movement, but in many ciliates also aid their feeding, sensation, and attachment to things. Rare, if not unique, among eukaryotes, they developed two nuclei, one to handle reproduction and the other to regulate various processes within the cell. Some ciliates evolved mouth-like openings that they use in feeding on bacteria, algae, other organic matter, and in some cases even on other ciliates. Still other ciliates absorb their food directly through their outer membranes. Ciliates can reproduce both sexually and asexually.

S*porozoa* are parasitic protists. As their name suggests, sporozoa have reproductive processes that typically involve the production of spores. Their reproduction generally proceeds in two phases, one in which the cell reproduces asexually and another in which it reproduces sexually. This often requires that they inhabit two different hosts to accomplish their reproduction. Sporozoa also lack cilia, flagella, or any other means by which they might move rapidly through their watery environment. As parasites they take their energy directly from their hosts and also often cause disease in humans. Malaria, for example, is caused by the sporozoan *Plasmodium*, which lives within the saliva of the female *Anopheles* mosquito until the

6. See the Geologic Time Scale in the appendix.

insect bites its victim, whereupon it moves from the saliva into the victim's bloodstream and thence to the liver, where it usually, in the absence of medical treatment, leads to death.

Protophyta, meaning "first plants," is, like the term "protozoa," a misnomer, as the protophyta are not true plants (which by definition are all multicellular entities with distinct types of specialized tissues). Predominant are the single-celled algae. As we have noted, the oldest algae fossils date to roughly 2.1 bya, but it seems that a significant increase in the diversity and number of algae in the oceans did not occur until more than a billion years later, in the late Proterozoic Eon (1 bya to 541 mya). Interestingly, these early fossil algae are quite similar in form to algae that are still found today, including *diatoms*, *dinoflagellates*, *calcareous nanoplankton*, and several types of multicellular algae. Some evidence suggests that algal scum had already formed on the land 1.2 bya.

Diatoms are single-celled algae that, like radiolarians, secrete exoskeletons of silicon dioxide, with the difference that diatom shells exist as two pieces that fit together, rather like a shallow canister and its lid. They live either at the bottom of shallow pools of water or float near the surface, where they can receive sufficient light for photosynthesis, which provides them their food. *Dinoflagellates* are floating algae characterized by the presence of two flagella that aid their motion through the water. When conditions turn unfavorable, dinoflagellates armor themselves with a tough organic outer covering, existing in this dormant, cyst state until conditions are favorable again. *Calcareous nanoplankton* (or *coccoliths*) are small, spherically-shaped algae that secrete exoskeletons of calcium carbonate (like forams), though in the shape of rounded plates. These serve to protect them as they float in the ocean. When they die, their shells sink to the bottom of the ocean where, under pressure from the thick layers of sediment that eventually cover them, they become the fine limestone we call chalk. *Multicellular algae*, mostly seaweeds, typically attach to the bottom in shallow bodies of water and grow upward toward the surface and the Sun.

The importance of the protophyta cannot be overstated. Algae lie at the base of the food chain in the oceans, constituting a large part of the plankton that float in the water, and so are critical for the continuation of nearly every form of life in the seas. Algae are also are responsible for roughly half of the oxygen production in the world today, having surpassed the amount contributed by cyanobacteria sometime after 2 bya. For both these reasons it is clear that without algae there would be no complex life either on land or in the sea.

Finally, there are the *fungi*. Fungi are plantlike in that they are typically rooted to the ground and reproduce by means of spores, as did the earliest land plants. Yet they are even more like animals, from which they diverged after plants had gone their own way some millions of years earlier, probably around 950 mya. Early fossils with fungi-like features reach back to more than 1.4 bya. Like animals, fungi do not have chlorophyll and so do not manufacture their own food. Rather, they typically feed on dead matter, both plant and animal. Their genetic signatures are also more animal-like than plant-like. Most fungi are so small that they go unnoticed and typically live in moist environments rooted by means of fine filaments into wet soil.

It is also important to note that during the Proterozoic Eon new mountain ranges came into existence on the continents. These mountain-building events, termed orogenies, occur when tectonic forces caused by the collision of Earth's plates cause the crust along the margins of the plates to crumple, sending parts of the crust upward. The earliest known orogeny during this eon is the *Wopmay Orogeny*, which occurred between 1.97 and 1.84 bya in what today is the arctic region of the northwestern part of the Canadian Shield. Later, circa 1250–980 mya, the *Grenville Orogeny* occurred, raising mountains that stretch from what today is Scotland down through Newfoundland, the eastern and south-central United States, and into Mexico. It was caused by the collision of plates that combined to form the supercontinent Rodinia. From these times onward, mountains would be part of the terrestrial landscape.

THE LATE PROTEROZOIC EON: THE ORIGINS OF PLANTS, ANIMALS, AND FUNGI

By about 1.1 bya, eukaryotes had begun to diversify greatly, possibly as the result of the development of sexual reproduction. This diversity seems to have set the stage for the appearance of the first complex multicellular creatures. The most common view among paleontologists today is that plants, animals, and fungi all descended from a single common protist ancestor. While some estimates put this divergence as early as 1.6 or 1.5 bya, the more common scientific estimate today seems to be around 900 mya to 1 bya. Others more conservatively place the divergences closer to the end of the Paleozoic, perhaps somewhat before 600 mya. Most paleontologists are agreed that the first group to evolve from this original protist ancestor were the plants, with fungi and animals continuing to share a common ancestor for several tens of millions of years after this first divergence.

Several lines of evidence point to this common ancestor of animals and fungi having been a kind of colonial protist called a *choanoflagellate*. A choanoflagellate had an ovoid- to spherically-shaped cell body a few micrometers long and a single flagellum. It has been suggested that the frond-like choanoflagellate *Diplotheca* might have been this last common ancestor of animals and fungi, which perhaps around 965 mya or so then diverged into their own separate evolutionary lines. Thereafter plants, animals, and fungi would continue their divergent evolutionary trails, eventually producing the myriad species within each kingdom that exist today. Some molecular evidence suggests fungi might have colonized the land as early as 1 bya, with the first plants following by perhaps 700 mya. Both would have lived near the edges of bodies of water.

The evolution of plants, animals, and fungi during the next several hundred million years—all taking place within the oceans—is largely unknown, as fossils of new creatures do not begin to appear until around 600 mya. It is believed that plants and animals diversified greatly during this period but that their soft bodies left few traces behind. It has been suggested that among the animals the *protostomes* (animals in which the first cavity formed in the embryo is the mouth) diverged from the *deuterostomes* (animals in which the first body cavity formed in the embryo is the anus, with the mouth forming later from a second cavity) around 670 mya (though in the scenario in which plants and animals diverged some 1.5 bya, that date lies somewhere after 1.5 bya). It seems that multicellular green and red algae became even more abundant during this period, as did *acritarchs*.

Acritarchs are small (50 micrometer-plus), spherical, unicellular microfossils with tough organic walls that appear in the fossil record as far back as 3.2 bya, though well-preserved forms began to appear only around 1.6 bya. Paleontologists are not entirely certain what produced them. They may be fossils from a range of species from all three domains of life, including dormant cysts of different kinds of algae and the egg cases of larger animals from late in the Proterozoic Eon. Acritarchs became more abundant, complex, and diverse in shape, size, and character beginning about 1 bya, the same time as the longevity of the species that produced them declined. This, combined with the fact that significantly spinier acritarchs began to appear around the same time, suggests increased predation by larger animal-like protists. Acritarchs reached their maximum abundance and diversity around 850 mya, were substantially reduced during the Sturtian and Marinoan glaciations between about 750 and 580 mya (see below), and gradually declined thereafter.

In the midst of this early *Neoproterozoic Era* (the last phase of the Proterozoic Eon), several significant developments occurred that might have affected the evolution of the first true multicellular entities. One was the formation and subsequent dissolution of the supercontinent *Rodinia*. Rodinia probably began to form sometime around 1.1 or 1 bya and persisted for perhaps 300 million years. Rodinia sat near the equator, and when it broke up around 750 mya a major glaciation, the *Sturtian glaciation,* soon followed and lasted for some 50 million years. The cause of this glaciation event uncertain, but it appears that the formation of supercontinents generally has led to phases of planet-wide cooling. It may be that the formation of supercontinents led to increased chemical weathering that pulled carbon dioxide from the air, thereby reducing its warming effect on the earth.

As glaciers began to form at higher latitudes, their whiteness reflected more of the Sun's energy back into space than the dark ocean waters, causing further cooling, which caused more glaciers, which caused more solar energy to be reflected back into space, creating a feedback loop, the final result of which was severe planet-wide cooling. Ice sheets are generally believed to have covered at least the three continents formed by Rodinia's break-up: Proto-Laurasia, Proto-Gondawana, and the Congo craton. Others think that ice sheets covered the oceans as well, perhaps leading to a second Snowball Earth in the planet's history. Most seem to believe that at least some stretches of ice-free ocean must have continued to exist or the entire population of photosynthesizing ocean life would have become extinct. There is evidence of a very small, disk-shaped animal that lived after the Sturtian glaciation, but it appears to have died off, along with much other marine life, during the *Marinoan* (or *Varangian*) *glaciation* that followed soon (as geological time goes) thereafter, lasting from roughly 625 to 580 mya.

These glaciations occurred some 100 million years later and may have created another Snowball Earth condition. This cooling event seems to have just preceded, or perhaps was roughly coterminous with, the formation of the next supercontinent, called *Pannotia* (or, occasionally, *Greater Gondwanaland*). Pannotia formed around 600 mya by the reconnection of the three continents that resulted from Rodinia's break-up. This time, however, they assembled not at the equator but near the South Pole. The collisions evidently involved more glancing blows, so that Pannotia already experienced some rifting at the same time it was forming. This is probably the reason the supercontinent lasted a relatively short 60 million years. The Marinoan glaciation also seems to have resulted in ice sheets covering most of the world, forming even at the equator. Many geologists think this was the most severe glaciation in Earth's history, ending finally around 580 mya. It appears that these two major glaciations, the Sturtian and the Marinoan, caused most forms of life in the oceans to become extinct. In this way ecological niches were created for the evolution of new forms of life once the temperature had warmed again in the *Ediacaran Period* at the end of the Proterozoic Eon.

THE EDIACARAN PERIOD (630–542 MILLION YEARS AGO)

The Ediacaran marks the beginning of the formation of Pannotia and conclusion of the Marinoan glaciation. During this period the average oxygen content of the atmosphere was roughly 8 percent of the total atmosphere, or about 40 percent of the current level, while the carbon dioxide content was some sixteen times its level in modern times (before its gradual rise since the industrial revolution). Existing evidence indicates that throughout most of the Ediacaran the temperature ranged from warm to hot at middle to low latitudes and was warm even further south of the equator. Only late in the Ediacaran did the temperature cool and glaciations appear at the poles.

The Ediacaran Period constitutes a major hinge-point in the history of life on Earth. For the previous three-plus billion years, or upwards of 80 percent of its history, life had been almost exclusively microscopic single-celled archaea, bacteria, and protists. Now, suddenly, much larger animals (or *metazoans*) appeared, ranging generally from several cm (about an inch) to almost a meter (roughly a yard) in length, and on the whole growing larger and more complex throughout the period. The fact that the oxygen levels had risen quickly over the preceding several hundred million years would have favored the evolution of these larger animals. In addition, they were evidently the first multicellular organisms to have distinct tissues. More advanced plant and fungal life also appeared in the Ediacaran, even if less spectacular in the range of their forms than the animals that evolved during this period.

The new metazoans were astonishing in the novelty and variety of their types, most of them bearing little resemblance to modern life-forms. More than one hundred different genera of Ediacaran animals have been discovered to date, though the nature and lifestyle of many of them remain mysterious, and few, if any, can be definitely related to the animal life that succeeded them in the Cambrian Period. Some of them were the earliest animals known to build shells of calcium carbonate, though most lacked such defenses and so their soft bodies remained exposed to the new metazoan predators that arose during the period. Some Ediacaran animals apparently lived deep in the oceans, while others inhabited the shallow seafloor. Some appear to have been flat disks while others were pillow-like in form. Some were smooth-surfaced and others quilted and fan-like. Some had heads and others appear to have had neither heads nor tails. Indeed, so strange was much of the Ediacaran *fauna* (animal life) that paleontologists are uncertain what many of the animals were, how they should be classified, and which, if any, evolved into the subsequent, more recognizable animal forms of the

Phanerozoic Eon. This "Ediacaran explosion" of new life-forms became especially rich after about 580 mya.

Fossils of these new animals have three general types. The first type consists of *fossil impressions* left by the soft bodies of these animals in the seafloor sediment. Fossils of such soft-bodied beings are rather rare, as they do not fossilize as easily as shells do, and so these impressions, along with the impressions made as some of them moved across the seafloor, are the only evidence that these creatures existed. Some of these impressions appear to have been left by animals similar to today's *jellyfish* (which float freely in the oceans) and *sea pens* (which are anchored by stalks to the seafloor and resemble plants more than our typical notions of what animals look like). Along with the more recently evolved sea anemones and corals, these animals appear to constitute the members of the phylum *Cnidaria*, though whether these Ediacaran animals are ancestors of the later cnidarians continues to be debated.

If they were ancestral to today's jellyfish and sea pens, they would have been similarly simple animals with quite rudimentary nervous systems. Their bodies would have been composed of outer and inner layers, the latter of which lined the stomach cavity, with a jelly-like substance in between. They also would have had stinging cells on the ends of their numerous tentacles with which they immobilized their planktonic prey and then drew them through their mouths and into their interior digestive cavities. They would also have had the same type of *radial symmetry* (which, as with a spoked wheel, looks the same in every direction as it radiates out from the center) as do jellyfish today. The sea-pen-like creatures, on the other hand, such as the frond-like *Charnia*, would have had the same *bilateral symmetry* (in which one side of the animal was essentially the mirror image of the other) that is characteristic of most animals, including humans, today.

Other Ediacaran animals that left seafloor impressions include what appear to be flat, segmented, worm-like creatures that might have been ancestors of today's *annelids* (segmented flatworms and roundworms). One such example is the three-foot-long, bilaterally symmetrical *Dickensonia*, which, though seemingly wormlike, had an oval, fan-like shape. Having no visible means of locomotion, it may have simply lain on the shallow seafloor, absorbing nutrients that passed by in the currents, or floated above it. Other seafloor fossils appear to have been made by animals with jointed legs and external skeletons that look like modern *arthropods*, a group which includes crabs, lobsters, spiders, insects, centipedes, and millipedes. One example is the segmented *Spriggina*. While some think Spriggina was more closely related to the annelid family, others think its plated body and curved head make it a good candidate as an ancestor of the arthropod trilobites that would appear some 50 million years later in the early Cambrian Period. Still other fossil seafloor impressions resemble those left today by marine snails, members of the *mollusk* (also spelled *mollusc*) family that also includes octopi, squid, sea slugs, and cuttlefish.

The second type of Ediacaran animal fossils are called *trace fossils*, impressions left by animals as they moved across the seafloor or burrowed down into it. Many of these have been found, especially from rock dating from 570 mya and later, during which period they demonstrate increasing complexity and diversity. These would have been made by tubeworms, which, by burrowing beneath the seafloor sediments, found a new, untapped ecological niche in the ancient oceans.

The third type of Ediacaran fossils are remains of the *external skeletons of small animals*, especially in the form of shells. Key to the Ediacaran explosion was the evolution of the ability to produce the protein *collagen*, a strong, flexible compound that is essential for the creation of shells and other skeletons. Shells of certain animals, such as the *Cloudinia*, constitute one example of this type of fossil. *Cloudinia* had small (some 1–2 mm/fraction of an inch-across) tube-like shells seemingly made of calcium carbonate (*calcite*) and looking as though a number of hollow tubes had been jammed into each other, the whole looking rather like a stack of paper cups.

Silica *spicules* (tiny pieces) of mineralized *sponges*, which are among the very simplest of animals, are another type of Ediacaran exoskeletal remnant. Lacking nervous, circulatory, and digestive systems, sponges get their food by extracting it from the water flowing through their bodies. These currents also carry away their waste products. Recent genetic evidence makes a strong case that certain very primitive types of sponges may have been the first multicellular animals and so the ancestors of all succeeding animal types, humans included.

Defying any classification whatsoever are various other animals whose unique body shapes tell us almost nothing about their structure and habits. These include the small disk-shaped and tri-armed *Tribrachidium*, the similar *Arkarua* with its five "arms," and a host of others. All these animals appear to have been evolutionary dead ends that failed to survive the Ediacaran Period.

For a diorama of the Ediacaran seafloor at the National Museum of Natural History, Smithsonian Institution, see https://astrobioloblog.files.wordpress.com/2011/01/ediacara-reconstitution.jpg.

It is likely that the evolution of terrestrial fungi, perhaps from water molds, had occurred by the end of the Ediacaran Period, though the fossil record provides no hard evidence for it. The movement onto land must have involved particularly the loss of the flagellum used to move the fungal cell through the water and the evolution of branching fungal filaments. The first terrestrial fungal cells would likely have been part of the scum-like mats composed also of algae, evidently the first and earliest colonizers of the land during the Proterozoic Eon.

As we have noted, the connection between these Ediacaran animals and later *cnidarians*, *annelids*, *arthropods*, *sponges*, and *mollusks* is far from established. The subsequent fossil record of the early *Phanerozoic ("visible life") Eon*, termed the *Cambrian Period*, evidences virtually none of these Ediacaran forms. The most commonly held explanation for this sudden disappearance of virtually all soft-bodied and shelled animals, as well as the acritarchs, from the fossil record is a mass extinction—though what might have caused it remains unknown. Perhaps the beginnings of the break-up of the supercontinent Pannotia, which ended the Proterozoic Eon and brought on the Phanerozoic Eon, caused different chemical or temperature conditions in the oceans to which most Ediacaran fauna could not adapt. The geological record indicates an increase in black shale formation consistent with a lack of oxygen in the oceans, which could have been caused by changes in oceanic circulation patterns.

Whatever the cause or causes may have been, the fossil record points to a sudden disappearance of Ediacaran fossils, then a noticeable gap in all kinds of life forms around 542 mya, and finally the gradual rise of very different types of fauna which arose to fill the ecological niches vacated by the Ediacaran fauna. Clearly there must have been some survivors from the Ediacaran Period that evolved into these new creatures of the Cambrian Period, or life would have had to begin again in rudimentary forms far more primitive than those that appear early in the Cambrian. Perhaps these ancestral animals were small and soft-bodied and so left no trace in the fossil record. Precisely which earlier animals might have evolved into what later creatures, however, may never be known.

15

From the Cambrian Explosion to the Permian Extinction

THE CAMBRIAN PERIOD (542–488 MILLION YEARS AGO)

THE BEGINNING OF THE Cambrian Period is marked by the break-up of the supercontinent Pannotia around 540 mya into the three relatively small continents of *Laurentia* (which would eventually form part of the core of North America), *Baltica*, and *Siberia*, and the supercontinent *Gondwana* (also sometimes called *Gondwanaland*) surrounding the smaller bodies on the west, south, and east (see the QR code below). The bodies lay mostly at or below the equator, with Gondwana stretching all the way to the South Pole. The climate was quite warm, more arid near the equator and more humid near the poles. There was no ice covering either pole. The mean surface temperature for the entire planet was roughly 21° C (about 70° F) throughout the period, or about 7° C/13° F higher than it was at the beginning of the Industrial Revolution. The warm temperature caused sea levels to rise to between 30 and 90 meters (or roughly 100–300 feet) higher than they are today. This resulted in late Cambrian continents being largely submerged, which created large, shallow seas between the continents into which life could expand and diversify. During the Cambrian Period oxygen increased to about 12.5 percent of the atmosphere (about 63 percent of its current level), while carbon dioxide stood at roughly sixteen times its preindustrial amount.

For an image of the Early Cambrian (Mollewide) Globe (by Ron Blakey and Colorado Plateau Geosystems Inc.), see http://cpgeosystems.com/540moll.jpg.

In rocks of the Cambrian Period, fossils become common for the first time. They witness to the evolution of a profusion of new animal forms. Why this occurred following the decline of the Ediacaran fauna is unknown, but a variety of explanations have been proposed.

The rifting of the ocean floor associated with the break-up of Pannotia had several important consequences. First, it released more calcium and phosphate into the oceans, where they would have been available for the formation of the shells of small animals that are the characteristic fossils of the Early Cambrian. Second, the rifting of the crustal plates that caused the supercontinent to break up created massive underwater chains of volcanoes spewing lava across the ocean floor, causing the water level to rise and in the process flooding the low coastal areas of the continents. This flooding throughout the Cambrian introduced still more chemicals necessary to life's flourishing into the seas, further aiding evolutionary surge throughout the Cambrian Period.

It may also have been that the increasing oxygenation of the atmosphere caused Earth's ozone layer to thicken sufficiently by this time to provide the new animals with greater shielding from harmful solar radiation. The development of shells might have had the same effect. The concomitant increase of oxygen in the oceans provided the necessary conditions for the evolution of the larger creatures that appear as the Cambrian progresses. As creatures grow in size, their oxygen needs increase; thus the increase in the oceans' oxygen during this period would have permitted larger beings to evolve. The warm climate through virtually the entire period was also conducive to the evolution of new plant and animal types.

It is also the case that small animals that grazed on algal mats appeared in the early Cambrian, after which these algal mats began to decline. This would have cleared spaces in shallow coastal waters for other new kinds of animals to thrive. It is also possible that the break-up of Pannotia separated species populations, permitting each to follow its own evolutionary path and so enhancing the rise of new species. Probably some combination of all these causes explains the explosive rise of new animals and their steady growth in average size throughout the Cambrian Period. This so-called

"*Cambrian Explosion*" constitutes the richest period of the evolution of new creaturely forms in the entire history of the planet.

Included within this "explosion" of life were the first unambiguous representatives of most modern phyla, that is, of the different types of body plans that exist today. Where the Ediacaran fauna are difficult if not impossible to assign with confidence to any phylum, the Cambrian fauna clearly contain ancestors of nearly all of today's invertebrates as well as the first vertebrates. Thus we find *sponges*, *brachiopods*, *arthropods*, *echinoderms*, and *chordates* (see below)—though it is also the case that some creatures seem to have had a foot in one modern phylum and another foot in a different one and so defy precise categorization. Still other Cambrian animals, however, reflect their Ediacaran forebears in the bizarreness of their bodily forms. Nearly all of these oddities went extinct, which is the other side of the Cambrian explosion. Extinction rates were evidently high throughout the period, with several mass extinctions near the end of the Cambrian drastically pruning the varied fauna that had evolved. In essence the Cambrian was a period of wild experimentation that produced a profusion of differing body plans, which were gradually winnowed to leave only the major animal types that continue to exist today. No fundamentally different animal body plan has arisen since.

The early Cambrian fossil record indicates that most of the newly evolved animals were still relatively small. Trace fossils indicate a gradual increase in the number and diversity of burrowing animals. Wormholes appear that reached several centimeters beneath the seafloor, indicating that these new worms had evolved the stiffer, rounded bodies necessary to digging down into the seafloor mud. These worms helped destroy the microbial mats that were an important feature of the Ediacaran seafloor, leading to the extinction of organisms dependent on the mats for their survival but opening up space for new ecosystems to evolve. A large variety of other invertebrates without protective shells—many having incredibly bizarre forms—also existed during the Cambrian. Many well-preserved fossils of these strange and as yet unidentified creatures were discovered in the now famous *Burgess Shale*, which was laid down during the Middle Cambrian.

But the Early Cambrian animals of greatest importance to the subsequent evolution of life on Earth were the large variety of new soft-bodied creatures with shells, and these evidently evolved within the first 10–12 million years of the period. Nearly all of these new shells were constructed of calcium carbonate. Calcium is a natural byproduct of animal metabolism, and it may have been that some small animal or animals developed the ability to secrete this excess calcium through their skin. When in the presence of sufficient oxygen—the amount of which continued to increase in the atmosphere and oceans throughout the period—calcium combines with it to form the calcium carbonate of which most modern shells are made. Increased phosphate in the ocean also resulted in shells made of calcium phosphate; and still others were made of silicon dioxide (silica).

Shells were important in that they provided support so that animals might evolve into larger forms. They also provided protection against the elements, such as sediments encroaching on space within and around the animal needed for digestion and other biological processes. They also provided a means of defense against the larger, more complex animal predators that began to arise in the Cambrian. Animal predation makes sense biologically in that animals are more nutritious than plants, and eating them therefore allows an animal to grow larger faster. As animals grow larger, they have more room within themselves to store excess food as fat. This fat can be used later as a food source, particularly during times when other sources are unavailable, and thus is an important survival mechanism.

This rise of animals that preyed on other animals created a whole new level in the food chain in the oceans, initiating the metazoan "arms race" that would encompass the rest of evolutionary history. Thus, as predators arose, prey animals developed defenses to protect themselves from them, following which the predators evolved ways to counter these new defenses, causing the prey to evolve yet more effective defenses, and so on, back and forth through the succeeding eons. The evolution of eyes during this period among both predators and prey also testifies to this intensifying arms race. Thus, the gentler life of prior times gave way to a new, increasingly ferocious battle for survival in the oceans. In this way the Cambrian oceans set the stage for today's ecosystems, with grazers, hunters, and filter feeders interconnected in complex ecologies.

One type of shelled prey animal that appeared early in the Cambrian was the *Hyolith*. Hyoliths had cone-shaped shells with pairs of small spikes on either side of the shell to help stabilize it on the seafloor. They also had something like a trap door at the wide end of their shells, behind which the small animals could retreat at the approach of predators. Lacking appendages that would have permitted them to move, they likely remained in one place throughout their lives unless moved by some other force, such as a brisk current. Their food was therefore likely organic material supplied by dying creatures, which was then deposited on the seafloor around them.

Another shelled animal was the heavily armored *Halkieria*. This animal appears to have crawled across the ocean floor by undulating its soft, flattened body, rather like a modern slug. Surrounding it on all sides was a complex calcium carbonate coat of armor. On top its scales were hollow and pointed, on its sides they were thinner and more knife-like, and along the bottom edges they were sharp and sickle-shaped. When a halkieria died, its armored scales separated from each other, leaving a host of small shells that, until a complete fossil specimen was finally discovered, paleontologists thought were the remains of a host of different, much smaller creatures. Hyoliths and Halkierids were perhaps related to later shelled mollusks, other types of which probably also existed in the early phase of the Cambrian.

Somewhat later, probably between 10 and 25 million years after the dawn of the Cambrian, other, still generally small animals began to appear which can be categorized with greater certainty as sponges, cnidarians, brachiopods, echinoderms, annelids, mollusks, arthropods, chordates, and vertebrates.

Sponges (phylum *Porifera*) of the early Cambrian shared significant traits with their Ediacaran forerunners. The earliest Cambrian sponges, dating to about 540 mya, were also generally vase- or bowl-like in shape and had tiny silica spicules woven together to make up their hard, glass-like skeletons. Their spicules came in four- or six-pointed varieties. Some ten million years later other sponges evolved whose skeletons were made of calcium carbonate rather than silica. The soft-bodied sponges most of us are familiar with, constructed of the organic material *spongin* rather than silica or calcite, only appeared much later. Bacteria were the main food of these early sponges, which they strained from the seawater as it passed through the maze of chambers within them. There protist-like cells with flagella captured small bits of food, drawing them within their small soft bodies and digesting them.

Cambrian sponges were also the creators of the first reefs in Earth's history in the shallow continental seas. One particular kind of early Cambrian creature, the *Archaeocyathid*, was probably a kind of sponge, given its simple body plan, and was one of the earliest, if not the earliest, reef builder. Reefs were important in that they created new ecological niches for Cambrian life-forms to inhabit. Archaeocyathids quickly branched into hundreds of species yet went extinct by the end of the early Cambrian.

Cnidarians, predatory animals including jellyfish, corals, sea pens, and sea anemones, were present in some forms in the early Cambrian, with jellyfish and sea pens particularly resembling, like the sponges, their probable Ediacaran ancestors. The largest (*medusae*) jellyfish reached two feet or more in diameter, making them the largest jellyfish ever to have existed according to the fossil record. Like all cnidarians today, they had *radial tentacles* (tentacles reaching out from their bodies). They were one of the largest predators in the Cambrian ocean and may have fed even on certain soft-bodied arthropods lacking *exoskeletons* (skeletons on the outside of their bodies, such as shells). *Corals* and *sea anemones* were other cnidarians that also made their first appearance in the early Cambrian, though, unlike most corals today, these early forms were solitary individuals rather than the communal types responsible for so much of the reef building in the oceans today.

Brachiopods are marine creatures with shells that are divided into two halves tied together by a hinge. Brachiopods look like *bivalve mollusks* but have different shell orientations, their shells being on the top and bottom of the soft animal within, whereas the mollusks have their shells on the right and left. They also have quite different internal anatomies and feeding habits, with the brachiopods employing a frilly loop-shaped organ called a *lophophore* that pumps water and strains small food particles from it. Most attach themselves to hard surfaces by means of a fleshy stalk. Abundant—second only to the arthropods in the Cambrian—and diverse already during the early Cambrian, brachiopods contributed significantly, along with the sponges, to the reef building that occurred during the period.

Echinoderms ("spiny skin"), members of the phylum *Echinodermata*, include starfish, sand dollars, sea urchins, sea cucumbers, and crinoids. *Crinoids* are vase-shaped creatures that are more commonly known as *sea lilies* (which are fixed to the seafloor by short stalks) and *feather stars* (which swim or float in the sea). Echinoderms typically have small tube-like feet with suction cups in order to attach more firmly to seafloor structures (and, with starfish, their prey). Most also have internal skeletons of calcite plates, yet some, such as starfish, are flexible. Echinoderms feed in a variety of different ways: some, such as crinoids, filter their food from the water passing by; others, such as sea urchins, are grazers; still others, such as sea cucumbers, are scavengers that eat decaying matter on the seafloor; and yet others, such as starfish, are hunters. Cambrian echinoderms include *Helicocoplacus*, a tubular creature covered with calcium carbonate (calcite) plates that stood upright on the seafloor but lacked the typical five-fold radial symmetry of other members of the group. Early *eocrinoids*, ancestors of the later sea lilies, also appeared during the Cambrian. They too were covered with calcium carbonate skeletal plates like most echinoderms, though some lacked the five-lobed radial symmetry typical of other members of the phylum. Later crinoids were larger and more complex creatures with silica spicules.

Annelids (worms) were a diverse group in the Cambrian and reach back at least to the early part of the period. They are mainly known only from their representatives found in the Burgess Shale; otherwise their fossil record in this period is poor. They seem typically to have been predators that either swam or crawled along the seafloor. Their heads had two good-sized tentacles protruding from them and their bodies were covered with bristles. One type found in the Burgess shale, *Canadia*, was 2.4–4.8 cm (1–2 in) long. Another type found elsewhere grew to 2.4 cm (1 in) in width and was more than 8 cm (3.3 in) long.

The evolutionary history of *mollusks* is heavily debated. The earliest Cambrian types, if indeed they were mollusks, were the small, snail-like, single-plated *helcionelloids* that crept along the seafloor on a wide fleshy "foot," grazing on algae. They have been dated to some 540 mya. Other, larger heicionelloids have also been found with cap-like bodies up to several cm (1+ in) long. In fact, Cambrian mollusks possessed a wide range of body forms, from snails to bivalves to some early forms of *cephalopods* ("head foots"), which are mollusks whose foot has evolved into a group of tentacles surrounding a beak-like mouth at the base of a large, bilaterally symmetrical head, as with modern octopi and squid. The precise character of these first cephalopods remains uncertain.

The most common mollusk feature is a *mantle* (or *mantle skirt*), a muscular organ that surrounds the animal's *mantle cavity*, protecting the animal's internal organs within the cavity. The mantle also serves a variety of other functions, the specific function depending on the type of mollusk. In shelled forms the mantle is what secretes the shell and causes it to become progressively larger and thicker as the animal grows. The mantle can also serve as the foot that moves the mollusk along the seafloor or that helps with feeding. In cephalopods the mantle surrounds a tubular funnel into which water is drawn, then squeezes the funnel forcefully, expelling the water and so propelling the animal in the opposite direction.

All mollusks, other than bivalves, also have a tongue-like *radula* that aids in feeding. In herbivore species, minute "teeth" along its outer edge scrape algae and diatoms off rock surfaces and mats; in carnivorous species, they aid in shredding the captured prey. Others mollusks, such as some carnivorous snails, have a specialized radula like a harpoon with which it stabs and poisons its victims

Arthropods were the overwhelmingly dominant animal population numerically during the Cambrian Period. Early on there was the monstrous *Anomalocaris* ("abnormal shrimp"), a fierce predator that grew to about 2 m (6 ½ ft) in length and is deemed by most paleontologists to be an arthropod (or at least something very closely related to arthropods). In addition to the segmented bodies which are a hallmark of all arthropods, anomalocaris also had two jointed appendages, one on either side of its segmented, disk-shaped mouth, with longish barbs facing inwards by means of which it impaled its prey. Rather than crawling along the seafloor, like most arthropods, the legless anomalocaris swam through the Cambrian seas by means of flaps composed of a series of overlapping lobes on either side of its body.

Another arthropod oddity was the small (4–7 cm/1.5–2.7 in) *Opabinia*. Opabinia's five eyes reached upward from its head on stalks, giving it an excellent view above and around it. Even stranger, it had a long, flexible proboscis with spines used to capture prey and drag it to the mouth underneath and behind it—somewhat like an elephant. Other Cambrian arthropods, such as the *Euthycarcinoids* and/or the *Eurypterids* (sea scorpions), are especially important in that they are likely candidates for the first animals ever to leave the seas and walk on land. Their body plans, and particularly their appendages, match up well with some apparently Cambrian terrestrial fossil imprints known as *protichnites*. This monumental first step toward animal colonization of the land occurred perhaps around 500 mya.

The iconic Cambrian arthropods, however, are without question the *trilobites*. Trilobites first appear in the fossil record from about 530 mya, though by then they were already quite diverse and widespread and sported their trademark calcium-carbonate exoskeletons. These facts suggest that they descended from earlier arthropods lacking such exoskeletons. In all over 100 families and 600 species of trilobites are known from the Cambrian Period alone, with more than 5,000 genera and 17,000 species known in total from the succeeding periods down to their final disappearance during the massive Permian extinction about 250 mya. Trilobites ranged in size from about a millimeter to more than 0.6 m (2 ft) in length. Most crawled along the seafloor, while others created burrows in the mud; the smallest floated along with the plankton. Most ate organic material from the seafloor, but some were carnivores, preying on smaller animals such as worms. Some trilobites inhabited the shallow continental shelves while others lived in the depths of the ocean. Early trilobites seem not to have protective shells nor the ability to curl up into a ball to protect themselves; later species had shells and could curl up in an attempt to thwart predators.

As their name suggests, trilobites were divided into three main lobes that ran the entire length of their bodies. At the head of the trilobite was a single head plate (*cephalon*) that often had long spines at the edges curling back toward the rear. This was followed by the three-lobed body (*thorax*), the lobes running parallel along the length of the body, with a single fused tail-plate (*pygidium*) at the end. The central lobe was taller and rounder than the two flattened lateral lobes but all three were equally segmented from the front to the back of the trilobite. The number of segments varied with the species, ranging from two to sixty-one, with most species having between two and sixteen.

The entire upper body was enclosed within an exoskeleton made of calcium carbonate and calcium phosphate. Beneath and stretching out to the sides of the exoskeleton were a varying number of unarmored legs, the number proportional to the body length or number of thorax segments. Each leg was composed of either six or seven segments, like other early arthropods, and some of these bore features used for functions other than walking, such as hairs on the segments evidently used for feeding and/or sensation as well as spines or claws on the last segment used to tear food apart.

The cephalon contained the mouth and stomach, with the rest of the trilobite digestive system extending through the thorax and pygidium to the anus. The cephalon also contained the trilobite's eyes, which were typically compound structures made up of from 1 to more than 15,000 calcite rods. Each rod was shaped like an elongated prism, and together the bundle of these rods in each eye would have produced a mosaic-like image.

The first *chordates* and *vertebrates* also appear in the Cambrian Period. Chordates are animals with a *notochord*, which is a flexible pipe-like structure that runs the length of the animal's body. Notochords exist in all higher animals in their early life, including humans, but develop into the *vertebral* or *spinal column* in vertebrate animals. The spinal cord sits next to the notochord in invertebrates and the spinal column in vertebrates. Paleontologists believe that the first vertebrates, the fish, evolved during the Cambrian, and that the first of these was the tiny *Myllokunmingia*. Only one fossil of this 28 mm (1.2 in) long and 6 mm (¼ in) tall creature has been found, dating to about 524 mya. It apparently had a skull and skeleton made of cartilage, like that of modern sharks, but lacked movable jaws and so is termed, with others of its kind, as *agnathans* ("no jaws"). It was likely a filter-feeder, drawing in water through its fixed mouth and straining out its food.

Similar in size, appearance, and probably in internal structure was another early agnathan fish named *Haikouichthys*. *Conodonts* form another example of this animal group. They ranged in size from perhaps 1 to a quite large 40 cm (16.7 in) in length and had teeth-like structures in their round mouths. *Pikaia gracilens* was yet another small agnathan, averaging some 5 cm (2 in) in length. Likely also a filter-feeder, it evidently swam above the seafloor with help from its wide tailfin. Agnathans persisted for some 150 million years alongside the jawed fish that evolved from them until the extinction late in the Devonian Period, about 364 mya, from which they never recovered. Today's lampreys and hagfish are the only existent agnathans.

Finally, the Cambrian evidently also saw the rise of the first metazoic *parasites*. Parasites are animals that exist in a symbiotic relationship with another, usually larger organism and benefit from that relationship while the host animal suffers from it. Parasites plague all forms of higher life, from plants to fungi to animals. Most gain their food from the host, but some, such as viruses, can only reproduce using the cellular reproductive mechanism of the host. Since a parasite needs a host in order to survive, it is clear that parasites originated after their hosts and continued to evolve in response to the defenses that evolved among those hosts. It is probable that parasites evolved from nonparasitic ancestors. Cambrian parasites are primarily known from the *pentastomids*, several genera of worm-like invertebrates sometimes called tongue worms, given their visual similarity to a human tongue. Whether these pentastomids were actually parasites is debated, as is the identity of their possible hosts. Conodonts have been suggested, and the deformed trilobites that have been found might well have been another animal plagued by these early parasites.

The Cambrian oceanic *plant* record, by contrast, shows little in the way of the evolution of new types. Algae and seaweed continued to be the predominant forms. Planktonic algae continued to evolve and to populate the upper layer of the oceans, and some animals developed larvae that lived among and consumed these algae. Algal acritarchs still existed as a major component of the ocean's phytoplankton, and algal-bacterial stromatolite mats persisted throughout the period. As the Cambrian progressed, however, these stromatolites began to diminish in abundance and became confined primarily to coastal intertidal zones. This was probably due to the rise of the new animals that grazed on the mats beneath the water—especially, it seems, mollusks. Intertidal zones evidently provided some respite from these grazers, though the presence of numerous holes reaching down through the stromolites that survived suggests the rise of other burrowing animals that tunneled through them. Also, red and brown seaweed existed along shallower seacoasts. *Radiolaria*, *udoteacean green algae*, and *calcareous red algae* may have evolved during the Cambrian, too.

For a painting of the Cambrian Seafloor, by D. W. Miller, see http://paleobiology.si.edu/burgess/cambrianWorld.html.

But just as the first arthropods were perhaps making their way tentatively onto the land, so were the earliest plants and fungi. While the fossil record indicates that an algae scum may have inhabited the land along the margins of bodies of water or perpetually wet places as far back as 1.2 bya, during the Cambrian it appears that a "richer" scum, this one a mix of algae, fungi, and perhaps also lichen, lived on the land. But even more importantly, a recent fossil find in Siberia has indicated the existence during the Cambrian of what is possibly the earliest true land plant, *Aldanophyton*. Incomplete fossil specimens indicate the plant was possibly a *lycopod*, a plant like modern *club mosses* that reproduces by spores and lives near wet places, sending up stems containing vascular bundles to carry water and nutrients upward in the plant.

The fossil evidence indicates a plant with a length of over 8 cm (3 in) and a width of 1.3 cm (½ in) with its stems covered by thin leaf-like structures. Other kinds of spores have also been found in Cambrian rock from diverse places, indicating perhaps the presence of other early plants for which we have no fossil evidence. However, the fact that only partial remains of *Aldanophyton* from a single location have been found and the possibility that other spores in Cambrian rocks may be the result of contamination by later spores means that some paleontologists are unconvinced that true plants had indeed moved landward during the Cambrian.

Paleontologist Richard Fortey brings this Cambrian world to life by offering a picture of how it would have looked to someone who was there to observe it:

> I can imagine standing upon a Cambrian shore in the evening . . . Where the sea meets the land there is a patch of slightly sticky, rounded stromatolite pillows, survivors from the vast groves of the Precambrian. The wind is whistling across the red plains behind me, where nothing visible lives, and I can feel the sharp sting of wind-blown sand on the back of my legs. But in the muddy sand at my feet I can see worm casts, little curled wiggles that look familiar. I can see trails of dimpled impressions left by the scuttling of crustacean-like animals. On the strand line a whole range of shells glistens—washed up by the last storm, I suppose—some of them mother-of-pearl, others darkly shining, made of calcium phosphate. At the edge of the sea a dead sponge washes back and forth in the waves, tumbling over and over in the foam. There are heaps of seaweed, red and brown, and several stranded jellyfish, one, partly submerged, still feebly pulsing. Apart from the whistle of the breeze and the crash and suck of the breakers, it is completely silent, and nothing cries in the wind. I wade out into a rock pool. In the clear water I can see several creatures that could fit into the palm of my hand crawling or gliding very slowly along the bottom. Some of them carry an armor of plates on their backs. I can recognize a chiton, but others are unfamiliar. In the sand there are shy tubeworms. A trilobite the size of a crab has caught one of them and is shredding it with its limbs. Another one crawls across my foot, and I can feel the tickle of its numerous legs on my bare flesh—but wait, it is not a trilobite, but a different kind of arthropod with eyes on stalks at the front and delicate grasping "hands." Now that I look out to sea I can see a swarm of similar arthropods

> sculling together in the bright surface water—and can that dark shape with glistening eyes be *Anomalocaris* in pursuit? Yes, for the top of its body briefly breaks the surface, and I can glimpse its fierce arms for an instant. Where the water breaks it shines luminously for a while in the dying light—the seawater must be full of light-producing plankton—and I have to imagine millions more microscopic organisms in the shimmering sea.[1]

The Cambrian Period was a time of vast experimentation that produced a cornucopia of new animals and animal forms that competed for survival, primarily in the seas. Extinction rates were evidently high during the whole of the Cambrian. By the end of the period, however, the experimentation had ended and a severe winnowing of this diversity of life forms occurred. Only those animal types that were members of the animal plyla that still exist today remained.

A major cause of the diminished diversity was certainly the series of mass extinctions that occurred near the end of the period. There appear to have been extinctions around 517 mya and another around 502 mya, with a third occurring around 488 mya at the boundary of the Cambrian and Ordovician.[2] It has been noted that every time in the past 500 million years that supercontinents have formed, the global temperature has dropped. Perhaps the formation of Gondwana and its movement to the South Pole caused the cooling that devastated life in the seas. Perhaps the oxygen levels in the seas dropped as well. Whatever caused these extinctions, the damage to life in the seas was massive. Many brachiopods, conodonts, and other more unusual life-forms disappeared. The trilobites, dominant in the later Cambrian seas, were substantially reduced, with a number of species vanishing altogether. The reef-building archaeocyathids had also disappeared by the end of the period. These extinctions opened up the Ordovician seas to the evolution of new life-forms that more closely resemble life on Earth today.

THE ORDOVICIAN PERIOD (488–444 MILLION YEARS AGO)

The Ordovician Period through most of its history continued trends in the oceans and air begun during the Cambrian. The oxygen level increased to about 13.5 percent of the air (about 68 percent of its amount before the

1. Fortey, *Life*, 106.
2. See the Geologic Time Scale in the appendix.

Industrial Revolution), and carbon dioxide declined somewhat to about fifteen times its early modern level. The average temperature during the period was roughly 16° C (61° F), or about 2° (nearly 14° F) warmer than pre-industrial Earth. Sea levels ranged mostly between 140 and 220 m (460–722 ft) above current sea levels, with some geoscientists placing its maximum elevation at as much as 600 m, or nearly 2,000 ft above current levels. This meant that, through most of the Ordovician, the continents were largely flooded, creating vast shallow seas into which life might expand and form diverse new ecosystems.

For an image of the Middle Ordovician Globe by Ron Blakey and Colorado Plateau Geosystems Inc., see http://cpgeosystems.com/470moll.jpg.

Gondwana, comprising what would later become the continents of Africa, South America, Australia, and Antarctica, continued to drift southward during the period, finally centering on, though still extending northward from, the South Pole. The smaller continents of Laurentia, Siberia, and Baltica remained separated and gradually shifted somewhat eastward and northward toward and beyond the equator. Throughout most of the period the warming trend continued, though late in the Ordovician the temperature cooled significantly, and glaciers formed at the South Pole, initiating an ice age that lasted some twenty million years. With seawater now held in glaciers, the sea levels gradually declined and with them significant parts of the shallow seas that had supported most of the planet's life. The result was a catastrophic extinction that ended the period.

Throughout most of the Ordovician, however, life-forms expanded into the ever-increasing shallow seas around and between the continents and diversified greatly as they populated these new ecosystems. The total number of orders of animals doubled, the number of families tripled, and the number of genera increased fourfold in the seas. The number of animals with calcium carbonate shells increased dramatically as well, and some animals that dominated the early Ordovician seas were gradually replaced by other, newer kinds of animals. This "Ordovician radiation" began slowly but increased significantly by the middle of the period.

The early Ordovician radiation was modest. *Trilobites* recovered from the Cambrian extinction and evolved a host of new forms well-adapted to the conditions of the new period. They continued to be the chief seafloor scavengers throughout the Ordovician. Some developed shovel-like heads for digging into the seafloor mud to find their food. Others became swimmers and so exploited a new ecological niche. Some evolved with no eyes, while others developed eyes at the end of long stalks that assisted them in locating the new and more fearful predators that feasted on trilobites throughout the era, gradually decreasing the seafloor dominance they had enjoyed during the Cambrian. This continuing arms race also resulted in some trilobites evolving spiny shells for protection and others the ability to curl up into a ball, with nothing but their armored plating exposed.

The chief new predators of the Ordovician, the ones that would sit atop the shallow-water food, were the *nautiloids*. Nautiloids were *cephalopods* with the typical features: tentacles surrounding a beak set at the base of its large head, like modern squid and octopi. The unique feature of the nautiloid was its long, tapering, protective shell surrounding most of its body, from which its head and tentacles extended. The nautiloid's shell became larger as the animal grew and added new chambers to inhabit. Throughout the Ordovician, nautiloids radiated into a variety of ecological niches and diversified accordingly, with some reaching some 2 meters (more than six feet) in length. Later nautiloids would develop coiled shells as well, looking essentially like the modern nautilus, which still survives in deeper waters, generally between 100 and 600 feet beneath the ocean surface. The other important developments within nautiloids were the ability to regulate the amount of gas in their abandoned chambers to control buoyancy and to take in and expel jets of water rapidly, allowing them to move at a greater rate through the water. Upon locating the smaller animals that were their prey, they would grasp the animal with their tentacles and draw it toward their beaked mouths, which would tear the animal apart, followed by the ingesting of its soft parts. The trilobites seem to have been the chief prey of the nautiloids throughout the Ordovician. Other early Ordovician seafloor animals include a variety of new kinds of brachiopods, bryozoans, mollusks, echinoderms (including the first brittle stars and sea stars), and also planktonic *graptolites* (colonial animals, many of which resemble plants).

For a painting of the Ordovician seafloor, by Fritz Geller-Grimm (for a diorama at the National Museum of Natural History, Smithsonian Institution), see http://commons.wikimedia.org/wiki/File:Nmnh_fg09.jpg.

Life on land was primitive, evidently composed of fungi, bacteria, and algae that lived near sources of water. Fossilized spores from the late Ordovician suggest that some more advanced plants might have arisen at the time, and the presence of small burrows means that some invertebrates, such as millipedes, might also have adapted to conditions on land and fed on these first spore-bearing plants. No conclusive evidence for the existence of land animals exists, however, until the Silurian Period.

Many of these new life-forms were wiped out in the great extinction that occurred during the long ice age that ended the Ordovician Period—the second greatest extinction event in the history of Earth. It began around 450 mya and continued into the early Silurian, causing nearly half of all plant life genera and vast numbers of warm-water animals to disappear.

THE SILURIAN PERIOD (444–416 MILLION YEARS AGO)

During the Silurian Period,[3] the major continent, Gondwana, continued to inch southward, surrounding the South Pole, while the three land masses of Laurentia, Siberia, and Baltica drew closer together around the equator, where Laurentia and Baltica would eventually unite into a second large continent scientists have named Euramerica (or Laurussia). This smaller continent would later divide to form the foundational cratons of North America, northern Europe, Russia, and Greenland.

For an image of the Silurian Globe by Ron Blakey and Colorado Plateau Geosystems Inc., see http://jan.ucc.nau.edu/~rcb7/430moll.jpg.

Early in the Silurian, temperatures began to rise, averaging about 17° C (63° F) during the period—significantly warmer than today—resulting in a major shrinking of the ice caps that had covered the poles. Sea levels rose to some 180 meters (590 feet) above the current level, again flooding the continental margins and so creating the warm, shallow seas ideal for the rapid evolution of life.

The Silurian Period saw the recovery of sea-dwelling plants and animals following the end of the mass extinction resulting from the long ice age at the end of the Ordovician, and with the recovery a significant diversification of many life-forms. This occurred especially in the warm, shallow seas around and between the equatorial continents. Several important innovations also arose there. Corals appeared, joining prior reef-builders such as bryozoans and certain sponges to produce massive coral reefs in the tropical shallow seas. These small animals feasted on the abundant plankton in the warm waters around them. Graptolites rebounded only somewhat following the Ordovician Ice Age and virtually disappeared by the end of the Silurian.

3. See the Geologic Time Scale in the appendix.

Brachiopods rebounded especially well, with the older, thin-shelled and flatter types replaced by ones with thicker shells and deeper bodies. They far outnumbered the bivalves, gastropods, and trilobites that also inhabited the Silurian seas. Trilobites were relatively scarce during the period, perhaps because of the great number of predators. These included the nautiloids, which remained major predators despite beginning to decline during the period. The top predators, however, were the sea scorpions, which attained their greatest numbers and range in the Silurian.[4] While most were less than 25 cm (10 in), one species reached nearly 2.5 m (8 ft) in length. Sea scorpions had claws on their front appendages for capturing and tearing apart their prey and paddle-like rear appendages for moving efficiently through the water. Some sea-scorpion species began to inhabit brackish bodies of water on the edges of the seas and even freshwater streams and swamps. Leeches also made their first appearance during the Silurian

For an image of the Silurian-Carboniferous seafloor, by Keiji Terakoshi, see http://www.terakoshi.com/item/dino&extinct/extinctEn-html/84-08081En.html.

Probably the most significant innovation in the seas was the rise, during the late Silurian, of new kinds of fish (see the QR code above). One type were the jawless fish that developed armored bodies and faces, termed *placoderms*. They lacked movable fins on the sides of their bodies, suggesting that they were not very maneuverable swimmers and most likely bottom-feeders. They also featured slit-like mouths with sharp plates rather than teeth, with which they captured and shredded their food, probably mollusks such as trilobites. It may be that other placoderms were filter feeders. Even more important was the appearance of the first fish with movable jaws, including the *acanthodians*, which had bony skeletons, strong jaw muscles, large eyes, and a row of spiny fins along its sides. These fins would have made them good swimmers, and they were likely quite successful predators. All later bony fish were evidently descended from these

4. Editor's note: In the final stage of editing Jarvis's manuscript, an MSN news feed (Geggel, "'Bizarre,' Human-Size Sea Scorpion Found in Ancient Meteorite Crater") caught my eye. An Iowa Geological Survey team researching a "meteorite impact crater" created 470 mya discovered the fossil of a eurypterid species that lived about 460 mya (in the middle Ordovician period), about 9 million years earlier than scientists previously thought. The team named it "Pentecopterus decorahensis," based on a Greek word for a "five-oared warship" (*pentēconteros*) and a Greek word for "wing" (*pteron*) and the location of the discovery near Decorah, Iowa. This species, which is related to "modern arachnids and horseshoe crabs," had earlier Cambrian representatives. For the original study, see Lamsdell, "The oldest described eurypterid."

acanthodians. The first known freshwater fish also appeared during the late Silurian.

The first definitive evidence for the presence of land plants and animals appears in rock strata from the Silurian. First came the creeping, spore-bearing plants we call *bryophytes*, descendants of earlier multicellular green (chlorophyll-bearing) algae and the ancestors of all modern mosses and liverworts. The first plants that colonized the land faced several important challenges. One was how to stay moist while exposed to the air. Bryophytes accomplished this by living only in wet environments, taking up water from the ground through small, hair-like structures called *rhizoids* that reached only a short distance into the soil, and dispersing the water throughout the plant by diffusion from cell to cell. This rather inefficient transport of water limited the bryophytes' vertical growth to less than 25 cm (10 in). When the groundwater periodically dried up, so did the plant, ceasing its metabolism until the water returned, at which point it rapidly recovered.

The small rhizoids of the bryophytes also answered another problem for land plants: how to stay anchored to the ground given the current unpredictable climatological conditions. A third novel problem bryophytes faced was how to reproduce on land. These early mosses and liverworts solved this, as did their terrestrial ancestors, by producing spores that were spread by the wind. The development of stems that rose higher above the ground made for more efficient dispersion of the spores and gave those bryophytes that evolved this capacity a survival advantage over other species. Finally, there was the problem of ultraviolet radiation from the Sun, which might have killed them. The gradual accumulation of an ozone layer evidently protected them and the other terrestrial plants and animals of that time and thereafter. Bryophytes dominated the Silurian Period, extending widely along bodies of water and other moist locales. So-called "moss forests" that have been found in rock strata indicate their wide distribution near bodies of water and across alluvial plains. Bryophytes have changed very little since they first evolved; today's mosses and liverworts remain much like their Silurian progenitors.

A second type of land plant that developed during the Middle Silurian, around 430 mya, was the *vascular plant*, and with it came a set of new answers to the problems created by terrestrial living. These new answers would set the course for the development of the more complex plants that would follow and that would come to dominate the earth. We have noted that bryophytes had a rather inefficient way of remaining moist despite their exposure to the air. Vascular plants developed several improvements that raised the efficiency of the process. First, they developed roots that could reach deeper into the ground, which allowed them to access to water unavailable to bryophytes and so to continue functioning long after the early mosses had desiccated and gone dormant during dry spells. Second, they developed a waterproof *cuticle* that kept the plant's moisture from passing through its skin and evaporating.

This development, however, created a new problem: how to draw in the carbon dioxide necessary for photosynthesis to produce the energy the plant needed to survive and how to expel its waste product, oxygen. Bryophytes simply absorbed their CO_2 from the air around them and expelled their O_2 by the same diffusion process, but the vascular plant's impervious cuticle prevented the absorption and expulsion of gases in this way. Thus, along with the cuticle there quickly arose *stomata*, small mouth-like openings that could open and close to take in or expel gases, and these stomata were spread throughout the plant's outer skin.

Vascular plants also developed another crucial ability that bryophytes lacked—the ability to move water and nutrients more efficiently by means of small tubes spread throughout the plant—hence the term *vascular* plant. Those tubes that carry nutrients and water upward in the plant via *osmosis*[5] are known as *xylem*, and those tubes that move the food created by photosynthesis, especially sucrose, down to feed the lower parts of the plant are known as *phloem*. Not only was this a more efficient way of moving food and water throughout the plant, it also enabled vascular plants to develop rigid stems and so grow far taller than bryophytes. Spore packets, or *sporangia*, soon developed atop these tall stems, enabling the plants to disperse their spores more widely and giving them a reproductive advantage.

The earliest known vascular plants are *Cooksonia* and *Baragwanathia*. Cooksonia dates from around 425 mya and is characterized by its branching, leafless stems atop which were spore-producing structures. A small plant, Cooksonia was only several centimeters in height. Baragwanathia was larger and more complex, growing as much as 11 inches (28 cm) tall. Perhaps the first vascular plant to have leaves along its stem, it is the ancestor of all club mosses. Some algae evolved protective waxy coatings and began to live on land, too, while terrestrial fungi also appear for the first time during the Silurian.

For an image of Cooksonia, by Danilo M. Bonanni, see http://commons.wikimedia.org/wiki/File:Cooksonia_pertoni.png.

5. As water is lost through leaves, water from the soil is pulled upward through the thin xylem tubes to replace it.

For an image of Baragwanathia, see http://commons.wikim org/wiki/File:Baragwanathia_artificial_model.jpg.

The emergence of land plants had several important ecological effects. First was the production of copious amounts of oxygen that went directly into the atmosphere rather than into the water, where only some of it had been diffusing into the air. This would make possible the evolution of more complex animal life in the periods that followed. Second, once the oxygen concentration in the air reached 13 percent in the early Silurian, wildfires became not just a possibility but a reality, as the fossil record shows. From that time on, charcoal has been present in strata from every period down to the present.

Animals, too, emerged from the waters and began their conquest of the land during the Silurian Period. All the first land-dwellers were arthropods, and the earliest of these, according to our current knowledge, was the *scorpion*, an arachnid relative of those scorpions that were dominating the food chain in the seas. It has been suggested that the first of these scorpions lived mainly in the water but came ashore when it molted its shell. Other land-dwelling Silurian arthropods included millipedes, centipedes, spiders, and some primitive insects.

Toward the end of the Silurian there was a series of relatively minor extinctions, probably caused either by impacts of objects from space or by climate change.

THE DEVONIAN PERIOD (416–359 MILLION YEARS AGO)

For an image of the Late Devonian Globe by Ron Blakey and Colorado Plateau Geosystems Inc., see https://www2.nau.edu/rcb7/370moll.jpg.

During the Devonian Period,[6] all the continents were moving northward, with Siberia in the northern latitudes, Eurasia straddling the equator, and Gondwana stretching from the South Pole to the equator (see the QR code above). Temperatures continued the rise begun in the Silurian, averaging some 20° C (68° F) or about 6° C (11° F) above today's average global temperature. Earlier in the period the sea level remained at some 180 meters (590 feet) above its current level, reflecting the almost total absence of ice at

6. See the Geologic Time Scale in the appendix.

the poles and elsewhere, but then gradually fell, along with the temperature, later in the Devonian, finally reaching only 120 m (395 ft) above current levels. Earth's oxygen content rose to about 15 percent of the atmosphere, or about three-quarters of its modern level, while the average carbon dioxide content of the air halved from its Silurian level to about eight times its modern, preindustrial level. This reflected the CO_2 absorption by the rapidly-spreading plant cover on land and the increased land area in tropical to temperate zones.

In the seas, invertebrate life continued to evolve, especially in the numerous warm seas surrounding the tropical parts of Eurasia and Gondwana. Plankton continued to flourish, as did the brachiopods, bryozoa, and crinoids, while corals continued their extensive reef-building. Trilobites remained common seafloor dwellers, with several novel species emerging during the period. Some featured protruding spiky tails, while others bore bulbous compound eyes. Perhaps the most interesting of these latter types was the *Phacops*, which had evolved the ability to roll itself up like a sow bug for protection against the numerous predators of the Devonian. Straight-shelled nautiloids declined, being gradually replaced by the curved-shelled ammonoids, their descendants. These mollusks began their lives surrounded by small chambers. As they grew, they secreted new, larger chambers in a pattern that curved back along the older chambers, giving them, when seen from the side, a spiral appearance. In this process, each older and smaller vacated chamber was walled off from the new one, becoming air-filled and so aiding the ammonite's buoyancy as it moved through the water.

The great Devonian innovation was the rise of new types of fish, hence the other name often used for this period, "The Age of Fish." The early Devonian was dominated by the armored jawless fish that appeared in the Silurian, such as the placoderms, while some evolved strange forms with shield-like or horseshoe-like heads. Some were more lightly armored and flattened, like modern rays and skates, while one giant, *Dunkleosteus* (also known as *Dinichthys*), grew to as much as 12 m (40 ft). But these heavily armored and jawless fish would become extinct by the end of the period, likely due to the rise of the new fish with jaws that evolved during the period. The earlier jawed Silurian acanthodians continued into the Devonian but also went extinct by the end of the period.

There arose two main types of jawed fish, the *bony fish* and the *cartilaginous fish*. Bony fish, as their name suggests, had bony skeletons. They lacked the bony plating of armored, jawless fish, being instead covered in scales, making them more flexible and so more maneuverable. They can be classified as belonging to one of two main groups, the *lobe-finned* and the *ray-finned*.

Cartilaginous fish of the Devonian featured skeletons made of cartilage and included the first true sharks. They had long, flat skulls; sharp, replaceable teeth; small, rough scales all over their bodies; and fixed fins, including large pectoral and two dorsal fins, with some species featuring spines in front of the dorsal fins. Devonian sharks were small by today's standards, the largest reaching only about 2 m (6 ½ ft).

Ray-finned fish featured bony spines in their maneuverable fins and had gas-filled swim bladders that enabled them to control their buoyancy. They are the ancestors of 99 percent of today's fish. Lobe-finned fish had rounded, fleshy fins with thicker bones in them, which they used to maneuver along the seafloor, and some of them even on land. They also had internal nostrils, which became useful as the fish developed lungs, enabling them to survive both in and out of the water. They evolved into three main groups: the *lungfish*, species of which persist today; the *coelacanths*, thought to have gone extinct long ago until a fisherman caught one off the southeast coast of Africa in 1937; and those that eventually developed into the first *amphibians*, ancestors of all subsequent land vertebrates.

One of the earliest of the proto-amphibians, or *tetrapods* (literally, "four feet") is the recently found *Tiktaalik*. It featured a long head that tapered to its snout, in which were its nostrils; a neck; stronger ribs (to aid breathing while lying on land) which enabled movement on land; and strong bony fins, which it used like legs to move through the shallows and perhaps briefly on land. Other species were *Elginerpeton*, a tetrapod sharing many of Tiktaalik's features, which grew to perhaps 1.5 m (5 ft), and *ichthyostega*, with its fish-like tail, long jaw, and seven toes—five was the norm—on splayed legs with which it waddled along. All these proto amphibians probably had finned tails, breathed through their skin and primitive lungs, and likely lived most of their lives in shallow water, perhaps crawling onto the land to escape predators or to eat.

Life on land was evolving rapidly, hence the phrase "the Devonian Explosion." During the early Devonian, plant life remained much as it had been during the later Silurian, dominated by low, creeping, spore-bearing bryophytes. Typical was Rhynia, with its cuticle-covered, leafless stalk and sporangia on the top, which seldom grew to more than half a meter (18 in) in height. Interestingly, the tallest early Devonian land-dweller was not a plant but a fungus, *Prototaxites*, with a fruiting body that rose in excess of 8 m (26 ft) above the ground.

The middle Devonian saw the appearance of shrub-filled expanses of *lycophytes* (or *lycopsids*, *lycopods*), which were the first major group of vascular plants, possibly the descendants of Cooksonia. Most were small, creeping forms that lived near water, but some developed stronger vascular bundles and strong roots that enabled them to reach 1 m (39.4 in) or more. They also featured leaves that bore the plant's spores and grew directly on the stout stems in a spiral formation. One lycophyte species, *Wattieza*, dated to about 385 mya, rose to an amazing 10 m (34 ft), towering above all other plant-life on the planet. As it grew, it shed its frond-like, spore bearing leaves, leaving diamond-shaped scars on its trunk. By the late Devonian, other tree species had evolved with strong, woody trunks that could support the weight of branches with leaves, some of which evidently rose 30 m (98 ft) above the ground. They and other plants with deeper roots were the first to be able to survive on dry land, and as they spread out these earliest trees began to cover the landscape to form the first true forests.

The late Devonian also saw ferns, with their complex, branching leaves and sporangia beneath them, appear in moist areas. Even more significant, the period saw the advent of the first seed-bearing plants, such as the *seed ferns*. Seeds have the advantage over spores in that, where the latter must be kept constantly moist so that they do not dry out from contact with the air, seeds have their own waterproof cover and food inside, and so can colonize environments where moisture is more sporadic.

All these rooted plants began to create the first soils on the land, which provided fertile spaces for their offspring as well as homes for the myriad mites, scorpions, millipedes, and other arthropods that were evolving in the Devonian. With the major increase in plant life during the period, the stage was set for marine vertebrates to seek out these ready food sources on the land.

The Late Devonian witnessed another major extinction, in which an estimated 70 percent of marine invertebrate species were destroyed, mostly in the warm waters of the tropics. Polar marine life seems to have been virtually unaffected. Tropical reefs were devastated as the reef-creating creatures died off to such an extent that they were extremely scarce throughout the rest of the Paleozoic Era. Their habitat was soon taken over by glass sponges, which do well in the cooler waters that evidently destroyed the corals and other former inhabitants. But reef-builders were not the only hard-hit life-forms. Some 85 percent of brachiopods were destroyed, all armored jawless fish declined severely, while placoderms became extinct, microplanktonic acritarchs nearly suffered the same fate, ammonoids were severely reduced in numbers, and trilobites and gastropods also declined. Land plants and animals were barely affected.

For an image of the Devonian seafloor, see http://encyclopediaurantia.org/images/pra248.jpg.

What caused the extinction is still a matter of debate, though some theories are better supported than others. The reappearance of polar ice indicates that the planet was cooling, perhaps due to the greening of the land, which brought a significant reduction in carbon dioxide, a greenhouse gas. There is some evidence that cold, non-oxygenated water from deep in the oceans was brought to the surface, further cooling the warmer equatorial waters where most life flourished. It has also been suggested that impacts from extraterrestrial sources may have raised enough debris into the air to cool the planet, though the evidence here is weak, since no sufficiently large impact crater has been found to explain the millions of years that the cooling event lasted. But marine life had dearly suffered, whatever the cause of the cooling period, and it was only as the temperature warmed in the Carboniferous Period[7] that marine life staged a significant comeback.

THE CARBONIFEROUS PERIOD (359–299 MILLION YEARS AGO)

Throughout this era, the Euramerican continent continued to straddle the equator, while late in the period the parts of Gondwana that would eventually become Africa and South America fused and collided with the eastern part of Euramerica, raising the Appalachian Mountain range to heights currently seen in the Himalayas. Siberia continued to creep northward while the main mass of Gondwana reversed its direction and began to move southward once again. Several microcontinents fused with southern Siberia to form the nucleus of what would become China. By the end of the period all the major landmasses were in the process of merging with each other to form the most recent supercontinent in Earth's history, *Pangaea* ("all land"). To the east of the continent the Paleo-Tethys Ocean formed, while to the west lay the gigantic ocean named *Panthalassa* ("all sea").

7. See the Geologic Time Scale in the appendix.

For an image of the Late Carboniferous Globe by Ron Blakey and Colorado Plateau Geosystems Inc., see http://cpgeosystems.com/images/300moll.jpg.

The average temperature of Earth during the Carboniferous Period was about 14° C (57° F), essentially what it was in modern times before the start of the Industrial Revolution. The period began warmer but cooled later, leaving southern Gondwana sheathed in ice. Sea level fell during this earlier period, reaching roughly its present level, but rose again later, reaching about 80 m (262 ft) above its present level by the end of the Carboniferous. The extensive swamp forests of the period absorbed up massive amounts of CO_2, reducing the atmospheric quantity to a mere three times modern levels, and pumped our copious O_2, which rose to about 32.5 percent of the air, or about 63 percent above its present level.

During the first half of the Carboniferous, life in the warm, shallow seas around and between the low-latitude continents flourished. *Foraminifers*, amoeba-like protozoans, became abundant during the period, as did some corals, lacy bryozoa, crinoid echinoderms, bivalve mollusks, and brachiopods of a wide range of types. Amminoids recovered and became the dominant invertebrate predator, while trilobites became very rare. As the waters turned cooler and receded from the land later in the period, the shallows where many of these groups lived disappeared, so many of these groups declined, some becoming extinct. Some, such as clams, adapted to life in freshwater streams and estuaries. Marine fungi continued to evolve.

Among the bony fish, the lobe-finned types continued to inhabit the oceans while some evolved into freshwater types. Most notable, perhaps, of these latter were the *rhizodonts* that were the dominant freshwater predators of their age, with one species reaching some 7 m (23 ft) in length. In general, the ray-finned fish thrived throughout the period. Sharks spread widely in the seas and developed a large variety of tooth structures and sometimes bizarre forms. There were sharks with the now-typical sharp, piercing teeth, others with cycloid teeth, and still others with flattish, crushing teeth designed to grind away the shells of various invertebrates. One type had a scythe-like head that resembled the head plate of some trilobites, and another had a brush-like dorsal fin with tooth-like structures on top. Some sharks also adapted to the freshwater environments of swamps and streams and reached as much as 2.1 m (7 ft) long.

The main innovations during the Carboniferous occurred on land. Dropping sea levels during the period opened up vast swathes of new land for colonization by plants and animals, including, importantly, swamps near the coasts of the more equatorial parts of the continents. These swamps were home to lycopod species that continued to thrive in the warm, humid environment, including a number of giants, as first seen in the Devonian. Tree ferns also rose into the forest canopy, as did other tree species with strap-like leaves. On the forest floor, ferns continued to occupy an important place, as did horsetails, liverworts, and mosses. In the absence of flowering plants, colorful fungi provided the only relief from the greens and browns that otherwise dominated terrestrial life in the Carboniferous.

Toward the end of the period, cycads and *gymnosperm* trees that would be ancestral to subsequent conifers arose. Gymnosperms are characterized by having their seeds exposed to the elements, such as on the scales of cones (*conifers*) or at the tips of stalks (*gingkoes*). Some trees developed deeper roots, which enabled them to begin to colonize the enormous stretches of land beyond the moist coastal areas. Spore-bearing and seed ferns grew in abundance under the swamp forest canopy. The vast numbers of trees in the swamps, when they died, were submerged, gradually decomposed into peat, and were eventually covered with layers of soil and compressed, finally becoming coal—hence the name Carboniferous ("coal-bearing") Period.

The historically immense amounts of oxygen produced by all these plants and trees had a significant effect on the animals as they evolved throughout the period, especially insects and other arthropods. These take in oxygen through holes in their exoskeletons, and with so much oxygen available in the air, by the end of the Carboniferous they had become gigantic. Some millipedes reached over 1 m (39.4 in) long, some centipedes attained double that length, and one millipede-like creature, *Arthropleura*, grew to 2.6 m (8 ½ ft), making it the largest known invertebrate in Earth's history. Spiders and scorpions—one species grew to more than 1 m (39.4 in)—also flourished, and snails had evidently emerged from the water and adapted to life on land.

Several insects appeared for the first time during the Carboniferous, including the first flying insects, such as mayflies, which sported wingspans comparable to those of medium-sized birds today. Cockroaches also appeared for the first time, some growing to 10 cm (4 in), and were so numerous that the late Carboniferous has sometimes been termed the "Age of Cockroaches." But the kings of the air, at least in terms of size, were the dragonflies, with stiff wings like mayflies that in some species spanned as much as 75 cm (30 in). They were the largest flying insects ever to exist on Earth. By the end of the period, insects with folded wings appeared, including beetles and the progenitors of grasshoppers and crickets.

For an image of a Carboniferous "swamp forest landscape with Hylonomus, Eryops and arthropod, front view," by Dorling Kindersley/Getty Images, see http://science.nationalgeographic.com/wallpaper/science/photos/carboniferous-period/carboniferous-swamp.

Amphibians diversified, grew larger and more numerous, and became less fish-like during the Carboniferous (see the QR code above). Some were aquatic, living in rivers; some apparently were semi-aquatic; and still others were terrestrial. They were also the largest predators during the period, with certain species reaching some 6 m (20 ft) in length. Others species ranged from about 15 cm (6 in) to as much as 2+ m (3 ½ ft). Many tended to look rather like alligators, with flattened heads, long snouts filled with sharp teeth, short legs, and some even with eyes located on the top of their heads, suggesting that they floated submerged in the water like their modern reptilian (crocodile and alligator) counterparts. Some had bony plates covering their heads, and others developed thick, scaly skins, which helped them retain moisture when they moved about on land. Another adaptation to more terrestrial living was the amniotic egg, which first appeared in the Carboniferous Period. Amniotic eggs feature protective outer coverings that keep in moisture while also allowing air to enter and exit. Within was a sack-like structure, the yolk, which provided food for the developing embryo. The first "shells" were leathery in texture, but eventually hard shells would evolve in other terrestrial vertebrates.

Toward the end of the Carboniferous the first reptiles appeared, all of which were small, had amphibian-like heads but lacked amphibian webbing between their toes, and laid amniotic eggs. The earliest was *Hylonimus*, a small, lizard-like creature from which would evolve all subsequent reptiles and birds. Some descendants of *Hylonimus* would return to the water, such as turtles, sea snakes, and crocodiles. Another early reptile was *Archaeothyris*, a primitive, more mammal-like creature (or *synapsid*) from which all mammals would eventually descend. By the end of the Carboniferous, reptiles had greatly diversified and spread, including into drier environments where amphibious vertebrates could not survive.

THE PERMIAN PERIOD (299–252 MILLION YEARS AGO)

For an image of the Late Permian Globe by Ron Blakey and Colorado Plateau Geosystems Inc., see http://cpgeosystems.com/260moll.jpg.

During the early Permian Period,[8] the continents continued to coalesce to form the supercontinent Pangaea, which stretched from pole to pole (see the QR code above). North of the equator and east of the main body of the continent the large Paleo-Tethys Ocean shrank, while south of that the Tethys Ocean formed and grew larger through the period. The existence of a single supercontinent had major ramifications for the Permian climate, which was marked by wide variations across the continent and relative instability as the period progressed. The south polar region was largely failing and would go extinct, along with many sea creatures. In the mass extinction that occurred at the end of the period came a wide dispersion of terrestrial life in all directions and with it a serious competition for survival in virtually all habitats. This would ultimately lead to a reduction in the diversity of species. Much of this winnowing came during the middle of the Permian, when temperatures dropped. The Carboniferous flora, which had continued to mark the early Permian with lush swamp forests, began to be replaced by more drought-resistant seed-bearing plants and trees. Lycopods were gradually replaced by *gymnosperms*, especially conifers, gingkoes, and cycads, which could also survive the more desert-like dry areas of Pangaea's interior. Large forests formed in many areas, with seed ferns rather than spore-bearing ferns covering the forest floor. Some tropical forests continued, featuring lycopods, huge horsetails and tree ferns, but by the end of the Permian these were relegated mainly to a few equatorial islands in the Paleotethys Sea.

Several new types of insects arose during the Permian: beetles, weevils, flies, and bees. Dragonflies remained the chief aerial predators, dining primarily on other insects but perhaps, due to their immense size, on smaller vertebrates as well. But the dominant insects, constituting some 90 percent of all insects, were the primitive relatives of cockroaches. They had the advantages of speed on the ground, efficient folding wings, good sensory faculties, protective exoskeletons of *chitin* (a long-chain polymer used in the outer shells of many arthropods), efficient mouth parts, and a primitive gizzard, a combination that made them formidable competitors in the struggle for life among insects.

8. See the Geologic Time Scale in the appendix.

The most significant developments among Permian land animals, however, occurred among the vertebrates. The early Permian was dominated by primitive reptilian *synapsids* (often termed *pelycosaurs*) and small amphibians. Reptiles, with their thicker, moisture-retaining skins, did especially well, given the drier climate and interior deserts of Pangaea. The primitive synapsids of the Carboniferous continued into the Permian and grew to become the first large herbivores and carnivores. Some developed large sail-like structures on their backs that might have helped to heat the cold-blooded animals in the early morning and so get them going. The largest grew to 3 m (10 ft) in length. One notable example, *Dimetrodon*, reached almost 2 m (7 ft) in length and some 100 kg (220 pounds/lbs) in weight and had sharp canines in the front of its mouth and flatter teeth behind. These large carnivores stood at the top of the early Permian food chain, feasting on herbivorous pelycosaurs, amphibians, and other reptiles, which in turn ate fish, insects, and plants. This has remained the basic structure of the fool chain ever since. The first proto-frogs also appeared in this period.

For an image of Early Permian life, showing an Edaphosaurus, by Interfoto Pressebildagentur/Alamy, see http://science.nationalgeographic.com/wallpaper/science/photos/permian-period/dimetrodon.

During the middle Permian, primitive *therapsids* appeared and would come to dominate the second half of the period. Therapsids were more mammal-like reptiles than the earlier synapsids/pelycosaurs. They quickly developed into a myriad of types, ranging from herbivores to carnivores and from the rat-sized *Robertia* to the hippo-sized herbivorous giant *Moschops*, which weighed more than a ton and featured a beaked mouth well suited for cutting vegetation.

In the late Permian, amphibians with large, flat bodies continued to exist in the more confined wetland areas, and salamanders made their first appearance. Synapsids, chiefly in the form of more advanced therapsids, evolved and ruled the terrestrial landscape. These therapsids were even more mammal-like than earlier species, with more mammalian jaws, skulls, and teeth (canines in front, multi-cusped molars in back). Some may have been the first warm-blooded animals, with hair and/or fur to help keep their body temperatures constant. Also, small *diapsids* began to appear—true reptiles that would eventually give rise to the dinosaurs, pterosaurs, lizards, and crocodiles.

The Permian Period ended with the greatest mass extinction in Earth's history, with an estimated 90–95 percent of sea-dwelling life-forms and 70 percent of terrestrial forms going extinct. Trilobites went extinct after some 250 million years of existence. So did archaic mollusks, including nautiloids, and almost all ammonite species. All corals also died out, and only two species of brachiopods survived. Echinoderms were devastated, with many types of sea urchins dying out and only one family of crinoids left. Many types of clams and snails also disappeared, and very few fish families survived. On land, significant reshaping occurred, with lycopods replacing gymnosperms, then gymnosperms recovering once again. This pattern seems to have repeated several times during the protracted extinction event. Terrestrial invertebrate life was also greatly affected, suffering the only mass die-off of insects known in Earth's history. Most types of amphibians and reptiles went extinct, including therapsids, which came close to disappearing altogether; other synapsids, however, fared somewhat better. These massive die-offs led to the significant reordering of life-forms and ecologies that followed, including the rise of, among other things, the dinosaurs.

The causes of this devastating extinction have been widely debated, with many scenarios proposed. Part of the problem is that the event seems to have lasted for quite a long time, though how long is unclear. Some have seen several phases of extinction, extending over several million years. Currently, global climate change from several causes seems to be the leading theory among paleontologists. We noted above that the colder middle Permian was followed by rapid warming at the end of the period. At that time, massive volcanic eruptions in Siberia began and continued for thousands of years, leaving much of the land covered in basalt. These eruptions would have hurled massive amounts of ash into the air, blocking out sunlight for a time and so cooling the planet. This might have disrupted plant photosynthesis, causing die-offs on up the food chain. But the increased CO_2 would have increased temperatures again over a longer period, with the greenhouse effect eventually raising global temperatures as much as 5° C (9° F). This rapid heating might have been more than a number of species could adjust to, leading to their extinction.

This process would have been exacerbated by warming of the oceans, which might have thawed frozen methane deposits on the ocean floors, sending that potent greenhouse gas up through the waters and eventually into the atmosphere, where it might have added another 5° C (9° F) to the global temperature. This would likely have triggered even more extinctions. Other possible scenarios that have been suggested, but have not garnered as much support, include a massive meteor strike, increased salinity in the oceans, decreased shallow marine environments as ocean levels dropped, decreased O_2 concentrations in the oceans, and several others. Whatever the cause, however, this extinction would thoroughly reshape Earth's varied ecosystems on land and in the seas, leading to a vastly altered, new biological world.

16

From the Triassic to the Quaternary

THE TRIASSIC PERIOD (251–201.6 MILLION YEARS AGO)

The Permian-Triassic boundary marks the end of the Paleozoic Era and the beginning of the Mesozoic Era,[1] often termed "the Age of Reptiles." Owing to the mass extinction of the late Permian Period, the fauna and flora that characterized life in the seas and on land during the new era was drastically changed, evolving into the more modern kinds of ecosystems that continue to exist today.

For an image of the Late Triassic Globe by Ron Blakey and Colorado Plateau Geosystems Inc., see http://cpgeosystems.com/220moll.jpg.

During the Triassic there continued to be the one supercontinent, Pangaea, which stretched from pole to pole. The Tethys Ocean continued to expand at the expense of the shrinking Paleo-Tethys Ocean. Long strings of volcanic islands arose off the western coast of Pangaea, while islands surrounded by reefs were concentrated in the tropical Tethys. The climate was largely hot and dry, and particularly in the vast deserts that covered much of the interior of Pangaea. Seasonal monsoons were evidently common along the tropical eastern coasts, leading to the formation of coal swamps there.

1. See the Geologic Time Scale in the appendix.

The poles were ice-free, quite moderate in temperature, and moist, making them good environments for reptilian creatures. Overall global temperatures for the Triassic increased to 17° C (62.6° F), or about 3° C (5° F) above modern levels. Oxygen levels dropped to about 16 percent, or some 80 percent of preindustrial times, while CO_2 levels increased to about six times the current amount. Sea levels rose and reached a maximum late in the period, causing the Tethys to flood stretches of northern Pangaea. Late in the Triassic, however, Pangaea began to break up, eventually to form, once again albeit in a somewhat different shape, the continents of Laurasia in the north and Gondwana in the south.

Recovery from the mass extinction at the Permian-Triassic boundary was slow, requiring about half of the period to reestablish diverse and complex ecosystems. The life-forms that did survive gradually spread and evolved into a host of new types, changing markedly the nature of life on Earth.

In the seas, the relatively placid life of seafloor creatures during the Permian gave rise to a host of new predators that initiated an arms race between the hunted and the hunters. At the bottom of the food chain, new types of phytoplankton and algae arose, including some varieties that are still present today. The first modern corals appeared with their altered internal structures, and they, along with other reef-building animals, created modest-sized reefs along the tropical coasts. Brachiopods suddenly became minor players in the seafloor environment, far outnumbered by the similar-looking mollusks and certain new forms of echinoderms, such as sea urchins. New types of filter-feeding bryozoans also appeared, and coiled-shell ammonites were present in large quantities and a wide variety of species. The late Triassic saw the appearance of the first crustaceans in the form of the first crabs and lobsters, which constituted yet other new predators in the seafloor communities. The abundance of shell-crushing predators that developed during the period caused animals with shells to evolve progressively stronger shells or other defensive strategies, such as burrowing or locomotion, or face extinction.

Fish were low in diversity and abundance, reflecting how few fish species survived the late Permian extinction. Ray-finned fish recovered well during the Triassic, while certain freshwater varieties adapted to life in the seas. Some of these developed new kinds of teeth that allowed them to prey on shelled creatures on the seafloor. Primitive sharks and rays, also hard hit in the extinction, were, like the fish, also reduced in diversity.

The most interesting life-forms in the oceans were the reptiles. New forms included the *placodonts*, which resembled today's marine iguanas but were much larger and feasted on mollusks, crushing them with their powerful jaws and peg-like teeth; *thalattosaurs,* appearing rather like lizards but with long, flattened tails for moving through the sea and growing to as much as 4 m (13 ft); *nothosaurs,* whose crocodile-like bodies grew to thalattosaur size, with sharp teeth in their elongated jaws ideal for catching fish and other marine animals; and *pachypleurosaurs,* which also resembled lizards but with long necks and tails, small heads, and paddle-like legs that also allowed them to live on the land. Pachypleurosaurs were relatively small, growing to a maximum of about 1 m (39.4 in) in length. The first turtles appeared later in the period.

The largest sea predators were the *ichthyosaurs* and the earliest *plesiosaurs,* which evidently descended from the nothosaurs. *Ichthyosaurs* ("fish lizards") appeared in the middle Triassic, resembling modern dolphins but with elongated jaws for catching fish in their sharp teeth, generally large eyes, and shark-like tails and powerful fins for moving quickly through the water. They began small but some species grew quite large in the later Triassic, reaching as much as 21 m (69 ft) long. *Plesiosaurs* had broad bodies, two pairs of large paddle-like flippers and short tails, and they varied greatly in size according to their species. Smaller plesiosaurs had short necks and reached more than 2 m (6 ft) long while the largest had long necks and grew to as much as 20 m (66 ft) in length. Both giants of the Triassic marine world remained dominant predators throughout the Jurassic Period and beyond.

Land plants were generally less affected by the Permian extinction than were terrestrial animals. The warmth and aridity of the period limited the location and extent of forests. Seed-bearing plants and trees (gymnosperms) came to dominate the period, with conifers, cycads, and ginkgoes the major tree forms (some ranging to 20 m/65 ft in height), though certain spore-bearing lycopods continued in moister areas throughout the period. Advanced seed ferns were the predominant species on the ground and evidently provided the main food consumed by herbivores; they were joined by moss, liverworts, and horsetails in moister environments. Overall, terrestrial flora were quite uniform through the more arid early and middle parts of the period, while the moister, monsoonal climate in the late Triassic enabled the expansion of lycopods and other water-loving plants. Fungi maintained a high degree of diversity.

Land animals fared variously, depending on their type. Insects flourished, with centipedes, millipedes, spiders, and scorpions present in abundance. In general, the arid climate and reduced wetlands and swamps through much of the Triassic did not favor amphibians as it did reptiles. Some primitive amphibian species that had survived the Permian extinction did not last very long in the Triassic, while others flourished throughout the period. Most amphibians were large, having large heads, short legs, and tails that varied in length, with different groups being more or less terrestrial. Proto-frogs continued to evolve.

It was the reptiles that truly flourished and diversified into a wide variety of forms during the Triassic. Some mammal-like therapsids survived the Permian and continued throughout the first half of the Triassic and beyond, but they began to be supplanted by the *archosaurs* and were nearly extinct by the period's end. Archosaurs ("ruling reptiles") were the reptilian group from which dinosaurs, pterosaurs, and crocodiles evolved in the middle to late Triassic. Dinosaurs began to appear in the middle of the period and were of two major types. The first type, the *saurischian*, is subdivided into groups, the carnivorous *therapods* and the herbivorous *sauropods*. Therapods tended to move swiftly on two legs due to their novel upright posture with legs located directly beneath their hips, giving them an advantage over slower-moving predators and prey. Their front legs became equivalent to arms. *Coelophysis* is a good early example of a theropod, standing as much as 2.7 m (9 ft) tall, weighing perhaps 45 kg (100 lbs), and feeding probably on amphibians and smaller reptiles.

Sauropods started out generally small but would give rise to huge herbivores such as the Brontosaurus in later periods. Sauropods featured long necks ending in relatively small heads, long tails, and stout, thick legs. They could stand on their two hind legs, giving them access to foliage from the tree canopy that other herbivores could not reach. *Plateosaurus* was the largest of the early sauropods, growing to as much as 10 m (33 ft) and 4,000 kg (8,800 lbs). But the most common herbivores during the period were the *rhynchosaurs*, pig-sized reptiles with beaks and, behind these, grinding plates covered in small teeth used to break down the tough seed ferns on which they likely fed. Claws on their toes would have also allowed them to forage beneath the ground for roots or tubers.

The other dinosaur type was the *ornithischians*. These were all beaked and herbivorous, the beaks useful in cutting plant stalks. Some are notable for their armored bodies, horns, and/or spiked tails, and others for their flat, duck-like bills. They evidently lived in herds, were generally smaller than sauropods, and were prey for therapods. The first ornithischians could walk on either two or four legs.

Pterosaurs ("winged lizards") arose late in the Triassic, perhaps from lizard-like archosaurs that glided short distances through the air by means of stretched skin between their legs (as with flying squirrels today) or from other reptiles with elongated scales that might have permitted gliding. Later in the period active flyers appeared with wings consisting of a membrane of skin covering front legs and feet that had much longer fourth digits. Early pterosaurs had teeth and long tails, while later species had much shorter tails and, in many cases, no teeth. All these early pterosaurs became extinct by the end of the period, replaced by the newer pterosaur types of the Jurassic Period. Still other archosaurs were lizard-like insectivores. The first crocodiles—small lizard-like reptiles—also first appeared in the later Triassic.

A second group of terrestrial reptiles also emerged which would lead to true lizards, the first of which appeared in the Triassic, and to snakes. Together all these varying reptilian forms would rule the earth for the next 150 million years.

Therapsids persisted into the late Triassic, though in diminishing numbers. At some point the first true mammals evolved from some unknown therapsid species. These were generally mouse-sized, shrew-like creatures that most likely were insectivores active mainly after dark. In this way, they carved out an ecological niche that was not in direct competition with the reigning archosaurs. Mammals generally gave birth to live young that had gestated in the mother's womb until reaching full development, though some species laid eggs like their reptilian forebears. Mammals were unique in feeding their young on milk and giving them extended care as they gradually became capable of fending for themselves. They were warm-blooded, with high metabolisms to keep their body temperatures high and constant. They also featured specialized jaws and teeth and had the greatest brain to body ratio of any animal of the period. Their nocturnal lifestyle may have promoted the evolution of fur. One of the earliest mammals, and a relative giant at some 1 m (39.4 in) in length, was *Eozostrodon*. It had a long snout filled with typical mammalian premolars and molars, four legs with eight-toed feet and claws, and a long, hairy tail. It laid eggs and fed its young milk from mammary glands.

For an image of Triassic dinosaurs, by Highlights for Children/Oxford Scientific/JupiterImages, see http://science.nationalgeographic.com/wallpaper/science/photos/triassic-period/triassic-dinosaurs.

The Triassic Period closed with yet another mass extinction, though this one was far less severe than the Permian extinction that preceded it. Indeed, current thinking is that there were probably three extinction events during the period, caused perhaps by climatic changes, and there is debate also on whether the end-of-the-period extinction was a single or multi-phased event. The end-Triassic extinction was likely caused by massive volcanic eruptions, occurring as Pangaea began to break up, which altered the climate significantly. Many terrestrial groups died out, including large amphibious herbivores, certain reptile types, and most synapsids, thereby opening ecological niches, which a wide variety of dinosaurs would quickly fill. The greatest devastation was in the oceans, where gastropods, mollusks, cephalopods, and brachiopods were greatly reduced in number and diversity. Sponges and bivalves were less affected, while all marine reptiles other than ichthyosaurs and plesiosaurs went extinct.

THE JURASSIC PERIOD (201.6–145.5 MILLION YEARS AGO)

The break-up of Pangaea that began late in the Triassic Period continued and accelerated throughout the Jurassic Period.[2] Laurasia in the north and Gondwana in the south continued to split apart, creating the growing equatorial Tethyan Ocean between them. In the second half of the period the two major continents began to rift further, with Laurasia splitting into what would become North America and Eurasia and Gondwana dividing into what would become Africa and South America in the west and Antarctica, Madagascar, India and Australia in the southeast. Later still, the northwestern tip of the future Africa would begin to break off from North America, creating the beginnings of the Atlantic Ocean in between. Later still South America, Africa, and eastern Gondwana began to divide. By the end of the period all the continents familiar to us today were beginning to assume their modern appearances.

For an image of the Late Jurassic Globe by Ron Blakey and Colorado Plateau Geosystems Inc., see http://cpgeosystems.com/150moll.jpg.

This rifting, and especially the creation of the Tethyan Ocean, caused major changes in ocean circulation and so the terrestrial climate. The hot and arid conditions typical of Triassic Pangaea gave way to the warm and moist conditions typical of the Jurassic. Lush jungles replaced deserts in many places, though some interior locations far from the oceans were arid. The poles were free of ice, and tropical conditions existed well into the northern latitudes. Overall, the average temperature for the period was about 16.5° C (61½° F), or some 3°C (4½° F) above modern levels.

Continental rifting meant that volcanic activity in various places was substantial, with mountains rising on the ocean floor and contributing to the rise of sea levels throughout most of the period before declining at the

2. See the Geologic Time Scale in the appendix.

end. Up to a quarter of the land area was covered at the high point of the oceans' rise, creating massive new marine environments in which sea life could thrive. This major volcanic activity was perhaps also responsible for the increase in atmospheric CO_2 to about seven times the preindustrial level. The increase in vegetation caused global O_2 levels to rise to about 26 percent of the air, or some 30 percent higher than today.

Many forms of life thrived during the Jurassic. At the bottom of the oceanic food chain were the plankton, which diversified significantly, with new kinds of foraminiferans (forams), dinoflagellates (protists), and micro-diatoms (algae) appearing throughout the period. On the shallow seafloors invertebrates fared well, with coral reefs proliferating, and sponges, snails, and mollusks thriving. Bivalves were dominant in most marine seafloor environments. Ammonites experienced a rapid evolution, developing into as many as ninety families from the single family that survived into the beginning of the period. Some of the new species grew to enormous size, with one late Jurassic ammonite, *Parapuzosia*, featuring a 3 m (10 ft) diameter shell and an animal within that was roughly the size of a modern giant squid.

Among marine vertebrates, fish and reptiles dominated. Primitive forms of ray-finned fish with their thick scales continued to dominate, though by the end of the period *teleost fish* appeared with their more flexible jaws that aided feeding and an air-filled swim bladder for buoyancy. A variety of shark species existed in two main types: those that preyed on swimming cephalopods and those that captured and crushed seafloor bivalves. Various types of rays also hunted in the warm, shallow seas.

Marine reptiles existed in profusion, with the dolphin-sized ichthyosaurs reaching their greatest diversity in the Jurassic. Large, long-necked plesiosaurs and shorter-necked, massive-jawed pliosaurs were formidable predators, along with huge marine crocodiles. Turtles were also present in significant numbers, as were a variety of smaller marine reptiles.

On land, seed ferns and horsetails continued to cover the forest floors, and larger tree ferns were also present in some areas. A variety of gymnosperms constituted the majority of Jurassic forests, with conifers especially widespread and cycads not far behind. In the northern latitudes ginkgoes were also present in abundance, while in the south *podocarps*—a type of conifer still indigenous in southern latitudes—proliferated. These widespread forests created massive amounts of organic matter, some of which eventually formed coal deposits. Small earthworm-like amphibians, the *caecilians*, seem to have arisen during the Jurassic, as well as the first salamanders.

The major story on land during this period is about the dinosaurs. So complete was their dominance that the Jurassic (and succeeding Cretaceous) Period has been termed "the Age of Dinosaurs." New habitats created by the breakup of Pangaea gave dinosaurs opportunities to diversify, which they did on a grand scale. The tendency throughout the period was for dinosaurs of most types to grow larger, both predators and prey, some reaching mammoth proportions.

The herbivorous sauropods were the ultimate exemplars of this trend, evolving to become the largest land animals that ever existed. *Apatosaurus* grew to some 23 m (75 ft) in length and weighed an estimated 27,000 kg (60,000 lbs); *Brachiosaurus* reached some 26 m (85 ft) long and 16 m (52 ft) tall; *Diplodocus* stretched more than 27 m (90 ft); and the recently found *Seismosaurus* reached some 36 m (110 ft) and weighed nearly 90,000 kg (200,000 lbs). The amount of vegetation these giants consumed, including ferns and the tops of various trees, which the long necks of these giants could reach, must have been prodigious, running in the tens of tons per day. Sauropods did not live in the water, as often depicted, but mostly on plains and in upland environments, and they likely moved in herds.

Predatory theropods ranged from small but swift *coelurosaurs*, from which birds would eventually evolve, through the medium-sized *ceratosaurs*, ca. 6 m (20 ft) long with horns located on their snouts, up to the giant of the era, the *Allosaurus*, which reached some 12 m (40 ft) in length. Some featured smaller forelimbs but all had powerful, muscular hind limbs, which allowed them to run in pursuit of their prey. They were apparently intelligent and may have hunted in groups, especially when trying to kill the larger sauropods. Some paleontologists have suggested that they were warm-blood, which would have supported their energetic lifestyles, but this claim remains controversial.

Herbivorous ornithischian dinosaurs also continued to evolve, though they were not so prominent as the sauropods and therapods. They existed in small- to medium-sized varieties and were natural prey for theropod predators. The best known of these dinosaurs is the armor-plated *stegosaurus*, which grew to some 9 m (30 ft) in length, 4 m (14 ft) in height, and weighed some 2,700 kg (6,000 lbs). The double-row of hard plates that extended upward along its spine were filled with canals for blood vessels and probably helped to regulate the animal's temperature as well as providing some measure of defense against predators. The stegosaurus also featured two pairs of long spikes on its tail for defense and the beaked mouth so characteristic of ornithischian dinosaurs.

Flying pterosaurs were common during the period, dominating the skies, but the late Jurassic possibly saw the first appearance of another type of flying animal: birds. As noted above, bird-like dinosaurs evolved from some species of early feathered coelurosaur, the feathers likely functioning to insulate the cold-blooded animal's body, keeping its heat in. *Archaeopteryx* has long been classified as a bird, but recent analyses suggest that it may better be classified as a transitional form between dinosaurs and birds. Archaeopteryx continued to share many therapod traits, including a bony tail, three fingers with talons (constituting the supports for the wings), teeth in the jaws, and certain other skeletal features. Yet, like birds, it was small, only about 50 cm (1.6 ft) long, and had wide, feathered wings. Whether it could actively fly or mostly glided is unclear. Other bird-like creatures also arose in the later Jurassic, most of them only recently discovered.

For a painting of a Mesozoic landscape, Jurassic period (by Publiphoto/Photo Researchers Inc.), including, e.g., stegosaurs, brachiosaurs, and allosaurs, see http://science.nationalgeographic.com/wallpaper/science/photos/jurassic-period/jurassic-landscape.

Mammals continued to be small creatures that lived in the undergrowth by day and were predominantly active after dark. They were a diverse group, with some eating only insects while others likely fed on a broader variety of foods.

THE CRETACEOUS PERIOD (145–66 MILLION YEARS AGO)

At the beginning of the Cretaceous Period,[3] Laurasia and Gondwana were already separated from each other by the Tethys Ocean and were in the early stages of breaking up into the smaller continents we know today. This rifting process continued throughout the period, with North America and Eurasia moving apart in the north, and South America and Africa doing the same in the south. The widening gulf between these continents gave rise to the Atlantic Ocean. As the continents moved apart east and west, Panthalassa grew smaller, yet would eventually become the Pacific Ocean. India, Australia, and Antarctica also rifted away from Africa and each other, and by the end of the period Antarctica was moving to its south polar position, Australia was still minimally joined to the eastern side of Antarctica, and India was a northward- and eastward-moving island in what would become the Indian Ocean. These continental movements would continue over the succeeding periods until the major landmasses reached the positions they occupy today.

3. See the Geologic Time Scale in the appendix.

For an image of the Late Cretaceous Globe by Ron Blakey and Colorado Plateau Geosystems Inc., see http://cpgeosystems.com/90moll.jpg.

The climate during the Cretaceous was generally significantly warmer than today. The global temperature for the period averaged about 18° C (64.4° F), or about 4° C (7.2° F) higher than modern levels. There was also only about half the temperature difference between the poles and the equator as has been typical in Earth's history. But temperatures were by no means uniform throughout the period. Early on, the climate continued the cooling trend seen at the end of the Jurassic, and snowfall was evidently seen in the higher latitudes, with some glaciers existing on mountains there.

As the period continued, Earth warmed considerably, until by the mid-Cretaceous temperatures were some of the highest in the planet's history. This was apparently caused by significant volcanic activity, owing at least in part to the shifting of the continents and the tectonic plates on which they rode. This volcanism produced massive amounts of CO_2, which filled the atmosphere and created greenhouse conditions. The temperature of the surface waters at the equator may have reached a stunning 42° C (108° F)—17° C (31° F) above modern levels—and averaged perhaps 37° C (99° F). Tropical to sub-tropical conditions existed well into the temperate latitudes, and dinosaurs roamed the forested areas and plains near the poles. At the end of the Cretaceous a new cooling trend began, though even then temperatures worldwide were noticeably warmer than today.

The combination of warming oceans and undersea volcanic activity caused sea levels to rise so much that a significant amount of the continental lowlands were turned into warm, shallow seas where life could flourish. As the continents split apart, more coastline was formed, further increasing the amount of shallow environments available for life's proliferation. The vast stretches of tropical to temperate climates meant that there were vast areas of vegetation, and this raised oxygen levels rose to about 30 percent of the air—about half again as much as today—and carbon dioxide dropped to some 6 times its pre-industrial amount.

The Cretaceous is notable for being a period in which many life-forms began to develop with more modern features. This is true of the many types of marine plankton that filled the seas and oceans, including *diatoms*, which made their first appearance. Particularly notable were the *cocolithophores*, algae that secreted *calcareous* (containing calcium, calcite, or lime) plates. So abundant were these small creatures in the late Cretaceous that their shells covered the shallow seafloors in many places, producing what would become deep layers of chalk, such as the White Cliffs of Dover in southern England. In fact, these chalk (Latin *creta*) deposits are what gave the Cretaceous its name.

The vast warm, shallow seas provided ample habitat for a variety of invertebrates. Clams and snails developed distinctly modern forms, while echinoderms such as sea urchins and sea stars (starfish) thrived. Coral reefs continued to expand, helped in part by new reef-building *rudist clams*. Ammonites were especially widespread and developed new forms in addition to the tightly coiled forms typical of the Mesozoic Era. Some spiraled outward to take on a more snail-shelled appearance, while the shells of other types straightened out in varying degrees. Some took on a J-shaped appearance, others looked like corkscrews or hairpins, and still others sported the kind of straight shells seen in more ancient nautiloids. They all ate other invertebrates that either floated freely or lived on the seafloor, and were in turn eaten by the larger vertebrate predators.

These predators included large numbers of sharks, rays, and teleost fish, of which hundreds of species existed, all of which took on modern features. The largest known of these ray-finned fish was *Xiphactinus*, a giant that reached as much as 4.5 m (15 ft) in length. Large marine reptiles, ichthyosaurs and plesiosaurs, continued at the top of the food chain throughout the Cretaceous, with some species reaching as much as 17 m (56 ft) in length and having long heads and jaws as large as 3.5 m (11 ½ ft). Late in the period these were joined by other large, formidable, vertebrate predators, the *mosasaurs*, which looked like monstrous snakes.

The early Cretaceous landscape looked much like that of the Jurassic and Triassic, with conifers, cycads, ginkgoes, and ferns dominating. The major novelty on land was the appearance of *flowering* plants and trees, or *angiosperms*, in the early part of the period and their progressive takeover of the land throughout the later phases. The chief difference between flowering plants and gymnosperms is their means of reproduction. Where gymnosperms mostly dropped their seeds, flowers achieved wider distribution by enlisting certain insects, such as bees, moths, wasps, ants, flies, and beetles, and some birds and bats, (and, to a lesser degree, the wind) to help them. Bees, for example, visit flowers and inadvertently pick up sperm-containing pollen grains, which stick to their heads, legs, and bodies. As the bee visits other flowers of the species, the pollen may rub off from the bee and attach to the flower, where it can fertilize the female ovary. This results in a seed containing the plant embryo, food to feed the growing plant, and a tough outer coat to protect it. In some plants and trees these seeds are contained within nuts or fruits, which are eaten by certain animals, which in turn excrete the seeds in another location.

Given the symbiotic relationship between insects, birds, bats, and flowering plants and trees, it is easy to see why these two different types of life *coevolved* (evolved different forms together). The plants produced various forms, flower colors, and scents to attract the insects, birds, and bats, and the insects, birds, and bats evolved bodily forms that enabled them to feed on the flowers' nectar. Today most of the 200,000 or so pollinators are insects. Insects appearing for the first time in the Cretaceous include ants, termites, some early moth and butterfly forms, aphids, grasshoppers, and certain wasps.

Flowers also have the useful ability to recover quickly from damage or being eaten. Early angiosperms were weed-like and relatively scarce, but by the middle of the period they were diverse and common, and by the late Cretaceous they constituted the majority of terrestrial plants and trees, with the latter replacing conifers as the predominant forest canopy tree. Leafy angiosperm trees included primitive forms of willows, magnolias, poplars, sycamores, figs, birches, maples, walnuts, and oaks. Some of these were the first *deciduous* trees—those that drop their leaves at certain times during the year. Palm trees also appeared late in the period.

Small terrestrial reptiles and some amphibians thrived during the Cretaceous, including snakes and lizards (which took on more modern characteristics), turtles, crocodiles, frogs, and salamanders. But it was the dinosaurs that continued to dominate all terrestrial vertebrate life and reached their greatest diversity in this period. Many dinosaur types evidence distinct regionalism. For example, sauropods were widespread on the southern continents but rare in the north, while certain therapods lived only in the north and others only in the south. Ornithischians lived on all continents except perhaps Antarctica, though not all species existed everywhere. The greatest diversity of dinosaurs appears to have developed late in the period.

Among therapods, the late Cretaceous *Tyrannosaurus Rex* was the largest form that inhabited the north, attaining a height of up to 4 m (13 ft), a length of up to 12.8 m (42 ft), and a weight probably between 5,500 and 6,800 kg (12,125 and 14,991 lbs). In the south, *Spinosaurus* was the dominant predator, at least in size, and was perhaps the largest therapod dinosaur ever to have lived, reaching as much as 18 m (59 ft) in length and as much as 20,900 kg (46,000 lbs) in weight. It had a long, crocodilian skull (which suggests that fish formed its main diet) and spines extending up perhaps 1.5 m (5 ft) from its vertebrae. These were most likely connected by skin, as with the Permian reptile *Dimetrodon*, to form a sail-like structure that was useful for regulation of body temperature and perhaps for mating displays. Smaller therapods likely scavenged food left behind by these monsters. One of these was perhaps *Velociraptor*, a small, late Cretaceous, feathered therapod that, while its overall length could reach to more than 2 m (6 ½ ft), stood only about as tall as a human's hips and weighed about 15 kg (33 lbs). Thus, they were in reality quite unlike the much larger and nonfeathered "velociraptors" pictured in the film *Jurassic Park*. They had relatively long claws on their fingers, birdlike arms, and a large, sickle-like claw on their feet for disemboweling prey. Hunting was thus clearly a part of their lifestyle, as was most likely scavenging.

Ornithischian dinosaurs existed in a variety of forms during the Cretaceous Period. *Iguanodon*, a large herbivore reaching as much as 13 m (43 ft) in length, arose during the early part of the period. It had arms roughly three-quarters of the length of its legs and hands with three relatively inflexible middle fingers, which permitted it to walk either on two legs or four. Instead of a thumb it had a conical spike, which presumably it used for defense, and its little finger was elongated and manipulable. Its feet had only three toes. Like most ornithischians, it had a beak for a mouth, behind which were numerous teeth, specialized for grinding down the vegetation that formed its diet.

Hadrosaurs were a highly diverse group of herbivorous dinosaurs that evolved from iguanodons and, like them, had long arms and so could move on either two feet or four. They are notable for having duck-like bills rather than beaks, as with most ornithiscian dinosaurs. Behind these bills were wide tooth plates with hundreds to thousands of small teeth to break down vegetation, and like iguanodons they had stiffened tails for balance. Most had elaborate crests on their heads, which were solid bone in some species and hollow and connected to their nasal passages in others. This latter feature would have enabling these hadrosaurs to produce loud, low-pitched calls. Most of them were around 9 m (30 ft) long, though some reached as much as 15 m (50 ft) and weighed perhaps 22,000 kg (48,500 lbs).

Another prominent and diverse ornithiscian group of the late Cretaceous was the *ceratopsians*, of which *triceratops* is representative. Triceratops walked on all fours and had the typical ornisthiscian beak, behind which were the typical plates filled with teeth specialized for grinding the low-lying vegetation they consumed. Their most notable features were the three horns on their faces, a relatively small one above the nose and two approximately 1 m (39.4 in) in length above the eyes. Triceratops had some of the largest skulls ever seen in land animals. They also had large, bony shields reaching outward from the tops of their heads. Other ceratopsians had different numbers and configurations of horns and different types of head shields, some with frilly edges. Triceratops grew to as much as 9 m (30 ft), and estimates of their weight range to as much as 12,000 kg (26,500 lbs).

Also interesting was another group of ornithiscians, the *ankylosaurs*. This was somewhat turtle-like in appearance, with thick armor knobs and plates made of bone that stretched in horizontal rows from its head to the tip of its tail. At the end of this tail was a massive bony club that was fixed at the end of seven fused vertebrae and may have had horizontal spikes protruding from it—a formidable weapon which it swung from side to side to crush bone and drive off predators. While standing only about as tall as an adult human, ankylosaurs grew to as much as 10 m (33 ft) long and weighed perhaps as much as 6,000 kg (13,000 lbs).

The Cretaceous Period saw the extinction of giant herbivorous sauropods such as brachiosaurus and diplodicus and their replacement by the even larger *titanosaurs*, which appear to have weighed as much as 100,000 kg (220,000 lbs). Featuring the typical long necks and tails and small heads of other sauropods, they were the dominant herbivores of their time, ranging across all continents.

Pterosaurs continued to rule the sky throughout the Cretaceous. One, *Quetzalcoatlus*, was a giant, with a wingspan estimated at perhaps 10–11 m (33–36 ft), making it one of the largest animals ever to fly. By the end of the Cretaceous, distinctly modern-looking birds were developing into a profusion of types, a number of which would be ancestral to the different varieties of birds known today. One example was *Hesperornis*, a flightless waterbird with small wings but strong legs and large feet that helped it maneuver in the water. Like modern cormorants, it dove for and captured fish, which it held in a beak at the end of its elongated jaws, aided by a row of teeth on its lower jaw that fit into sockets in its nearly toothless upper jaw.

For a painting of Cretaceous life, including maiasaurs (front left); tarbosaurs (top right), and pterosaurs (top center), the first flowering plants and one of the earliest mammal relatives (in the foreground), by Publiphoto/Photo Researchers, Inc., see http://science.nationalgeographic.com/wallpaper/science/photos/cretaceous-period/cretaceous-collection.

Mammals remained small and shrew-like in appearance and were minor players in the drama of life during the Cretaceous. Importantly, *marsupial mammals* appeared early in the period, differing from prior egg-laying mammals in that their young were born relatively undeveloped within the mother's pouch. The newborn had to crawl up across her abdomen to reach a nipple, where it would attach and complete its development before emerging into the world. Somewhat later in the period *placental mammals* appeared, the females carrying their young within a placenta until they reached full development, at which point they were born. It was these placental mammals that would become the dominant life-form on Earth following the extinction of the dinosaurs.

This mass extinction marked the end of the Cretaceous Period and Mesozoic Era and ushered in the Cenozoic Era,[4] which continues to the present.[5] This event is called the *Cretaceous-Paleogene (K-Pg) extinction* (formerly the *Cretaceous-Tertiary,* or *K-T, extinction*). It is believed to have been the third worst extinction event in Earth's history, killing some one-half of all life on Earth and up to 80 percent of animal species.

Many educated people are familiar with the theory that an asteroid crashing into Earth was responsible for the extinction of the dinosaurs, but the cause of the disappearance of so many life-forms is more complicated. The evidence for the asteroid theory is plentiful. Most paleontologists accept that an impact occurred, probably near the Yucatan Peninsula, where a crater 180 km (112 mi) wide dating to the end of the Cretaceous has been found. This impact would have sent massive amounts of dust and water vapor high into the atmosphere, blocking out sunlight for months, causing temperatures to drop significantly and plant photosynthesis to stall. As plants died, so did the herbivores that fed on them and the carnivores that ate the plant-eaters. In addition, widespread wildfires would have occurred (for which there is also evidence from soot deposited at the time), caused by the intense heat created by the impact, and this too would have taken a toll on plant and animal life. The energy released by the asteroid's collision would also have caused nitrogen and oxygen in the atmosphere to combine into nitrogen oxides, which, when combined with atmospheric water, would have formed nitric acid that rained down on the planet, causing yet another catastrophe for life on Earth. Once the particulate matter had finally settled out of the atmosphere, the water vapor that remained would have created a greenhouse effect, raising the planet's temperature quickly, further stressing whatever life remained.

Some half-million years before the asteroid impact, a massive volcanic event occurred in what would become India and Pakistan. It laid down some 10,000 cu km (2,400 cu mi) of lava and sent massive amounts of volcanic ash and CO_2 into the atmosphere, cooling the planet for some time and

4. The term "Cenozoic" (or "Cainozoic") is based on two Greek words, *kainos*, "new," and *zōē*, "life."

5. See the Geologic Time Scale in the appendix.

perhaps affecting ocean chemistry. Evidence of volcanic activity at roughly this same time, which would have exacerbated these effects, has also been found in southern Brazil.

The late Cretaceous also saw one of the worst oceanic regressions in the planet's history. As the sea levels dropped, the shallow seas that surrounded all the continents were replaced by dry land. This massive loss of prime habitat would have meant death to the wide variety of marine organisms that lived in these shallow seas. With the disappearance of these ecosystems, animals on land that consumed the marine creatures also would have disappeared.

Given these three occurrences, and also the fossil evidence that a large number of animals had already become extinct before the end of the period, most paleontologists today seem to favor a scenario in which environmental degradation had already caused the planet's life to begin declining and that the asteroid impact would have been the final blow in bringing about the massive extinction.

On land, the disaster wiped out not only the dinosaurs—herbivorous, carnivorous, and flying types—but most other herbivorous and carnivorous animals as well. A good amount of plant life survived, perhaps because their seeds could lie dormant until more favorable conditions returned. In the oceans, the damage was worst in tropical waters, where many varieties of plankton were substantially reduced, as were many seafloor inhabitants, especially bivalves such as mollusks and brachiopods, and certain corals and fish. Generally, those animals that lived on the seafloor did much better than those that lived in the waters above. Ammonites went extinct, as did all the large marine reptiles, including the ichthyosaurs, plesiosaurs, and mosasaurs. Those animals that survived, both on land and in the seas, were mostly those that fed on insects and rotting plant and animal matter or the predators that fed on those animals. Thus, mammals, birds, snakes, amphibians, turtles, and crocodiles survived largely intact and became the main repopulators of the devastated world. And as the Permian extinction meant the end of the dominance of amphibians and the beginning of the long reign of reptiles throughout the Mesozoic, the Cretaceous extinction brought to a close the reign of the reptiles and ushered in the still-continuing domination of the mammals, culminating in the rise of the most dominant of the mammals, humankind.

THE PALEOGENE PERIOD (65.5–23 MILLION YEARS AGO)

Throughout the Paleogene Period,[6] the continents continued their gradual movements toward their current locations. South America drew nearer to North America and both continued their westward movement as the Atlantic Ocean continued to widen at the rate of several centimeters per year. The inland seas covering North America dried up early in the period, unifying the continent. The island of India continued its northward movement toward Asia, finally colliding with that continent and raising the Himalayan Mountain range. Africa also moved northward toward Europe, finally connecting and forming the Mediterranean Sea between them. Australia completed its separation from Antarctica, with the former moving north toward Southeastern Asia and the latter moving south until it assumed its current position straddling the South Pole. All this movement meant that widespread volcanic activity continued during the period.

For an image of the Eocene (Paleogene) Globe by Ron Blakey and Colorado Plateau Geosystems Inc., see http://cpgeosystems.com/50moll.jpg.

Climate during the Paleogene varied wildly. At the beginning, temperatures warmed to Cretaceous levels, due to the large amount of carbon dioxide in the air and consequent greenhouse effect. Around 55 mya (10 million years into the period), temperatures on the surface of the seas rose an estimated 5–8° C (9–14° F) over a period of only a few thousand years—one of the most rapid temperature spikes in the history of the planet. Even in arctic latitudes, sea surface temperatures rose to an estimated 23° C (73° F), causing the extinction of a variety of sea and land animals and the appearance of new species as well as the growth of palm trees and guavas in the far north. Following this temperature spike, Earth began to cool and become drier, apparently because new currents formed in the Antarctic Ocean (also called the Southern Ocean) once Antarctica had broken away from Australia and moved southward. This cooled ocean temperatures significantly and trapped cold air over the South Pole. Temperatures worldwide dropped more than 10° C (18° F), causing ice sheets to form on Antarctica by 37 mya and vast, densely forested areas to become sparser woodlands and savannas. This spurred the evolution of new animals adapted to these new conditions. The temperature warmed somewhat over the next 20 million years or so

6. See the Geologic Time Scale in the appendix.

before cooling again. This time the cooler and drier climate remained, creating the conditions typical of our world today: permanent ice sheets at the poles and a wide variation in temperatures and moisture levels in between. It is estimated that the mean surface temperature for the Paleogene was about 14° C (57° F), or about what it is today. The oxygen content of the air was slightly higher than today (some 8 percent more) while carbon dioxide levels were roughly equal to those at the start of the Industrial Revolution.

Life required a substantial amount of time to rebound from the late Cretaceous / K-T extinction and fill the planet's open ecological niches. Initially, plant and tree forms continued the Cretaceous trends, due to the very warm global temperatures, with tropical and subtropical forms predominating. But once the cooling and drying trend began, these forms became common only in equatorial latitudes. Vast new temperate zones were cooler and drier and experienced greater seasonal temperature variations, causing, as already noted, dense tropical forests to disappear and more open woodlands and savannas and prairies to appear. In the process, deciduous shrubs and trees diversified greatly, filling the new cooler and drier environments. Flowers continued to evolve new forms and fill the various climatic niches, continuing their take-over of the planet. Notably, the first grasses evolved from flowers and quickly filled the dry prairies and savannas worldwide.

In the seas, microplankton recovered quickly, and more modern forms of those animals that survived the K-T extinction appeared, such as new corals, clams, sea snails, sea urchins, and bivalves. Teleost fish took over the niche previously occupied by ammonites, as did squid and other forms of cephalopods. Sharks expanded their predatorial dominance in the absence of the gigantic reptiles of the Mesozoic Era, with one form, *Carcharocles megalodon*, growing to as much as 17 m (55 ft) in length. Generally, though, life in the oceans evolved only modestly during the Paleogene.

The major exception was the evolution of new mammalian forms that returned to the sea and were significant predators. The first of these was the *whale*, with the earliest known proto-whale fossil, *Pakicetus* (which evidently shared a common ancestor with today's hippopotamus) appearing around 53 mya. The earliest fully aquatic whales probably arose a few million years later. They were toothed predators whose limbs had turned to flippers, with the hind flippers eventually disappearing and their limbs becoming internal structures no longer involved in swimming. They developed vertical tail flukes and thick layers of fat, or *blubber*, to insulate them from the increasingly cold waters later in the period.

Toward the end of the Paleogene, some whales developed a new, filter-feeding system in which their teeth were replaced by plates of bristle-like baleen. These plates vary greatly in length depending on the size of the whale species, reaching as much as 3.5 m (11 ft) in length and 90 kg (200 lbs) in weight. The whale feeds by opening its mouth and taking in water filled with small animals and then pushing the water back out through the baleen, capturing and then swallowing these small creatures.

At the very end of the Paleogene, other mammals, the *pinnipeds*, took to the sea, including seals, sea lions, and walruses. The earliest known pinniped was the bear-like *Puijila darwini*, which probably lived in the arctic. It had thick limbs with flattened hands and feet and probably webbing connecting its digits.

On land, as in the seas, most life-forms that survived the K-T extinction evolved slowly during the period. Reptiles were generally much smaller, their gigantic forebears having disappeared in the great extinction, and these included lizards, turtles, and snakes, with crocodiles being the largest form. Alligators appeared for the first time in North America. Amphibians and invertebrates continued to evolve. But the most significant changes arose within the bird and mammal lineages. The extinction of the pterosaurs left the skies vacant, and their cousins, the birds, were quick to diversify and take their place, assuming relatively modern forms by period's end.

Mammals began their terrestrial reign in the Paleogene—occupying, like birds, the niches formerly ruled by the dinosaurs, just as reptiles had earlier supplanted amphibian dominance during and after the Permian. As we have seen, some mammals took to the seas; others, bats in particular, evolved and took to the air in the middle of the period. But it was on land that the greatest diversification developed, and it did so early in the period. Not only did mammals occupy the new niches created by the changing climate in the Paleogene, but the final separation of the continents (and islands) from each other permitted evolution to take various routes in the independent ecosystems. Thus, for example, North America was dominated by placental mammals while in South America and Australia it was marsupial mammals that spread across the continent, with similar looking forms occupying similar niches in all three places.

As the climate cooled and temperate forests and plains appeared, modern hoofed herbivores appeared, such as horses, deer, camels, rhinoceroses, and elephants, which became more specialized over time as they adapted to the changing conditions. New mammalian predators likewise developed to feed on the herbivores, including the first bears and wolves with their better brains and eyesight. Interestingly, one of the largest predators was a bird that stood some 2.4 m (8 ft) tall. These predators spurred further adaptations over time in their prey—such as, among horses, longer legs for greater speed and much larger body size. The largest land mammal ever arose in the middle of the Paleogene, the herbivorous *hornless rhino*, which grew to 5.5 m (18 ft), dwarfing all elephant species then and to come. By the middle of the period most of the modern mammal orders had evolved. Around 55 mya the first primates appeared as lemur-like tree-dwellers in tropical jungles—the first forms of that mammalian lineage from which our human ancestors would ultimately evolve.

For an image of Paleogene-Eocene fauna, see http://i1-news.softpedia-static.com/images/news2/Neozoic-the-Era-of-the-Mammals-4.jpg. Credit: anthropology.si.edu.

THE NEOGENE PERIOD (23–2.58 MILLION YEARS AGO)

During the Neogene Period,[7] the continents continued their slow movement, and by the period's end were in essentially their current positions (where the motion that brought them to their new locales continues). The Atlantic Ocean grew while the Pacific Ocean shrank, and South America continued to inch toward North America, finally connecting to it through the formation of the Isthmus of Panama some 3.5 mya. Africa moved northward, merging with Europe and putting an end to the Tethys Sea while creating the Mediterranean. Italy pushed north into southern Europe, creating the Alps, and Spain's northward movement caused it to collide with France, creating the Pyrenees. India continued its northward movement into the Asian continent, pushing the Himalayas ever higher, as it still does today. Tectonic forces in South America created the Andes mountain chain and in North America the Rocky Mountains, Sierra Nevadas, and Cascades. The accumulation of snow and ice at the South Pole and atop these new mountain ranges removed water from the oceans, lowering sea levels and so exposing land bridges between Asia and Alaska and, in Western Europe, between North Africa and Spain. This led to the Mediterranean Sea drying up temporarily, refilling, and then repeating this cycle several times late in the Neogene.

For an image of the Neogene-Miocene Globe by Ron Blakey and Colorado Plateau Geosystems Inc., see http://cpgeosystems.com/20moll.jpg.

7. See the Geologic Time Scale in the appendix.

Climate during the Neogene began with a warming trend but soon began to cool, remaining relatively warm overall though with significant perturbations until the end of the period. Then temperatures dropped sharply, bringing an ice cap to the North Pole and initiating the ice ages that would continue into the Quarternary Period that followed.[8] This cooling, caused in part by the closing off of the equatorial seas between Africa and Europe and North and South America, along with the rise of new mountain chains, caused a drier and more seasonal continental climate to prevail in the latitudes north and south of the tropics, much as it does today. Grasslands underwent a significant expansion at the expense of forested land. The average temperature during the period was roughly what it is today, as were also the atmosphere's oxygen and carbon dioxide levels.

In general, one can say that life in the seas and on land during the Neogene was quite modern in form. In the oceans, brown algae, or kelp, appeared in cool, shallow places, creating an entirely new habitat that was populated by varieties of invertebrates, fish, and a new marine mammal, the otter. Some whales, such as the sperm whale, evolved into modern forms, and during this period whales attained their greatest diversity, one that is greatly reduced today. Pinnipeds, including seals, sea cows, sea lions, and walruses, became increasingly adapted to life in the seas. Overall, however, the variety of marine mammals was dwarfed by their terrestrial cousins. Sharks remained one of the top ocean predators, with the huge megalodon continuing as the largest and most fearsome of all.

On land, the patterns of plant life changed significantly, largely continuing trends begun in the Paleogene. Gradually the world's climate became cooler, drier, and more seasonal, causing tropical ecosystems to contract to a relatively narrow band on either side of the equator. In their place, deciduous and, especially in the northern latitudes, coniferous forests arose, with vast grassland areas on all continents except Antarctica, some deserts in Africa and Asia, and in the far north expanses of tundra. Grasses diversified greatly, with a host of new species appearing during the period and coevolving along with the diverse herbivores and grazers that consumed them. Toward the end of the period cycads began to evolve new forms after declining for long periods due to climate changes. By the end of the Neogene, some 95 percent of today's seed plant families were in existence.

Birds and mammals continued to diversify substantially on land, while other terrestrial creatures remained largely as they were in the Paleogene. Among reptiles, members of the crocodile family evolved several new forms, including giant crocodiles that appeared in India and South America, while

8. See the Geologic Time Scale in the appendix.

in Europe crocodiles and alligators died out as the climate cooled. New venomous snakes, including rattlesnakes, appeared along with the new forms of rodents and birds that were their prey. Giant tortoises continued to do well in North America, diversifying somewhat, while South America saw the rise of large piranhas in freshwater habitats.

The Neogene saw the appearance of most, if not all, modern species of birds that had not previously appeared. The new forms included ducks, owls, plovers, cockatoos, crows, grouses, and auks. Large birds that preyed on smaller ones encouraged the latter to evolve quick, agile forms in order for them to survive. Late in the period, after the Panamanian land bridge appeared, the enormous South American *Titanis* invaded North America. There also it was one of the top predators alongside the large carnivorous mammals. Titanis stood more than 2.5 m (8 ft) tall, weighed some 330 lbs (150 kg), and likely had a large skull and powerful beak for tearing apart its prey. Its small wings did not enable it to fly, but it has been estimated that titanis could run at speeds up to 40 mi/hr (65 km/hr).

Along with birds, mammals continued to be the dominant terrestrial vertebrates and, like the birds, evolved many new forms during the Neogene. As the cooling trend continued, more forests gave way to grassy plains and savannas. Just as grasses diversified, so did the herbivores that fed on them. *Ruminants* appeared—mammals that begin the digestion of these grasses by first chewing and swallowing them, then softening them further in the first of their four stomach compartments, and then regurgitating the semi-digested food (called *cud*) and chewing it once again (or ruminating) to further break down the grass. Eventually the food is passed through the second, third, and fourth parts of the stomach, where it is further broken down before passing into the intestines. Examples of Neogene ruminants include bison, sheep, antelopes, camels, and giraffes. Non-ruminant grazers, such as horses, continued to evolve new adaptations appropriate to life on the plains. Many traveled in herds for added protection for themselves and their young. *Mastodons* and *mammoths* also appeared in the later Neogene, as well as more modern forms of wolves, raccoons, beavers, and deer.

With the separation of the continents came the development of distinctive animals particular to each area over time, though the land bridges that joined North America to South America and Africa to Eurasia caused some intermingling of species. In the late Neogene North America, mastodons, opossums, and rodents did well, as did such carnivores as bears and weasels; but hoofed animals, including camels, deer, and horses, were in decline. Several animals, such as rhinos, certain canine species, and three-toed horses, went extinct. A number of grazers from the Paleogene likewise vanished. After the land bridges with Eurasia and South America formed, elephants and rhinos migrated into Europe from Asia, and armadillos, porcupines, certain rodents, and large ground sloths moved north from South America.

In Eurasia, rodents also flourished, as well as elephants and other elephant-like herbivores such as mammoths, cows, tapirs, rhinos, and antelopes. Horses declined in diversity, as did primates. New types of predators, such as early saber-toothed cats and hyenas, appeared alongside the canines, bears, and weasels already flourishing. Some camel species migrated in from North America, while elephants and apes came north from Africa.

In Africa, hoofed animals dominated. Elephants, antelopes, and cows diversified. Horses, giraffes, and modern forms of rhinos appeared, while camels came late in the period, having migrated from North America through Asia. Pigs, rhinos, saber-toothed cats, and rabbits migrated in from Eurasia. North American predators, including bears, canines, and weasels, also appeared, joining various cats, hyenas, and cat-like *civets* at the top of the food chain. Primates continued to evolve, and from them the early proto-humans.

Australia and South America, both islands isolated from the other continents, were the two places where animal life was widely divergent and marsupial mammals dominated the scene. In Australia, herbivores such as kangaroos and wombats were hunted by several cat-like and dog-like marsupial predators. Rodents arrived, evidently from Asia, and the platypus also appeared during the Neogene.

South America was also dominated by marsupials prior to the formation of the Panamanian land bridge. (Current theory suggests that Australian marsupials originally came from South America via Antarctica when these landmasses were joined in the supercontinent Gondwana.) Hoofed herbivores were plentiful, with marsupial equivalents of placental hippos, giraffes, camels, mastodons, antelopes, and horses roaming the plains. Marsupial predators resembled wolves, hyenas, and other placental types. Some placentals coexisted with these marsupials, including sloths, armadillos, and ancestors of anteaters. Once the land bridge formed, though, a number of North American placental species migrated south and eventually replaced many of the native marsupial and other mammalian forms (see the QR code below). Invading herbivores included camels, horses, deer, tapirs, and various rodent species. Immigrating carnivores included large members of the cat family, such as saber-toothed cats, pumas, and jaguars, along with bears, rodents, and members of the wolf, weasel, and raccoon families. One, the short-faced bear, *Arctotherium*, evolved into the largest known terrestrial mammal ever, weighing perhaps as much as 1,600 kg (3,500 lbs). Some members of the cat family likewise grew to enormous size in the

South American context, such as *Smilodon populator*, at some 500 kg (1100 lbs) the largest feline ever to roam the planet. These and other migrants quickly came to dominate their South American equivalents, driving most of them to extinction. South American fauna that survived were especially small marsupials and the very largest herbivores.

For Jay Matternes's painting of [Neogene-]Miocene fauna of North America, on a mural made for the Smithsonian Museum, showing Moropus, Daphoenodon Palaeocastor, Stenomylus, Parahippus, Daeodon, Merychyus, and Promerycochoerus, see http://commons.wikimedia.org/wiki/File:Miocene.jpg.

The Neogene was also the period of significant primate development, which would eventually lead to the rise of the great apes we know today and the earliest clearly identifiable forms of the lineage many think led to human beings, the *Australopithicines*. A more detailed description of the evolution of humans will be our focus in the chapters 18 and 19.

THE QUATERNARY PERIOD (2.6 MILLION YEARS AGO TO THE PRESENT)

During the short Quaternary Period,[9] the continents drifted perhaps 100 km (60 mi) along their previously-established trajectories, meaning that throughout the period the world has looked essentially as it does today. Geologists date the beginning of the Quaternary at the beginning of the recurring ice ages that persisted throughout most of the period. In fact, the Quaternary is divided by geologists into two epochs: the *Pleistocene Epoch*, which is the period of the ice ages and extends all the way down to 10,000 years ago; and the Holocene Epoch, which covers the period after the ice ages to the present and which is termed an *interglacial* time. The climate of the Quaternary was thus quite variable, switching back and forth between the cold ice ages and the warmer interglacial periods.

The cause of these ice ages seems to be the slight periodic shifts of Earth around its axis and the planet's elliptical orbit around the Sun, bringing it alternately closer to and further away from the Sun and its warmth. By about 800,000 years ago, a pattern had fallen into place in which each ice age lasted around 100,000 years and was followed by interglacial periods of roughly 10,000 to 15,000 years. During the ice ages, sheets of ice would

9. See the Geologic Time Scale in the appendix.

move out from the poles, reaching as far south as 40° latitude in North America, Europe, parts of Asia, and as far north as parts of South America. The greatest coverage was thus in the Northern Hemisphere. It has been estimated that at its farthest reach the ice sheets covered some 30 percent of Earth's surface and ranged between 1,500 and 3,000 m (about 5,000 and 10,000 ft)—that is, as much as two miles—thick. Stretches of permafrost up to a few hundred kilometers (several hundred miles) long reached outward from the base of the ice sheets, and below these large lakes formed from the run-off from the melting edges of the glaciers. Rivers also experienced substantial increases in flow.

The massive amount of ice covering Earth decreased global sea levels as much as 100 m (400 ft) or more, exposing shallow continental shelves and so temporarily increasing the size of the continents. Land bridges, such as the one presently submerged between North America and Asia, appeared, permitting animals and humans to migrate back and forth. Each change from glacial to interglacial periods brought changes of wind and water current patterns: with the colder global temperatures came a drier and, in places, windier climate, leading to an increase in the extent of deserts worldwide. At the end of the last ice age, roughly 10,000–12,000 years ago, sea levels rose once again, and the continents assumed their current appearance.

With each advance of the glaciers, earth and rock was pushed south, and with each retreat a new landscape emerged, including such features as the fjords of Scandinavia and the Great Lakes of North America. The effect of these changing global conditions obviously had an important impact on Earth's living things. As the glaciers expanded, parts of the tropics cooled and dried out, turning into seasonal forests or savannas, changing not only the plant life but also what animals could live there. During interglacials, the tropical forests expanded again, returning their flora and fauna to their former range.

Flowering plants, insects, and mammals continued their domination of the land. Few new animals evolved during the period, though some large mammals, such as mammoths, mastodons, oxen, and rhinos, grew larger and developed thick, hairy coats that allowed them to live close to the ice sheets. When the ice receded for the last time around 10,000–12,000 years ago, these large grazers went extinct, along with some predators, such as the saber-toothed cat. Whether this extinction was from the warming temperatures, predation by humans, or a combination of both remains an open question. Other large mammals survived in various places, such as elephants, rhinos, and hippos in Africa, while some disappeared from regions where they had previously flourished, such as horses, camels, and cheetahs in North America.

On the whole, life in the oceans was little affected, with whales and sharks at the top of the food chain, otters, seals, sea lions, walruses, squid, and certain fish below them as both predators and prey, and smaller crustaceans, sea urchins, plants, and microscopic plankton sitting toward or at the bottom.

The major evolutionary news during the Quaternary Period was the continuing evolution of primate species, from the Australopithicenes down to the *Homo sapiens*. In fact, the Quaternary is sometimes called the "Age of Humans." The evolution of humankind from its primate ancestors to our own species will be the focus of the next two chapters, our last concerning the history of the universe.

17

God's Interaction with the Universe and Its Creatures

My ultimate goal in the theological chapters in this book has been to lay the groundwork necessary to address what is most relevant to the topic of this book: how God can be understood to interact with the world and the way this interaction permits us to think of God as Creator. To do so it was first necessary to establish the fundamental Christian claim that God is Perfect Agapic Love and to note its implications for the other traditional Christian doctrines about God, including the nature of God's action in the world with regard to creatures, which is persuasive. We are now at the point where we can address more fully the precise nature of God's interaction with the universe, both with its creatures and with the conglomerate entities that constitute the rest of nature.

If God interacts with humans by means of *loving persuasion*, this must occur at some level of our mental lives, i.e., at either the conscious or subconscious level. As noted in chapter 3, I believe that this divine-human interaction is a subconscious process that goes on with all people, though some specially gifted people may be able to bring this interaction up to the conscious level, at least on occasion. This ability might, for example, be the means by which Jesus and others have been able to express their revelations from God to those of us who have not been able to bring our experiences of the divine into consciousness.

If most of our experience lies below the level of consciousness, then this is the probable place where our minds are aware in some fashion of the presence and communication of God as part of their experience. In understanding how this works more specifically I follow the analysis of human experience laid out by the twentieth-century philosophers Alfred North Whitehead (1861–1947) and Charles Hartshorne. In their understanding and in mine, in each moment of our existence we experience a myriad of influences upon us from everything in our environment, each of which exists as an option for focus, and thereby influence, in that moment. The greatest practical influence on us will be who we were in the previous moment and, by extension, who we have been in our previous moments of existence as we remember them subconsciously.

Our tendency, as noted in the first chapter, is to behave in the present as we have behaved in the past as long as that still works for us, i.e., as long as we do not suffer for that choice at some point. This, of course, is the power of habit. Nonetheless, all of the other influences we experience in that moment also offer behavioral options for us, and one of these is God. God offers in each moment the behavioral option that would be the most loving for ourselves and those around us and seeks, by the attractive power of such love, to persuade us to choose that option. Whether we choose God's option or some other is up to us in our freedom; and once we have made our decision and chosen what we will do, and so who we will be in that moment, that moment of our existence ends. Then it is on to the next moment, which repeats the process anew. Thus, each moment of our lives is involved in decision-making regarding what we will do and so the kind of person we will be (though most of these decisions will be subconscious), with God always seeking by means of love to persuade us to choose that one option which would be best for ourselves, other people, and all creatures.

It makes sense that we are influenced by the physical things around us, since we are physical beings ourselves, but physical things are not the only influences on us. Our *memories* of the past, which give us our sense of continuity, of being the same person throughout the many moments of our lives, also influence us, as do our *hopes* for the future regarding, ideally, who we might become and the kind of world we would like to help create. This suggests the means by which the Spirit that is the Loving God might communicate with us—from the divine Mind to our minds.

It is difficult to say how this might occur. Perhaps it is an inherent property of mind that it can communicate with other minds (via some sort of *telepathy*, which some people seem to accomplish to some extent but which is an ability we have yet to learn to employ universally and well; perhaps telepathy will be the direction toward which future human evolution moves), or at least it is an ability of the divine Mind/Spirit. Or perhaps a physical connection is required, one linking brain and mind.[1] Some scientist-theologians have suggested that God might communicate with us based on how God influences what happens at the quantum level in our brains. It might be that God in some way—perhaps by "observing" a particular quantum state in a certain atom of the brain—causes it to take a particular pathway that begins a cascade of causes which, according to the laws of chaos theory, are amplified to a point where they influence neurons to fire in a particular way so as to create a thought that is a communication from God. This then becomes, as we have seen, one influence among others for our subsequent decision as to how we will behave in that moment. It may be that there are other pathways that God might use in communicating with humans subconsciously that are unknown to us today. If we accept that it is appropriate to think of the world as in some sense God's body, as I do, then it seems reasonable to believe that as we come to understand the mind-brain/body connection in ourselves, we will find a likely analog for the way in which God communicates with us and the other creatures in the universe.[2]

The next question that inevitably suggests itself is how far down the scale of physical being "mind" (and so freedom) in some form exists. The general tendency among scientists and laypersons alike in the modern era has been to underestimate what animals other than ourselves are capable of doing, assigning their behavior to *instinct*, by which they mean a simple efficient causation of stimulus and response. This has produced a strong sense of human uniqueness. But evolutionarily we cannot presume such impermeable lines; in evolution, each stage of whatever we are investigating arises out of some former stage. Nothing appears suddenly without precursors that have preparatory anticipations of what will later evolve. Thus, evolutionary thinking must argue that things such as freedom, creativity, interiority, and the like do not suddenly appear with humans, but that they exist in the

1. The issue of the nature of mind and its relation to the brain is complex and beyond the scope of our concern here. I simply affirm that I, like many, conceive of mind as an immaterial "something" that exists in a two-way relationship with the body, such that the mind is able to influence the body and the body is able to influence the mind (i.e., that both top-down and bottom-up causality exist). This should be obvious to anyone who has ever intentionally moved his or her body to do something, such as raise an arm (mind influencing body) or who has ever felt the mental disorientation that certain illnesses bring (body influencing brain). What this mind-body connection is and how it works no one currently knows, though it is an area of intense research and should at some time in the future give up its secrets.

2. There seems to be a progressive relationship between the complexity of a creature, the "amount of mind" it has, and the degree of freedom it can exercise, such that more complex creatures have more mind and more freedom. This will become clearer in what follows.

creatures from which we evolved, albeit to a lesser degree. Thus, it makes sense to look at our own abilities and make the judgment that these things also exist in the animals from which we evolved, though in attenuated form.

This line of thought is supported by the fact that things that were once considered uniquely human are for the most part no longer seen that way. *Primatologists* know that chimps and other primates have conscious minds (though whether they are self-conscious as we are remains uncertain). Anyone who has owned dogs or cats knows they are individuals possessed of their own distinct personalities, and personalities emerge from their possessing minds and the freedom to make certain choices in behavior where others of their kind typically make different choices. We also know that, like us, they remember things, such as when it is time to be fed, or walked, or where they hid food, and also that they enjoy certain experiences, such as play. Songbirds display creativity in their songs and seem to sing for, among other things, pleasure.

How far down the chain of being can we take this progression of attenuated human behaviors? If the relationship between complexity of creature, mind, interiority, creativity, and freedom holds, as I believe it does, then one might well say that these things are present all the way down to the simplest forms of life. Clearly, at that level of simplicity, e.g., single-celled creatures, there is nothing like a thinking organ (e.g., a brain), but this need not mean that the organisms do not experience their world and have some very attenuated sense of mind by which they react to their world with some degree of freedom. For example, even single-celled amoebas interact with their environments, knowing when they have been exposed to light and moving away from it into darkness. The way they move into the darkness is not the same in each instance, suggesting perhaps some freedom regarding how they choose to move toward darkness. This question of the degrees to which attenuated forms of human behavior might be found in simpler creatures is one that scientists might fruitfully begin to investigate.

We can press this question even further. Is it possible that freedom and so "mind" in some sense goes even further down the scale of creaturely being, into the realm of nonliving entities? The reason for even posing such a seemingly outlandish question is the fact that, on the subatomic scale, particles of matter do not behave deterministically, as we have seen in our study of quantum mechanics. Rather, things like electrons behave probabilistically—their motions can only be predicted according to probabilities but not known in advance, as can be done with conglomerate things such as rocks. Certainty exists in the realm of determinism; probability exists only in things possessing at least some degree of freedom.

As anyone who has studied psychology knows, human behavior is described by bell curves, in which the width of the bell's base represents all possible behaviors given a certain stimulus. Primates, too, could have their behaviors charted according to bell curves, as could dogs, cats, and other simpler animals. What one would see is that the less complex the creature—i.e., the less mind and freedom it possessed—the narrower the base of the bell curve would be. But it would still be a bell curve. So if the behavior of quantum particles is indeterministic, then it too can be graphed as a bell curve, though one with so narrow a base that at anything less than extreme magnification it would appear as a single perpendicular line (which would indicate determinism). Quantum particles would thus have extraordinarily little freedom, but not none at all. As Geoffrey Chew has suggested, we might think of the behavior of such subatomic particles as "localized impulses."[3] And if a small amount of freedom exists for such particles, then perhaps mind in some unimaginably attenuated form might also exist in them.

Would things such as rocks and hammers, then, have some amount of freedom and mind? No. Only individual entities possessing an integral being, some controlling center of experience, such as the subatomic particles we have been discussing, would have some freedom. Rocks and hammers—and seemingly plants—are only conglomerates of simpler particles, each with some freedom but operating individually, not in concert with one another as an ordered whole. If freedom is found at the level of the subatomic particle, then, as suggested above, it is here that God might have some influence, as well as at all the levels above them where an integrated whole, such as a cell, or an animal, or a human, exists. With regard to mere inanimate things such as rocks and hammers—or suns, planets, or galaxies—God would have no influence simply because there is no integral entity there to influence. Such things take the course they do based on deterministic physical laws alone.

What Whitehead and Hartshorne proposed is that every individual entity interacts with every other entity in its environment, seeking "enjoyment" or "satisfaction" in whatever form that is possible given its nature. For simpler creatures, that enjoyment is momentary gratification; for more complex creatures, that enjoyment can come in much more varied and less immediate ways, as with humans, where enjoyment can come from anticipations of future enjoyments and from abstract things, such as the flourishing of a cause we believe in, or reading a book, hearing music, or observing artistic creations. Creativity and the experience of maximum intensity and enjoyment are sought simply as ends in themselves. In this sense one could

3. See Eastman and Keeton, eds., *Physics and Whitehead*, 85.

say that aesthetic enjoyment is the fundamental goal of all action by all creatures.

In chapter 7, we noted that the Christian doctrine of creation (as well as that of Judaism and Islam) claims most fundamentally that it is God who causes the universe to exist, in the sense that without God's sustaining will and power all physical reality would descend into chaos. This is because it is God who establishes the ultimate laws of nature by which the universe exists and proceeds—including laws that determine how universes propagate, if that is how the universe operates—and holds those laws in being, so that there is cosmos, an ordered universe, and not chaos.[4] And it is because of the nature of God as Perfect Agapic Love, one who seeks to create creatures capable of enjoying life and engaging in loving relationships with other creatures, that the Divine Spirit engages in the creation of a cosmos at all.

At this point, and with this theological preparation, we can now begin to consider what God has been doing in the history of our universe and our planet. Given the nature of God's power as persuasive and the miniscule freedom that exists in subatomic particles, atoms, and the like, it would seem that through much of the early history of the universe there was little that God could accomplish. The natural laws would have determined almost entirely what transpired as subatomic particles formed with the cooling of the earliest universe and eventually combined into the atoms of hydrogen and helium that virtually alone constituted the universe for many millions of years. Following the births and violent deaths of the first stars, new elements were created which could then foster the more rapid birth of second- and, later, third-generation stars.

Around some, if not most, of these stars, planets formed and gradually stabilized, with some located in orbits that permitted liquid water to exist and life to appear. Here God would likely have worked to influence the chemical building blocks to draw together in unique new ways so that they might interact and proceed along the pathway that led to life. The laws of nature in our universe at least permit the development of life, which is obvious, but I understand them to do more than this: I believe they actively encourage life to exist and then to develop increasingly complex forms, and this not only because of the relative rapidity with which life began on

4. It may be that God determines the laws of nature in each universe individually (if more than our universe exists) or that God permits each universe to have different laws not directly created by God but spun out by the process of universe-creation, whose laws God established in the infinite past. If the latter were true, it would mirror God's action in evolution (which we will presently discuss), in which God does not *control* evolution but interacts persuasively with the creation—meaning that the biological world participates in its own creation—just as God does with the creation of individuals within the creation.

Earth but also because this was logically part of God's intention in creating a cosmos.[5]

Once life had begun on Earth, God desired the evolution of new and more complex forms of life, with every creature *enjoying* its life, whatever that might mean for simple organisms. God worked to persuade each one to behave in such a way as to make such enjoyment for self and others possible. But God has also sought, working through the process of evolution, to draw forth novel and ever more complex creatures, knowing that with greater complexity comes the possibility of greater enjoyment (though also greater sorrows) for the creature and for God. The miniscule freedom of single-celled organisms minimized God's influence early on, which is why it took so many millions of years before multi-celled organisms appeared. Once they did, God had more complex creatures with which to interact. God then worked with and through the seemingly random mutations[6] that occur genetically in creatures, working also with whatever freedom exists in each entity for its own benefit and for the benefit of others. Such divine influence would likely be undetectable—or at least is undetectable at present, given our current scientific capabilities.

The genius of evolution as a strategy for the creation and development of life should not be undervalued, as it is by those conservative religious people who deny it as God's means of creating. Because God does not control

5. Several decades ago, physicists came to the conclusion that we live in an *anthropic universe*—that is, one whose natural laws must be very much like they are or life could never have evolved anywhere in the universe. Indeed, some of the particles and forces we noted in chapter 4 have been found to have had to exist within a miniscule fraction of a percent of what they actually are or the universe would have been forever a barren place. With no apparent reason why these particles and forces must have had precisely the natures that they do, some have calculated the odds of all the constituents of our universe having precisely those values they do at trillions to one against it. Such long odds have suggested at least three possible options to those engaged in this work: either there are trillions of universes that exist and we happen to live in the one (or one of the few) anthropic universes among them; or that in the future science will discover that the laws of nature must have been just what they are; or that only our universe exists and, given the odds against its laws being anthropic, they must have been designed by God so that life would exist. Any of these answers seems possible to me, though with my belief that an infinite number of universes have existed and may currently exist, the value of our anthropic universe as an argument for the existence of God is not particularly strong. I do not, however, expect that there would exist trillions of sterile universes with only one or a few bearing life. I would expect that God, who established at the least the laws by which universes come to be and are propagated, would establish such laws that, as with worlds in our universe, so in these other universes life would typically be abundant.

6. Genetic evolution certainly appears random, but, as any knowledgeable mathematician will confirm, randomness cannot be proven.

things like large asteroids that occasionally pummel Earth, or massive and long-lasting volcanic eruptions that foul the air and shut out the life-giving light of the Sun, God uses evolution to heal their most destructive result: the depletion of life at all levels. As we have seen, life on Earth has on several occasions been wiped out to a startling degree, only to return in myriad new forms to repopulate the planet—all due to the remarkable healing power of evolution. Were it not for evolution, life on Earth would have remained a relatively insignificant form following its inception and would not have rebounded as it has from each major catastrophe the planet has experienced.

Thus, the history of Earth has been one in which the laws of nature, and especially evolution, have worked to create ever more complex creatures capable of increasingly complex forms of enjoyment, in accordance with God's Agapic design. While not God's primary goal, Agapic as God is, God also must obtain enjoyment from this process, sharing the creatures' more or less modest forms of enjoyment as God experiences, along with creatures, every moment of their existence. For as God offers the paths that will bring maximum enjoyment for all creatures, so God also experiences directly the choices of all creatures and the enjoyment—or suffering—they bring to themselves and others. In this way God rejoices with each creature as it rejoices in each moment or suffers with it as it suffers. Both the joy and suffering of God are beyond imagination yet are the natural and inevitable experiences of any being who truly loves agapically an infinity of others.

At this point we can address the question of natural evil and God's role in it, though the answer should already be fairly clear. God does not control conglomerate entities any more than God controls free creatures. Because God's power is persuasive, not coercive, God cannot control those things in nature that harm us. There are only a handful of natural laws to govern the entire universe, and these function primarily in the service of the production and sustenance of the natural order. But they can also have side effects that can produce evil for creatures. Bacteria, for example, can harm as well as support the life of more complex creatures, and this is a simple byproduct of the evolutionary development of life in general on our planet. God must be working to attempt to mitigate the negative effects of harmful microbes, including working to help scientists find ways of prevention and cure of the diseases they cause. Thus, just as most of us view our lives as wonderful gifts despite the times of suffering and sorrow we inevitably must face, so too should we view the natural evils of this world as the inevitable accompaniments of what is otherwise a wonderful place to live, rather than blaming God when God is not responsible for their occurrence.

This understanding of what God can and cannot do, and the way in which God acts in the world, also offers us a solution to the question of *miracles*. The first problem we face is how to define the term, since many definitions have been and continue to be used by different people. The term "miracle" is actually never used in the Bible; the two main terms often translated as miracle are "signs" and "wonders," by which the biblical authors seem to mean surprising occurrences that often indicated the power of those performing them. There was no sense that miracles were violations of the laws of nature, as they have often been understood in the modern era, for the simple reason that the idea of natural laws did not exist yet. Things were understood to be governed by God and expressive of the divine will, and the ancient Jews understood the rhythms of nature, such as the passing of the seasons, one into the next, to be expressions of the consistent ordering of things by the good and trustworthy God. Thus, signs and wonders were acts of power that were unusual and generally indicative of the power of the one who did them, and this was true whether that person and the acts performed were good or evil (cf. Rev 13:11–15).

This sense of the miraculous as something wondrous (usually the expression of something good) persists today among many even where such events are not understood to be a violation of natural law, such as when the birth of a child is termed a miracle. But the predominant use of the term in the modern era has seemed to be in reference to things that appear to violate the laws of nature, with the understanding being that God, who established the laws of nature, is also free to violate them. Such an understanding is difficult to maintain, however, for at least two reasons.

First, we do not understand all the laws of nature—and are far from it, in the view of many, myself included—and so we cannot say with any confidence what is a violation of the laws of nature or not. What was thought to be impossible only a few centuries ago, such as travel into space and to the Moon, is now a given, and this pattern will be repeated in the coming centuries. Even today, when someone claims a miraculous event, such as a sudden healing, the first inclination of most people, even people of faith, is to look for a natural explanation for the event rather than believe that the laws of nature have somehow been violated.

The second problem with this understanding of miracle is that it assumes a God who is coercively in control and so acts occasionally like a magician, suddenly, for example, healing someone who had been dying of cancer. That such surprising healings have happened is well known among physicians and has been reported in responsible medical journals. But to insist that God alone was responsible for the healing is simply wrong, given our understanding of the nature of God's power as persuasive, not coercive. Perhaps even more pointedly, it must also cause us to question the very goodness of God that calling such a healing a miracle intended to convey;

for if God miraculously and unilaterally heals some sick people, then why does not God heal all of them? If God has such power, why are any children born with horrific diseases, which permit them only short lives filled with suffering? Why does not God heal all the young people who contract cancer or any of the other fatal diseases that claim millions of children around the world each year? Must not their suffering mean that God does not love some children? Extend that logic beyond the realm of children to encompass adults and the aged, and you are presented with a very clear and stark choice, as we saw in our discussion of the problem of evil: either God is coercively omnipotent but not good and loving, or God is Perfect Agapic Love but lacks the coercive power to violate the laws of nature once those laws have been established. The choice really is that simple, despite all the attempts to complexify it by numerous theologians past and present. The evil of suffering and death is real, and either God is good or God is coercively omnipotent. Which of these Gods is worthy of worship—the God of love or the God of coercive power? I take my stand with Jesus: God is the God of Perfect Agapic Love.

These reflections raise also the question of *prayer*, and specifically *petitionary* prayer, in which people ask God to do good things for themselves or others. If one understands that God is Perfect Love, it makes no sense to ask God to do what is good because God is surely already doing whatever God can to bring about whatever good is possible for them. In this sense petitionary prayer is useless.

However, petitionary prayer is not useless with regard to its effect on us. Like all prayer, it helps us to live in the recognition that we exist as creatures of God and can relate to God by our words as well as by what we feel but can find no words to express. It helps us to live in faith, with our lives centered on God and so on love for others, as God desires. In this way it helps us to get what we really want and need, even though we may not know it—the love of others. The genius of Christianity (and all the great religions, I believe) is that it provides the only way to ultimate personal fulfillment by the very act of *not* seeking that fulfillment for ourselves but instead seeking it for others by loving them. In this way, in the end, everyone wins.

Here we return again to the question of why God desires the creation of more complex forms of life. Yes, God desires the existence of creatures such as ourselves so that we may experience richer forms of enjoyment and fulfillment for ourselves, but that is only part of the story. Like all other creatures in nature, humans are relational beings, affected by our relationships with others, including relationships with the earth on which we live, the multitude of creatures in it, and God. God wants all of us creatures to experience enjoyment to whatever degree is possible for us, again because God is Perfectly Loving.

What God gains with complex creatures like us are potential coworkers in God's attempt to facilitate the enjoyment of others as well as ourselves. Our minds have reached a level where we can conceive God's existence and relate to God consciously, can understand God's fundamental purpose for all creatures as their enjoyment/fulfillment, and can align ourselves with God in that task. Because love is what most fulfills human beings, our task is to love them—which means, even more than feeling love for them, working to bring about the most loving state of affairs for them. This means trying to see to it that they have the means to live in the world and the opportunity for personal fulfillment. Our love for others means that we seek to support them through good times and bad and to help them to learn that it is love that will truly fulfill them and nothing else. Other things—success, the respect of others, even money, properly used—can bring us some fulfillment, but they are only the frosting on the cake. Real fulfillment comes only from loving and being loved by others, and the greater the range and intimacy of that love, the greater the fulfillment.[7]

However, human fulfillment is not our only concern, nor is it God's. Like God, our love must encompass all of God's creatures by expressing love for those that can feel and are themselves at least partially fulfilled by love (larger-brained mammals, at least, including our mammalian pets), while seeking also to foster those conditions that will allow them to flourish. This includes especially caring for the earth on which we all live, and not only for ourselves and other creatures today but for those who will live in the future. The offer each of us is given, then, is to be cocreators with God in seeking to bring about a better world and better lives for all creatures, human and nonhuman alike.

7. As I write these words, I am some twenty-one months into my battle with pancreatic cancer, a battle the statistics say I will soon lose. I have learned this lesson about love and fulfillment personally, in the depths of my being, with an intensity I had never experienced before. It has been the greatest "silver lining" of this battle. The outpouring of love I have been shown by, and have been able to return to, my friends has allowed me to live in peace and with an overwhelming sense of fulfillment in life.

18

The Rise of Humankind

From Primates to the Australopithecines

As we have seen in our discussions of the origin of the universe and the origin of life on Earth, it is questions of origins that are among the most difficult and uncertain of all scientific inquiries. Thus, it should come as no surprise that the question of the origin of humanity is fraught with problems that also cannot be resolved, given our present knowledge. Fossil finds over the past century and more have provided *paleoanthropologists* (scientists who study the evolution of the human family) with a growing number of possible human ancestors. The nature of these beings, however, remains highly uncertain. Researchers, studying what are mostly scant fossil remains, find it difficult to determine their proper genus and species their dates, the relationships among the various species, and which were the ones from which we *Homo sapiens* might have evolved. These potential human ancestors are called *hominids* (or *hominins*, roughly equivalent terms that refer to those increasingly human-like primates that lived after the last common ancestor before the divergence of the great ape and human lineages).

We will begin with a discussion of primate evolution in greater detail than was done in chapter 16, after which we will review the various hominid fossils discovered to date and note whether they are currently considered to have been ancestral to humans or not.

PRIMATE EVOLUTION

Primate fossils begin to appear in rock from around 55–58 mya, or less than ten million years after the extinction of the dinosaurs opened up numerous ecological niches that quickly began to be filled by what had been small, mostly nocturnal, mammals. Genetic evidence pushes the origins of primates significantly earlier, to perhaps 85 mya. Most primates today continue to be mainly *arboreal* (tree dwelling), with some monkeys, apes, and humans being the exceptions.

Why primates evolved is the subject of a variety of theories, the most likely of which is that, in the absence of dinosaurs, some mammals took to the trees to feed on the insects that inhabited the lower levels of the canopy. Catching these mobile insects required both visual acuity and effective grasping, and so these traits were selected for evolutionarily over millions of years, giving rise to these two key primate characteristics. Visual acuity required forward-facing eyes for depth perception, and very large eyes capable of taking in greater amounts of light were important to the nocturnal lifestyles. Efficient grasping required separately moving digits (in contrast to those of dogs, lions, elephants, and other such mammals), opposable thumbs, and nails rather than claws on at least some fingers and toes. Most early primates also developed *prehensile* (able to grasp things) tails to aid them in moving through the trees. Each evolutionary step along the way to these innovations made primates more effective hunters.

Over time, more advanced primates developed relatively large brains, came to engage typically in more social behavior, usually gave birth to single offspring that required a relatively large amount of time to mature, and lived long lives. Many also became *sexually dimorphic* (different in form according to their sex), especially in terms of differences in body dimensions and the size of their canine teeth. Primates share the same four basic kinds of teeth—(from the front of the jaw to the rear) incisors, canines, premolars and molars—though the number and form of each varies among the different primate species. Humans normally have four incisors, two canines, four premolars and six molars total on their upper and lower jaws.

Scientists today divide primates into two major groups: *prosimians*, including lemurs, lorises, and tarsiers, and *anthropoids*, including monkeys, apes, and humans. Prosimians were the first primates to evolve and lived in the tropical and sub-tropical regions of the "Old World"—that is, Africa and Eurasia—as opposed to the "New World" of the Americas. They typically have very large eyes to aid their mostly nocturnal lifestyles; wet noses and longish snouts to give them a well-developed sense of smell; large, mobile ears to aid their hearing; prominent whiskers to enhance their sense of touch; nails at the ends of their digits except for a claw on the index toe for grooming; and small, forward-pointing teeth with nearly horizontal incisors in the lower jaw called *grooming combs* to aid their hygienic practice. In some species these grooming combs are also used to feed, such as by gouging into the bark of trees to get at the insects that are their primary food, or tree gum. Prosimians also live typically solitary lifestyles.

Monkeys include both Old World and New World varieties and evidence a wide range of sizes, social organizations, and behaviors. Monkeys are often arboreal, though some, such as baboons, live primarily on the ground. They vary widely in size, ranging from the pygmy marmoset, which is only 12–16 cm (5–6 in) in length (not counting the tail) and 120–140 grams (4–5 ounces) in weight, to the chacma baboon, which can be as much as 1.2 m (47 in) tall and weigh up to 40 kg (88 lbs). There are something like 260 known monkey species, most if not all of which are generally thought to be intelligent. Typical food sources include fruit, seeds, nuts, flowers, leaves, eggs, insects, spiders, and other small animals. There are, however, some notable distinctions between Old and New World monkeys, such as in their tails: New World monkeys have prehensile tails while Old World monkeys do not. In fact, some Old World monkeys have quite small tails, and others have no visible tails at all.

Apes are divided into two categories: the *lesser apes*, such as the smaller-bodied gibbons, and the *great apes*, including the larger-bodied orangutans, gorillas, chimps, and bonobos. Because the great apes are closer to us evolutionarily, we will examine them more closely. As we move forward in time, we will note the appearance of certain traits that become important points of continuity between them and our earliest human-like ancestors.

For an image of the evolution of the primate family tree, see http://genealogyreligion.net/wp-content/uploads/2013/03/Primate-clade-tree.jpg.

Orangutans are arboreal, typically less social than the other great apes (though mothers and their young remain together usually for two years), and, while they once lived in Malaysia as well, today they are found only in the rain forests on the islands of Borneo and Sumatra in Indonesia. Orangutans display substantial sexual dimorphism: males can grow to as much as 1.75 m (5 ft 9 in) in height and 118 kg (260 lbs) in weight, while females reach only about 1.25 m (4 ft 2 in) and 45.4 kg (100 lbs). Both can live for up to thirty years in the wild. Orangutans probably diverged from the rest of the great apes somewhere between 15.7 and 19.3 mya.

Gorillas are the most imposing primates, with the largest males, or *silverbacks*, reaching as much as 1.8 m (5 ft 9 in) in height and 230 kg (510 lbs) in weight in the wild, while females attain a significantly smaller average of 1.4 m (4 ft 7 in) and 100 kg (220 lbs). Gorillas are generally herbivorous and ground-dwelling, inhabiting the forests of central West African countries and, to the east, of the eastern Congo and Rwanda. They are primarily *knuckle walkers* (they walk on their feet but with bent knees, and the knuckles of their hands bear a significant part of their weight as they move), though they can walk upright for short distances when carrying food or warding off rivals or enemies. They are more similar to humans than are orangutans, having diverged from the chimp, bonobo, and human lineage some 7+ mya. They are only 1.6 percent different from humans in terms of the correspondence of their genes. There are, however, other differences, such as how many copies of each gene we and they have, that account for yet more genetic variability relative to us.

Chimpanzees and *bonobos* are quite similar in many respects and would be difficult for most people to tell apart; in fact, bonobos used to be called pygmy chimpanzees. Chimps are more *robust* (strongly built, muscular) in form while bonobos are more *gracile* (slim or slender)—a distinction we will note later in differentiating certain hominid species. Male adult chimps can grow to as much as 1.7 m (5.5 ft) in height and 70 kg (154 lbs) in weight, with females reaching as much as 1.3 m (4.25 ft) and 47 kg (100 lbs). Bonobos are somewhat smaller, with males averaging 83 cm (2.75 ft) in height and 39 kg (86 lbs) and females a slightly smaller 76 cm (2.5 ft) and 31 kg (68.3 lbs).

Chimps and bonobos are both arboreal and terrestrial, with knuckle walking their primary mode of locomotion on the ground, though, as with gorillas, they occasionally walk upright for short stretches, especially when carrying things. They have bare palms with opposable thumbs on both their hands and feet, which aid tree climbing. Like gorillas, they are active during the day and sleep at night. While bonobos spend more time in the trees and live only in forests, chimps live in a variety of habitats that range from swamp forests to open savannas. Fruit is the primary and preferred food for both chimps and bonobos, though other plant parts such as seeds, leaves, and flowers, plus eggs, insects, other invertebrates, and even small mammals supplement their diet. Bonobos have rarely been seen to hunt actively for meat, while in chimps the practice is more commonly noted, often involving other chimps to capture larger game, such as the red colobus monkey. Rarely, cannibalism has also been observed among chimps.

Chimps and bonobos are more closely related to humans than any other primates, sharing more than 98 percent of their genetic material with us. Gestation lasts eight months, and weaning begins at about three years, though chimps and bonobos usually remain close to their mothers for several additional years before reaching puberty at between eight and ten years of age. In the wild they have been known to live for up to forty years.

Chimps and bonobos are also social creatures, like us. They live in large communities of one hundred or more members but spend most of their time traveling and foraging in smaller groups ranging up to about fifteen members. Where chimp societies are male-dominated and evidence more aggressive male behavior, bonobos are more female-dominant and peaceful. Also like us, both chimps and bonobos create and use tools, such as modified sticks, grass, leaves, and stones, for acquiring food and water. They also use sounds, postures, and facial expressions to communicate with each other, and are quite intelligent. Chimps are known to display what appears to be altruistic behavior, such as sharing food with one another (and occasionally, it seems, with other species, such as turtles) and adopting young orphaned chimps, even, sometimes, those of other, unrelated groups. They also evidence other human-like behaviors, such as play, curiosity, mourning for friends that have died, and even, it seems, an appreciation of beauty. Chimps and bonobos shared a last common ancestor with humans some 6–7 mya and began to diverge from one another only about 1.5–2 mya, probably when the Congo River was formed, effectively isolating the two groups to evolve separately.

The fossil record is spotty and incomplete for the evolution of great apes, making it impossible at present to determine which of the evidently numerous apes that flourished 10–20 mya is the ancestor of the lineage that would eventually lead to humans. As a result, it is not currently possible to know exactly what our last common ancestor with chimpanzees and bonobos would have looked like, and this makes judging which early hominid features were novel and which were not relative to that common ancestor more speculative. Nonetheless, paleoanthropologists do note certain differences that exist between humans and great apes today, and it is these differences that have guided them in their attempt to discern ancient ape from ancient hominid fossils.

Chief among these differences is *bipedality*, the ability to walk upright on two legs rather than knuckle walking. As we have seen, great apes today are knuckle-walkers, though not all do so in precisely the same way. This may indicate that this form of walking developed after the last common ancestor between apes and humans. There is no evidence to date that the earliest hominids employed knuckle walking, though, again, the evidence is sketchy for this period.

Why bipedalism evolved remains a topic of debate among paleontologists, though most seem to favor the idea originally put forth by Darwin that it permitted hominids to carry things while moving, such as food and their young. That more upright walking would have also allowed hominids to see predators more easily, such as over tall grasses or shrubs, is another favored theory. It might also have helped them follow and kill game migrating across the savannas. These points are often tied to climate changes that were occurring during the time hominids were evolving.

As we have already noted, the Neogene Period saw Earth's climate generally cool while exhibiting a number of temperature fluctuations throughout the period. The downward global temperature trend became permanent after a major cooling occurred between 15 and 14 mya. This cooling meant that the atmosphere could not hold as much water as it had previously, leading to less rain and so a generally more arid climate in many places. The rain forests that had stretched from Spain to China diminished, creating deserts and scrub grasslands in various places; and many of the species of apes that had flourished in tropical forests disappeared along with them. Around 8 mya another significant cooling occurred, further drying the climate. Some 2 mya East Africa became drier as the eastern equatorial region of the Indian Ocean warmed and the western region cooled, converting more forested land to savannas. With the increasing openness of the landscape, most paleoanthropologists seem to think that bipedalism would have become increasingly selected for by providing the advantages in survival and reproductive success noted above. Sawyer and Deak offer the following plausible scenario:

> A drying trend, and just as important a tendency toward marked seasonality of rainfall, had begun to affect the kind of vegetation the continent supported. Most significantly, the once monolithic African rainforest began to break up and to become interspersed with patches of woodland and even grassland. Formerly diverse ape populations began to find themselves under stress as their habitat began to shrink, and some of them would have found themselves obliged to struggle along the forest edge rather than in deep forest, and even to move out into the more open woodlands that now lay adjacent. The postures that the ancestral apes had adopted in the forest setting would have varied considerably. Some of them would have moved through the trees on the tops of branches, using mainly four-legged postures. Others would have preferred to suspend themselves from branches and would habitually have held their bodies more vertically, especially while foraging. Some suspensory apes evidently found it

> easier to continue using vertical postures when habitat changes forced them at least part-time to the ground, either to forage there or to move from one clump of trees to another. These were the primates that ultimately became the ancestral hominids.[1]

Thus, it seems reasonable to think that the evolution of hominids had already begun in the tropical forests prior to significant drying out of the climate in Africa, so that a drier Africa may have encouraged this trend toward bipedalism but did not originate it. Most paleoanthropologists think that it was probably a combination of factors rather than any single one that favored the gradual evolution of bipedalism.

In addition to bipedality, several other traits evidenced by great apes today and fossil apes from the past are used to try to tease out which anthropoidal fossils belong to hominids and which belong to the ape lineage. One has to do with the shape of the palate: in apes the palate is flatter where in humans it is rounded. Another concerns the shape of teeth within the jaws. In apes the teeth form more of a U shape, with the molars in both jaws parallel to one another, whereas in humans the teeth in both jaws take on a more V-shaped arrangement. Further, ape canines are larger than human canines, typically interlock, and tend to have less enamel thickness (though this is not always the case and probably reflects the differing diets of the two). In addition, the *foramen magnum*, the hole at the base of the skull where the spinal column attaches, tends to be farther back in apes than in humans, reflecting their different walking postures.

Some paleoanthropologists, however, have challenged the usefulness of some of these criteria for distinguishing the ape and human lineages based on fossils of the early ape *Oreopithecus* that date to 7–9 mya. These fossils suggest that Oreopithecus was smallish, weighing some 30–35 kg (66–77 lbs), evidently lived in a swampy habitat, and ate mainly leafy foods. Its pelvis, lower back length, knee joint and forward foramen magnum have led some paleoanthropologists to think that Oreopithecus was bipedal in some sense. Its teeth are also more human-like in a number of ways than is typical for apes. This suggests to some scholars that the last common chimp-human ancestor might already have had these same features, making judgments about which fossil finds belong to the hominid lineage and which belong to the ancestors of modern apes even more uncertain.

We will encounter more such differences as we survey the various hominid fossils that have been discovered and their similarities to and differences from our ape-like ancestors. In so doing we will group the fossil finds under several headings, which will prove useful in drawing conclusions

1. Sawyer and Deak, *The Last Human*, 19–20.

about the probable path from the last common ancestor of apes and the human lineage and the steps from that ancestor to us *Homo sapiens*.

CANDIDATES FOR THE EARLIEST HOMINIDS

The earliest of what are thought to be possible hominid fossils come from the fossil-poor era of approximately 5–8 mya. A combination of fossil dating and molecular evidence, particularly of protein and DNA, places the last common ancestor of apes and humans in the range of 6–7 mya, and the first two of these four early hominid candidates are dated within that time frame. They are sometimes grouped, along with the *Australopithecus* species that appeared later, as *Australopithecines*, but this term is probably better used to refer only to those species of the *Australopithecus* genus.

Sahelanthropus tchadensis

A 2001 fossil find in Chad (central Africa) named *Sahelanthropus tchadensis* (often called *Toumai*, or "hope of life" in the local Goran language) has been estimated, on the basis of the sediments in which it was found and the extinct fauna discovered with it, to have lived approximately 7 mya, roughly the same time as the ape-human split. Its discoverers have claimed that it is the oldest known hominid. Others have disputed that judgment, asserting that *Sahelanthropus* probably lived before the last common ape-human ancestor and may have been an early member of the gorilla or chimpanzee family. If they are correct, then the "*anthropus*" ("human") in its name is a misnomer and should be changed.

The chief problem in determining on which side of the human-ape divide this primate should be placed is the paucity of its fossil remains, amounting so far to a single skull, some lower jaw fragments and several teeth. The skull features a combination of both ape-like and human-like traits, as would be expected in a creature intermediate between the two. Its more ape-like traits include an elongated skull, relatively small brain (average estimate: 350 cu cm [21.4 cu in], about the same size as that of as modern chimps but only a quarter that of modern humans), a large *browridge* (the protruding bony ridge above the eyes), and a robust lower jaw. Its more human-like traits include a relatively short and flat face (the area beneath the eyes does not protrude as it does in apes), a forward foramen magnum (indicating more upright walking), small canines with long roots and worn tips (suggesting they were not used for cutting, as in apes), generally human-like tooth proportions, lower premolars that do not rub against and therefore sharpen the upper canines, the lack of a gap between the lower premolar and canine, and molar enamel thickness intermediate between apes and the later *Australopithecines*. However, these human-like traits are also shared with *Oreopithecus*, clouding the picture significantly. The complete lack of skeletal fossils from the neck down makes claims for bipedality speculative.

Sahelanthropus lived near the massive prehistoric Lake Chad in a wet grassland and woodland habitat with seasonal floodplains, sand dunes, and more desert-like areas nearby, much like that region is today. Its diet would have been largely vegetarian, though it probably included insects and small vertebrates as well.

Currently, paleoanthropologists as a group cannot decide with any degree of certainty on which side of the ape-human divide *Sahelanthropus* stood. Only new fossil finds, particularly *postcranial* (below the head) skeletal remains, will help decide the matter, though whether such fossils will be found is anyone's guess. Given how few sites in Africa and beyond have been investigated by paleoanthropologists, it is rather amazing how many possible hominids have been discovered—and the pace of these discoveries has been accelerating in recent years. It is reasonable to hope, therefore, that more fossils predating and postdating the ape-human split—though not necessarily fossils of *Sahelanthropus*—will continue to be found in the future.

Orrorin tugenensis

Found in the year 2000 in Kenya (East Africa), *Orrorin tugenensis* is thought by most paleoanthropologists to date to about 6 mya (the age of the fossils has been estimated to be between 6.2 and 5.6 mya) and to have a stronger claim than *Sahelanthropus* to be the first true hominid. This claim remains controversial because of the limited nature of the *O. tugenensis* fossils. These include two jaw fragments with three molars and some other isolated teeth, a right arm bone shaft, several thighbone fragments, and perhaps an isolated finger (or possibly toe) fragment. These are believed to have come from five or more different individuals living as much as 400,000 years apart, and so may be evidence of changes within the species or even the coexistence of different species during that long time period.

The tooth fragments represent a mix of ape-like and more human-like traits, with the incisors, canines, and premolars being relatively large and more ape-like and the molars with their thick enamel and relatively smaller size being more human-like. Overall, the teeth are relatively small relative to *Orrorin's* body size and so apparently more like ours today, though the fact that they come from several individuals whose sex is unknown (female teeth often differ from those of males) and from various times makes this judgment less certain. Since there are currently no *Orrorin* skull fragments, nothing can be said about its brain size or facial characteristics.

Interpretation of the few skeletal bones is controversial, with some arguing that the *femur* (thighbone) in particular indicates bipedalism while others claim the evidence is insufficient to support that claim. The *humerus* (upper arm bone) suggests that *O. tugenensis* was well adapted to life in the trees. The size of these bones indicates that it was about the size of a female chimpanzee. While there is some disagreement on the point, its habitat was seemingly a mix of thick forests and woodlands but not savannas. Its teeth indicate a diet that likely consisted mainly of fruits supplemented by seeds and/or other harder vegetation.

If *Orrorin* was indeed bipedal, and so a hominid, it would mean that bipedality originated in the forests and not the savannas, as many paleoanthropologists suggest. More evidence will be required, however, to convince the majority of scientists that it was indeed bipedal. The discoverers of the *Orrorin* fossils have suggested that it was a direct ancestor of humans, but that claim is even more speculative than the claim of bipedality and so has garnered little support.

Ardipithecus kadabba

Fossils of this species, found in the east-central region of Ethiopia, date from about 5.8 to 5.2 mya, with 5.6 mya often given as a rough estimate. Skull fossils include a lower jaw fragment with an incisor, canine, premolar, and three molars from one site and nine additional isolated teeth from two other sites. The jaw is roughly the size of a chimp's and the teeth show a mix of ape-like traits, including upper canines that *honed* (sharpened) by rubbing against the lower canine when *A. kadabba* ate, and other traits more similar to the later *Ardipithecus* species, *A. ramidus*. Together, the teeth suggest that this species consumed a wide variety of vegetable matter.

A. kadabba's skeletal remains consist of part of a collarbone and two arm bone shafts along with partial finger and foot bones. Interpretation of these bones is controversial, as a toe bone that suggests bipedal walking was found some 25 km (15 mi) away from the other remains, in deposits as much as half a million years older. It has also been suggested that the finger bones might be toe bones and/or have belonged to monkeys or apes. Thus, very little can be said with any degree of confidence regarding *A. kadabba's* manner of locomotion. The skeletal bones do suggest, though, that it was somewhat smaller than *Orrorin*.

Fossil animal bones found together with *A. kadabba* suggest that it lived in an area of mixed woodlands, grasslands, and swamps, though which, if any, particular habitat it preferred cannot yet be determined. The climate of the region was probably similar to that of the area today, warm with seasonal rains. *A. kadabba* is believed by at least some paleoanthropologists to be the ancestor of the other *Ardipithecus* species, *A. ramidus*.

Ardipithecus ramidus

This species is represented by a substantially larger number of fossil remains—some 110, including a 45 percent complete female skeleton—from approximately the same area in Ethiopia as *A. kadabba*. The estimated age of the fossils ranges between 4.51 and 4.39 mya, with 4.4 mya the typical estimate. This makes *A. ramidus* more than a million years younger than *A. kadabba*. This age gap alone is sufficient, in the eyes of some paleoanthropologists, to justify classifying it as a separate species.

Fossil remains of *A. ramidus* include a skull that indicated it had a brain about the size of a modern bonobo or female chimp, or roughly a quarter the size of ours. Its shorter skull base is like that of later, upright-walking hominids, though its foramen magnum is too incomplete to confirm this upright stance. Its face is smaller, shorter, and less *prognathic* (the facial region below its eyes protrudes less) than those of apes. Its incisors are more like a gorilla's and its *milk molar* (or baby molar) is more chimp-like. On the other hand, and unlike apes, *A. ramidus* molars and canines are generally smaller and its upper canines, shaped more like incisors (like later hominids), do not project outward. They are about the same size in males and females, and do not hone against the lower teeth. Its teeth generally have enamel thicknesses intermediate between chimps and later hominids and lack the adaptations to heavy chewing seen in apes and some later hominids. The smaller and roughly similar size of *A. ramidus*' canines suggests to a number of paleoanthropologists that there was less male-on-male conflict than in chimps, increased bonding between males and females, and perhaps increased parental investment in raising their young.

The differences in the teeth of *A. kadabba* and *A. ramidus* are another major reason, along with the differences in age of their fossils already noted, that many paleoanthropologists cite for placing these two *ardipithecines* into different species. But a number of scholars have also claimed that if the features of *A. ramidus* noted above have been correctly reconstructed, then they call into question, at least to some degree, the use of modern chimps as analogs for early primates. Rather, they suggest, it may be that some of the traits of modern apes evolved after the last common ape-human ancestor and so were not characteristic of the earliest hominids. If so, then the ape and human specializations seen today were the products of very different evolutionary paths after their lineages split off from one another, meaning that the last ancestor they shared in common would have looked significantly different from either apes or humans today. If this proves to be the case, it will change significantly the way in which early, possibly hominid, finds are evaluated and classified. It seems that, again, only the discovery of new fossils from the period of the splitting off of the human lineage from that of the apes will help resolve this important question.

The kind of diet suggested by the teeth of *A. ramidus* is a matter of debate, with some scholars hypothesizing that it ate a minimally varied diet composed of less abrasive foods and perhaps more ripe fruits, immature leaves, and succulent plants. Other scholars suggest it was more *omnivorous* (eating all types of foods), with a diet of foods eaten both in the trees and on the ground. Most scholars, citing fossil plant and animal remains associated with the *Ardipithecus* finds, believe that it inhabited forests, or woodlands with some patches of forest, but not grasslands—essentially like modern chimps. This, again, would counter the traditional view that bipedalism originated when primates made the move out into the savannas.

The skeletal fossils of *A. ramidus* suggest that it stood about 1.2 m (4 ft) tall, weighed roughly 50 kg (110 lbs), and exhibited little difference between the slightly larger males and the females of the species. The pelvis, with its advanced-hominid-like bony protrusions that serve as sites where muscles attach, suggests bipedalism, while its grasping big toe indicates the ability to grasp branches in trees—though its feet also had certain adaptations for bipedalism. Together these features suggest that *A. ramidus* moved on all fours in trees yet was bipedal on the ground (though it would have been incapable of running or walking long distances). Some scholars, however, dispute this interpretation based on the scant evidence provided by the fossil remains.

From the evidence just presented regarding these four early hominid candidates, we must conclude that settling the question of which was the first hominid, or even which was a hominid at all, remains irresolvable at present. This is for two main reasons: first, there is currently no clear definition of what constitutes a hominid. The main criterion has been bipedality, but that has not been confirmed to the satisfaction of many paleoanthropologists for any of the four candidate hominids we have noted. Second, though related to the first, is the scantiness of the fossil record for key anatomical areas that would help determine bipedality and whatever other criteria might be deemed relevant in determining what constitutes a hominid. These problems will also plague discussions of the hominid status of the various *australopithecines*, as we will see in the following discussion.

THE AUSTRALOPITHECINES

The *Australopithecines* (also termed *Australopiths*) represent a number of species that lived roughly between 4 million and 1 million years ago. Australopithecines exist in two main forms: *gracile* and *robust*; and an increasing number of paleoanthropologists appear to believe that the phenotypical differences (differences in form)[2] between these two justify placing them into two or even three distinct genera: *Australopithecus* for the gracile types, *Kenyanthropus* for another seemingly gracile type, and *Paranthropus* for the robust types. The appropriate genus determinations for these hominids remain somewhat controversial, as do the interpretation of the fossil remains variously given for each species. In what follows we will follow a common practice among paleoanthropologists, which is to consider all the gracile and robust species under the general "australopithecine" heading.

Australopithecus anamensis

Dates for the *A. anamensis* fossils range from 3.77 to 4.32 mya, with a collection of nine fossils from one site in northern Kenya being older—some suggest by as much as 300,000 years—than the twelve fossils from the other northern Kenya site. As we will see, this has created significant difficulty in determining what the appropriate genus classification for these remains should be.

Only portions of the skull of *A. anamensis* have been found, including an upper and lower jaw, jaw joint, ear canal fragment, and a variety of teeth,

2. Characteristics of the gracile australopiths include a narrower, more well-defined forehead, lack (usually) of a *sagittal crest* (a bony ridge running across the top of the skull from above the forehead to the back of the head, to which powerful jaw muscles attach, the presence of which indicates a large jaw suitable for consuming a diet composed of hard substances), a weak browridge above the eyes, and differences in the shape of the teeth and palate. Robust types are characterized by the lack of a forehead, presence of a sagittal crest, a large browridge, and various palate and tooth differences. Several robust species also appear to have had somewhat more robust skeletal bones (and so musculature) than typical of gracile types, but their most prominent distinguishing characteristics remain the more robust skull, jaws, teeth, and other adaptations related to chewing tougher plant materials.

and these exhibit mostly primitive characteristics. Its face is prognathic, with a recessed nose and the incisors and canines protruding outward, and the lower jaw indicates a weak chin—all features seen in apes today. The jaws are narrow, the teeth rows U-shaped, and the jaw joint and ear canal indicate a more ape-like skull, one significantly different from later Australopithecines. *A. anamensis* also had somewhat longer canines with more robust roots than later species, yet its molars were smaller and its teeth had thicker enamel than the earlier *Ardipithecus* species. There is no evidence of the upper canines being honed by the lower premolars. In general, the teeth come in a wide variety of sizes, though this may be indicative of male-female differences or the presence of different species in the fossil remains. All these features indicate a skull that was largely similar to older ape fossils.

Skeletal fossils include a *humerus* (forearm) bone, which has variously been described as more chimp-like and more human-like, and a finger bone, which may or may not belong together with the humerus in the same species because they were found in deposits separated by some 300,000 years. Two *tibia* (the large lower leg bone) fragments have been interpreted by some to indicate bipedality, but of a type different from modern humans, while others believe more evidence is needed to substantiate that claim. It has been proposed that *A. anamensis* weighed between 47 and 55 kg (103–121 lbs), though this is a rather speculative estimate given the paucity and uncertainty of the fossils.

That there is no consensus among paleoanthropologists regarding *A. anamensis* should be clear from what we have just noted. Whether the two different fossil collections come from a single species, as proposed by their discoverer, or from multiple species (estimates among other paleoanthropologists range from two to five different species) continues to be debated. There is also disagreement regarding assigning these fossils to the *Australopithecus* genus, with some scholars believing they should be assigned instead to *Ardipithecus* while others believe they probably belong to some primate genus prior to the ape-human split. Once again, only further fossil discoveries will help clarify these issues.

Australopithecus afarensis

A. afarensis is by far the best documented of any early hominid species, with nearly five hundred fossils having been unearthed since the early 1970s from sites in Ethiopia, Kenya, and Tanzania. The two most significant finds have been *Lucy*, the fossil remains of whom constitute about 40 percent of her body, and the "*First Family*," a large trove of fossils found together that paleontologists estimate come from at least thirteen individuals ranging in age from young offspring to adults. This extensive fossil collection has allowed archaeologists to reconstruct and understand aspects of the behavior of *A. afarensis* to an unprecedented extent. It is estimated that this species appeared possibly as early as 3.9 mya and continued to exist for roughly a million years thereafter.

For a photo of a reconstruction of an Australopithecus afarensis dubbed "Lucy" (in CosmoCaixa, a science museum located in Barcelona, Spain), see http://en.wikipedia.org/wiki/File:A.afarensis.jpg.

Two relatively complete skulls, in addition to a wealth of other skull fragments and teeth, indicate that *A. afarensis's* head displayed a mix of more ape-like and more human-like features. Its skull and face are chimp-like, with relatively small brains ranging from about 380 to 550 cu cm (23.2–33.6 cu in)—slightly larger than chimps today—a sagittal crest atop its skull, essentially no forehead, a significant browridge, and a prognathic face lacking a chin. Its teeth however, were more humanlike, with smaller canines and molars than apes (though still somewhat larger than ours). Its jaw and teeth rows were intermediate between the U-shaped ape- form and the more V-shaped parabolic human form.

While evidencing certain ape-like features, *A. afarensis'* skeleton is markedly more human-like in certain features than earlier hominids. It is notably slender in build—i.e., gracile—making it a good candidate for being the ultimate progenitor of the later gracile australopithecines. Its lumbar curve, pelvis, knee, leg bones, ankles, foot bones, and somewhat shorter toes, along with big toes that stood closer to the rest of the toes (implying the loss of the ability to grasp things with its feet) are more modern features that together suggest to most paleoanthropologists that *A. afarensis* was bipedal—though not with the same gait that humans evidence—and likely spent most of its time on the ground. On the other hand, its curved fingers, toes bones, and shoulder joints were more ape-like, which would have aided climbing and grasping tree branches, suggesting that *A. afarensis* spent some of its time in the trees. It also had relatively long arms and short thighs compared with modern humans, which suggests to some that it used its forelimbs for locomotion at least some of the time, such as when rapid movement was required.

A set of footprints dated to approximately 3.6 mya was found at Laetoli, northern Tanzania, along with *A. afarensis* fossils. Given the geographical and temporal proximity to those fossils, most paleoanthropologists think the footprints were made by an *A. afarensis* individual, though some suspect they were made by some unknown early *Homo* forerunner. If the footprints were made by *A. afarensis*, they provide further evidence for bipedality in this species.

A marked sexual dimorphism, similar to that found in gorillas, is also noted in the *A. afarensis* fossils. Males attained heights up to 1.5 m (5 ft) and weighed some 45–50 kg (99–110 lbs). Females stood about 1 m (39 in) high and weighed roughly 29 kg (64 lbs). This suggests to some that social behavior in this species would have been similar to that of modern apes, meaning that they would have lived in small family groups of one male and several females. A recent study suggests that *A. afarensis* used primitive stone tools to carve animal carcasses. If so, it pushes back the use of stone tools by hominids another million years or so. Not surprisingly, this claim remains controversial.

Animal fossils found near the *A. afarensis* sites indicate that these hominids probably lived in an area of open grasslands and woodlands similar to that found in those areas of Africa today. Certain features of their teeth imply that the diet of *A. afarensis* likely consisted of fruits and leaves along with plant pith, seeds, and herbs. They probably stripped grittier plants with their protruding incisors.

Australopithecus bahrelghazali

In 1993, a partial lower jaw with one incisor, both canines, and all four premolars was found in the central African country of Chad by the same paleontologist who discovered *Sahelanthropus*. Dates assigned the specimen range from 3.8 to 3.0 mya. In some features the fossil resembles *A. afarensis* while in others it does not, such as its relatively thin tooth enamel, its premolars, and possibly more vertical face. It is debated whether these differences are indicative of variations within the *A. afarensis* species or are sufficient grounds, along with its geographical location some 2,500 km (1,500 mi) west of the *afarensis* finds, to indicate a new species. The fact that its discoverer, who assigned the new species name, has not permitted other paleontologists to examine the fossils has retarded attempts to determine whether *A. bahrelghazals* should be classified as a new australopith species or not.

Kenyanthropus platytops

K. platytops (literally, "flat-faced Kenyan human") is controversial for several reasons. Fossils for this hominid consist of one highly fragmented and distorted skull, with cracks filled by material from the surrounding stratum, and a fragmentary upper jaw, dated 3.57 to 3.4 mya. Other fossils, including a lower jaw, a skull fragment with its ear canal intact, and thirty-three other, mostly tooth and jaw fragments, come from another stratum dated 3.4 to 2.5 mya, making it uncertain whether these latter fossils belong to the same hominid as the former fossils. A toe-bone fragment, the only skeletal remain found, similarly may or may not belong to *K. platytops*. This, and the fact that no foramen magnum has been found, makes it difficult to say anything with certainty about its size or bipedality. Given these problems, it has also been a matter of significant debate whether this hominid belongs in the new genus *Kenyanthropus* or instead should be classified as another species within the *Australopithecus* genus.

The poor condition of the skull and questions about whether all the fossils belong to the same hominid has made drawing conclusions about *K. platytops* difficult. It appears to have had relatively large, high, and flat cheek bones and a flat bone plate beneath the nose, which together give the face a broad, flat appearance (hence its species name). This vertical face and its relatively small molars make *K. platytops* seem more human-like, which has led some scholars to associate it with the later *Homo rudolfensis* and/or *Homo habilis* and so assign it to the *Homo* genus. On the other hand, its ear canal, thickly enameled premolars and molars, and small brain size are similar to those of *A. ramidus* and *A. anamensis* or *afarensis*, which has led other paleoanthropologists to assign it to the *Australopithecus* genus, with some even placing it within the *afarensis* species.

Its small teeth and geographical location imply that *K. platytops* lived in a grassy woodland environment near streams and lakeshores and ate fruits, along with herbs, succulent plants, and leaves. Given its overlap in time and place with *A. afarensis*, it is likely the two lived alongside each other but probably did not compete for precisely the same foods. Such coexistence among different hominids was evidently common before the rise of our own species. No tools have been found associated with the *K. platytops* fossils.

Australopithecus africanus

A. africanus fossils come from four different locations in South Africa, each dated somewhat differently from the others. The earliest fossils are dated 3 to 2.6 mya (though some have dated them as much as 3.5 mya), while the latest fossils come from strata dated 2.4 to 2 mya. Some questions remain about *A. africanus*—including its proper species assignment and the date and significance of one find, the so-called *Taung child*—but, apart from the Taung child, most paleoanthropologists seem to understand the fossils from the various sites to belong to the same distinct hominid species. Some view the gracile australopithecine to be the South African, if somewhat more advanced, counterpart of the gracile East African *A. afarensis*.

The skull, face, and teeth of *A. africanus* display many human-like features, a number of which are not present in *A. afarensis*. Its face was relatively flat and short vertically, though its forehead was relatively high, and it had a relatively small browridge. Its foramen magnum was located forward and oriented downward, a trait associated with bipedality. Though lacking a chin, its jaw was relatively large, but its teeth rows were parabolic like those of modern humans. This larger jaw, combined with larger molars and premolars and certain skull adaptations, indicate that the diet of *A. africanus* involved chewing harder foods. Its incisors and canines were relatively small and non-protruding, like those of modern humans. Brain-size estimates range from 400 to 500 cu cm (24.5–30.5 cu in), with the fossil skulls averaging nearly 450 cu cm (27.5 cu in), making them slightly larger than those of modern chimps.

The many fossil skeletal parts of *A. africanus* evidence a mix of more advanced and primitive traits. The upper limbs appear to be larger than those of *A. afarensis*, and its lower limbs smaller, with an upper- to lower-limb ratio of about 85 percent (about the same as baboons today). Modern humans have a ratio of 76 percent. Its pelvis is longer than a human pelvis but better adapted for bipedalism than that of *A. afarensis*, though it was also suited, like its somewhat curved fingers, for tree climbing. Except for the finger curvature and slightly longer fingers, *A. africanus's* hands and fingers were much like ours. Though no tools have been found associated with the fossils, its human-like hands mean it was certainly capable of making, and likely did make, at least some tools. Like its fingers, *A. africanus's* toes were also relatively long. Its big toe stood somewhat apart from the rest of the foot, though less than with earlier hominids.

These skeletal aspects point to a creature that spent most of its time on the ground, often engaged in bipedal walking and standing, but that also spent some time in the trees. *A. africanus* was on average slightly larger than

A. afarensis, with significant, if less pronounced, sexual dimorphism. Males have been estimated to stand about 1.38 m (4.5 ft) tall and weighed some 41 kg (90.3 lbs), while females stood about 1.15 m (3.8 ft) tall and weighed approximately 30 kg (66.1 lbs). Such sexual dimorphism suggests a social arrangement in which one or two males lived with a number of females and their young.

The environment in which *A. africanus* lived was relatively harsher than that of *A. afarensis*, with seemingly higher temperatures and significant dry periods, as in the region today that made finding food more difficult. They likely lived in grasslands with woodlands that were close to streams and rivers. Their robust molars and jaws probably mean that they ate whatever foods were available at various times of the year, including large quantities of plants and plant parts, many of which were hard to chew. This likely meant that, as with primates in the region today, the staples of the *A afarensis* diet were leaves, fruits, flowers, nuts, seeds, stem pith, roots, tubers, bark, lichen, and honey, along with meat, insects, and insect larvae.

Paranthropus aethiopicus

With *P. aethiopicus* we move back northward to East Africa and encounter the first of the *robust australopithecines.* As already noted, *Paranthropus* is the new genus name that has been chosen by many paleoanthropologists to apply to all robust anthropithecines, though other paleoanthropologists consider them to be more appropriately assigned to the *Australopithecus* genus. The robustness referred to applies only to the skull; the bodies of *Paranthropus* individuals were very much like those of the gracile australopiths. The word "paranthropus" means "alongside humans," an indication that paleontologists believe the robust australopiths as a group were a dead-end lineage, not ancestral to modern humans.

The fossil record is somewhat controversial in that most of the remains are teeth and jaw fragments from different locales that seem, at least to some researchers, to differ in certain ways from the major fossil that has defined the species. That fossil is a nearly complete skull and jaw called the "Black Skull" due to its dark color, a product of manganese in the soil having leached into the bone. The Black Skull dates to about 2.5 mya, while the dates for the other, fragmentary fossils date to between 2.8 and 2.3 mya.

The Black Skull shows that *P. aethiopicus* had a small brain of about 410 cu cm (25 cu in), which is comparable to the lower range of other australopithecines. The rear of the skull is primitive like that of *A. afarensis*—as is its prognathic face and several other features. It is the marked skull and jaw differences that make this species "robust." It has the largest sagittal crest of any known hominid—comparable to that of gorillas—which would have anchored huge chewing muscles that connected on their other ends to massive jaws. *P. aethiopicus's* face was likewise large relative to its brain case, with heavy browridges, large cheekbones, and a thick palate. Its tooth rows were primitive, being U-shaped like early hominids. Also, although its front teeth were wide, they were relatively small compared with its massive premolars and molars.

All of these large-scale skull, facial, and dental features indicate that *P. aethiopicus* had a diet that consisted mainly of large quantities of low-nutrition plants and roughage. Fruits, tubers, and roots were likely eaten as well, when available. Its habitat was probably flooded grasslands.

While there are no skeletal fossils that are definitely attributable to *P. aethiopicus*, several such remains have been found in the same deposits as some of the tooth and jaw fragments and of a corresponding time. If these skeletal fossils do indeed belong to *P. aethiopicus*, then they can give us some hints about the nature of its body and lifestyle. In particular, the arm bones are chimp-like and were evidently used for supporting weight, allowing them to assist in locomotion both on the ground and perhaps in the trees as well. Foot bones indicate it had a flat arch, further supporting the idea that it spent a part of its time in the trees.

Later *Paranthropus* species have bodies much like that of *A. afarensis*, suggesting that *P. aethiopicus* did too. These various skull and body similarities to *A. afarensis* have led most paleoanthropologists to the conclusion that *P. aethiopicus* was descended from that gracile australopith, and that the differences in the skull reflect adaptations to *P. aethiopicus's* different diet. No tools have been found with the other fossils, but some found relatively nearby and from about the same time period that are different from those of other hominids may signify that, late in their existence, they acquired the ability to make and use tools.

Australopithecus garhi

A. garhi is another East African gracile australopithecine whose relatively few fossil remains were discovered in Ethiopia. The fact that its skull and teeth fragments were found separated from what might be its skeletal remains has raised questions about the proper classification of these fossils, with some suggesting they represent a late form of *A. afarensis*, *A. africanus*, a female *P. aethiopicus*, or perhaps the final link between the australopiths and the genus *Homo*. The fossils, both cranial and skeletal, date to around 2.5 mya.

The skull fossils include three small skull fragments, an upper jaw with teeth and associated skull fragments, a lower jaw with teeth, and two toothless lower jaw fragments. The skull remains suggest a cranial capacity of about 450 cu cm (27.5 cu in), which is about the same as the average for other australopithecines, and a prognathic face, which appears to be more similar to *A. afarensis* than *A. africanus*. *A. garhi's* incisors are relatively small, within the range of those seen in *A. africanus*, and its canines are strongly built and lie pretty much within the plane of the incisors. Its premolars and molars, however, are the largest of any gracile australopith and more like those of the robust *Paranthropus* species. These outsized molars are likely evolutionary adaptations to a diet of generally harder plant material. It is this odd mix of skull and teeth traits that has created problems in attempting to classify these remains.

The skeletal remains of *A. garhi* (if they do indeed belong to the same hominid as the skull and teeth fossils) also feature a mix of more primitive and modern traits. The arm bone fragments seem to suggest that its arms were longer, like earlier australopithecines and apes, and supported more weight than modern humans. Its toe bone is similar to that of *A. afarensis*, as well. Its femur (thighbone), however, appears longer, more like those of later humans than any australopiths. This suggests that *A. garhi* took longer strides in walking and probably spent significantly more time on the ground than in the trees. Probable height and weight estimates do not seem to exist.

Some stone tools have been found in nearby deposits of the same age as the *A. garhi* fossils, along with animal bones from relatively large animals. Some of these bones show cut marks, evidently from butchering, and others appear to have been broken open with stones, presumably to extract their marrow. If stone tools were made by *A. garhi*, then that hominid species was among the first to make stone tools. They might, however, only have used stones found with sharp edges. The animal bone finds may also be evidence for scavenging, which would make *A. garhi* the oldest known scavenger.

The animal remains associated with the *A. garhi* fossils suggest that it lived in an open area composed primarily of grasslands, scrub plants, and thin woodlands near rivers and lakes. The climate, as in the region today, was likely mild and seasonal, with long dry periods during certain times of the year. Its food was likely a mix of relatively hard plant material, with the occasional addition of scavenged meat.

Paranthropus boisei

Fossils of *P. boisei* have been found in three East African countries ranging from southern Ethiopia through Kenya to northern Tanzania, and also in the southeastern African country of Malawi. In some areas, *Homo* fossils have been found alongside those of *P. boisei*, making it difficult to know which remains belong to which hominid. Distinguishing *P. boisei* from *P. aethiopicus*, both present in the region and possibly representing the same species, only intensifies the classification problem. The fact that there are only a couple of skeletal bones that can be assigned with certainly to *P. boisei* makes reconstructing this hominid even more difficult. Dates of the fossils from the various sites indicate a maximum range of between 2.3 and 1.2 mya, with recent analyses of the carbon in the teeth of a number of individuals suggesting dates between 1.9 and 1.4 mya.

The skull of *P. boisei* indicates the typical robust characteristics that differentiate the members of the *Paranthropus* genus from those of the gracile australopithecines. These include large, gorilla-like sagittal crests, massive jaws, and molars that are the largest of any known hominid. These cheek teeth are four times as large as those of modern humans, and their chewing surface is more than twice the size of ours. Their enamel thickness is also the greatest of any known hominid. On the other hand, *P. boisei's* incisors and canines were small (similar in size to ours), its cheek bones wide, its eyes separated by a wide nose, and its face long and relatively flat. Its front teeth formed a straight line and were angled at nearly ninety degrees from its cheek teeth, forming an exaggerated U. *P. boisei's* braincase size ranges from some 475 to 550 cu cm (29–33.5 cu in), significantly larger than that of *P. aethiopicus* fossils.

As mentioned, skeletal remains of *P. boisei* are plagued by problems of attribution, particularly with regard to contemporaneous *Homo* bones found in the same deposits. Those who have assigned significant skeletal fossils to *P. boisei* suggest males stood about 1.37 m (4 ft 6 in) tall and weighed some 49 kg (108 lbs). Females were somewhat smaller, standing some 1.24 m (4 ft 1 in) tall and weighing about 34 kg (75 lbs). If these figures are correct, then *P. boisei* exhibited the least sexual dimorphism of any previous hominid. Also, comparing a male *P. boisei* with an *A. africanus* shows two species of approximately equal height but with substantially different weights, with the robust form outweighing the gracile one by about 8.2 kg (18 lbs). The (questionable) skeletal remains also show similarities to early *Homo* skeletons; and if this is true, it suggests the two shared a fairly recent common ancestor.

The habitat in which *P. boisei* lived seems to have been dominated by dry grasslands, but also with patches of forests featuring heavy underbrush, and wetland areas near rivers and lakes as part of its environment. While its massive facial features suggest a diet of tougher foods, such as nuts and tree leaves, analyses of the kinds of carbon present in *P. boisei's* teeth and the micro-wear patterns seen there indicate a diet primarily of grasses and other softer foods, perhaps including insects and meat from small animals. Harder foods may have been the fallback for *P. boisei* when other, preferred foods were unavailable.

Whether *P. boisei* made stone tools is debated. While some rather crude tools appear in the same deposits as *P. boisei*, *Homo* fossils, as we have noted, do as well, making it unclear which species was responsible for making the tools.

Paranthropus robustus/crassidens

Remains of *P. robustus* come from several sites just west of Johannesburg, South Africa. A specimen from one of these sites was originally named *P. crassidens* by its discoverer due to certain differences from other *P. robustus* fossils, but most paleoanthropologists now understand it to be a form of *P. robustus* that reflects differences that evolved over the significant time span between it and earlier fossils of the species. *Homo* fossils have also been found in the same sites, adding to the difficulties of interpreting the fossil record. Because the only means of dating the remains is by *biostratigraphy* (dating soil and rock strata by the fossils contained within them—as opposed to dating soils by the more precise radiometric means), the dates proposed for the *P. robustus* remains have varied significantly, with the maximum range set at about 2.5 to 1 mya.

Like other *Paranthropus* (robust australopithicene) species, *P. robustus* had large jaws, molars, and skull features, such as significant sagittal crests that supported these powerful chewing adaptations. Certain features of the cranium, such as a greater frontal lobe width and increased *cerebrum* height relative to the *cerebellum* (the *cerebrum* is the main part of the human brain that sits atop the more primitive and smaller *cerebellum*; its twin lobes are responsible for higher cognitive functions) have suggested, at least to some scholars, a more human-like brain. Most estimates of brain size for adult males are within the 515–530 cu cm (31.4–32.3 cu in) range. *P. robustus's* face was relatively large, flat, and vertically oriented, with heavy brow-ridges, essentially no forehead, and minimal prognathism, especially below the nose. Its more anterior foramen magnum was also more human-like, strongly suggesting a more erect posture and bipedalism. Its relatively small and non-protruding incisor and canine teeth, and large, flat-wearing and thickly enameled premolars and molars, are similar to those of other later *Paranthropus* species and nearly as large as those of *P. boisei*.

The relatively few skeletal fossils have made reconstruction of the body of *P. robustus* with any degree of certainty difficult. Estimates of the body size of males have varied significantly, ranging from about 35 to 54 kg (77–119 lbs) for males and from 25 to 40 kg (55–88 lbs) for females. Males may have been only 1.2 m (about 4 ft) and females slightly under a meter (3 ft 2 in) tall, though the lack of long bones make this estimate uncertain. The level of sexual dimorphism was probably similar to that of earlier *Paranthropus* species.' Its toes and fingers, including its thumbs, were relatively short and so more human-like, implying a mainly terrestrial lifestyle. This view is re-inforced by the relatively dry and treeless environment in which it lived. Its pelvis was also more like ours than previous species. Its backbone and thigh, however, suggest a nonhuman-like hip joint, meaning that *P. robustus* at times relied on all four limbs for movement, like previous australopiths of both robust and gracile types.

The large jaws and teeth of *P. robustus* suggest a diet consisting of tough foods, such as tubers, roots, nuts, seeds, and berries, though recent studies have indicated a more generalized diet, including fruits, insects, and small vertebrates approximately the same as that of baboons that live in the same area today. Harder foods were perhaps the foods of last resort when the more desirable foods were out of season or otherwise unobtainable. Some evidence indicates they consumed meat, supporting the omnivorous thesis, and also that they used fire. Other evidence suggests they used bone and stone tools, the density of such finds increasing through time, and that these were probably used for digging up tubers and other such foods. Whether P. *robustus* fashioned these tools or simply used stones found with sharp edges is unknown. Skull remains suggest that *P. robustus* served as food for the large carnivores with which they coexisted in the open grasslands.

Australopithecus sediba

In 2008, the remains of an adult female approximately thirty years old and a male about ten to thirteen years old, perhaps her son, were found in a cave in the Malapa region of northeastern South Africa. The two likely fell through a fissure in the roof of the cave and, unable to get out, starved to death. Other primitive animal remains were also found in the cave. Dated to 1.977 mya ago, the fossils of *A. sediba* have been described as the most human-like australopithecine ever found and one of the best examples of a mammalian transitional species in the entire fossil record.

The skull of *A. sediba* is one of the most telling of the fossils. The interior of the cranium shows the impressions left by the brain, which was still small at roughly 420 cu cm (25.6 cu in). But the impressions indicate that *A. sediba's* brain had undergone some reorganization into a more human-like form, particularly in the *orbitofrontal region* directly behind the eyes. These regions control reasoning, long-term planning, and the ability to multi-task. This upended the previously held notion that increase in brain size in hominids predated the reorganization of their brains. While the face of *A. sediba* was notably more human-like, certain facial and dental similarities to *A. africanus* are also seen, perhaps indicating that *A. sediba* descended from the earlier australopith.

The skeleton of *A. sediba* also exhibits primitive as well as more human-like characteristics. The female's hand fossils are more complete than any ever found for a hominid, with only a few bones missing. It shows that this creature had long thumbs and short fingers, much like us, which would have enabled it to grasp things using only its fingers and without the aid of its palms. Similarly, its pelvis is one of the most complete in the hominid fossil record and looks quite like that of a modern human, except that it is longer, like that of Neanderthals. *A. sediba's* shinbone, thin heel, and certain features of its anklebones are more ape-like, while its ankle and foot, particularly its arch, show more human-like traits. The pelvis, leg, ankle, and foot indicate that *A. sediba* was an upright bipedal walker, though it could also still climb trees. These bones also suggest that the young male stood some 1.3 m (4.25 ft) tall, while the adult female was somewhat taller.

A novel feature of the *A. sediba* fossil record is the possibility that it includes several pieces of skin. The remnants are being evaluated by various researchers to determine whether this is actually so. If they are indeed skin, it would be the oldest example of fossilized skin yet found.

Given the unique mosaic of primitive and modern aspects of the *A. sediba* fossil record, it is not surprising that it has engendered significant debate among the paleoanthropologists about its place in the hominid record. A number believe it is the link between the earlier australopiths, particularly *A. africanus*, and the later *Homo* genus, particularly *Homo habilis* or *Homo erectus*—the latter of which is thought by most paleoanthropologists to be an ancestor to *Homo sapiens*. Some others suggest that *A. sediba* might have been a side branch of the human family tree, one that eventually died out.

The most that researchers can say about *A. sediba's* habitat is that it included the sub-tropical alpine forest, mixed woodlands, and other types of forest that apparently were within short range of the cavern into which these two hominids fell. Their diet has not yet been established. While no tools have been found associated with the remains, the advanced brain and hands of *A. sediba* suggest it had the means to create them. Other hominids of this period were fashioning relatively primitive stone tools, which increases the likelihood that *A. sediba* did, too. Its advanced brain may also have permitted it to communicate nonverbally, such as by smiling—an ability unique to humans.

Given the preceding review of the various australopithecine species, what might we conclude about them generally? The fossils that we currently have appeared around 4.3 mya and continued to exist until perhaps as late as 1 mya. Paleoanthropologists believe that they likely appeared first in East Africa and later spread at least to the west and south and perhaps eventually came to fill the continent. The australopithecines generally were distinctive for their bipedality, though the paucity of skeletal remains for some species makes this claim probable but not certain. Their bipedal locomotion, however, was likely only occasional; most species still relied on four-legged movement at least part of the time, and they still spent some of their time in the trees. It seems that their bipedal locomotion increased and their arboreal habit decreased with the further evolution of the australopiths as they ventured out of the denser tropical forests in which they first appeared and into grasslands and sparser woodlands over the millions of years of their existence. Their particular types of bipedal motion also evolved with time, though none seems to have been quite like that of our species today.

The gracile types were apparently the first to appear and, according to the current fossil record, constituted the majority of australopithecine species. Whether the first australopith evolved from an ardipithecine, *A. ramidus* perhaps, is a topic of significant debate among paleoanthropologists. It is probable that one (or possibly more) gracile species eventually gave rise to the robust *Paranthropus* forms—perhaps *A. afarensis* to *P. aethiopicus* and *P. aethiopicus* to *P. boisei* and *P. robustus*—which began to appear around 3 mya, probably in response to the prevalence of harder foods available to them in the hotter and dryer environments in which they lived. Most paleoanthropologists today think it was a gracile australopithecine that eventually gave rise to the first member of the *Homo* genus—likely *Homo habilis*—while the *Paranthropus* genus died out, probably from competition with the more capable *Homo* forms with which they coexisted for at least half a million years. The recent discovery of the more human-like *A. sediba* has made this species a possible candidate for that transitional australopith, though here again there is no consensus. Other proposals for the original human progenitors include *A. afarensis, A. africanus,* and *K. Platytops* (the last of which, as noted above, is thought by some to be more correctly assigned to the genus *Homo*).

All australopithecines shared a mix of ape-like and human-like traits. Most species had brains one-quarter to one-third the size of ours, or roughly the same size as those of chimpanzees today. They were therefore likely to have been, on the whole, no more intelligent than chimps today. Their faces were generally prognathic, somewhat less so than those of apes today but more so than those of modern humans, though the degree of prognathism varied among the different species. They tended to have larger molars and premolars than humans, though their canines were smaller and so rather human-like. Most were significantly smaller than we are, ranging on average from 1.2 to 1.4 m (3 ft 11 in to 4 ft 7 in) in height and from 25 to 54 kg (55 to 119 lbs) in weight, and exhibited more—typically much more—sexual dimorphism than our ratio of about 15 percent.

The diet of most australopiths appears to have been omnivorous, consisting predominantly of plant materials, though in many species, if not most, probably augmented by the occasional meat when available. They likely used some organic tools, such as sticks and bones to aid them in obtaining food, probably as apes do today. They likely also used unmodified stones as well to dig for roots and tubers or to crack open bones to get at the marrow inside. Later australopiths perhaps made primitive stone tools or used sharp found stones to butcher meat. Being slower than the larger carnivores of the open grasslands into which they ventured, they were also their prey. Throughout their history, australopithecine species typically lived alongside others of their kind, with most species dying out over time and only an indeterminate few having been our direct progenitors.

19

The Genus *Homo*

As you might expect from what we have seen, there is no universally agreed-upon dividing line between australopithecines and the earliest *Homo* species or definition of the latter. Nonetheless, certain novel features do appear in the species ascribed to the *Homo* genus, such as increased brain size, smaller cheek teeth, longer femurs, enlarged hip joints, more human-like pelvises, and behaviors such as the fashioning of ever more sophisticated tools, hunting, the use of fire, and more complex forms of social organization. What follow are the various *Homo* species named to date in their estimated order of appearance, culminating in our own species, *Homo sapiens*.

INTERMEDIATE SPECIES

Homo habilis

Most paleoanthropologists understand *H. habilis* ("able" or "handy" human) to be an intermediate species between the australopiths and more advanced *Homo* species, but there are significant problems in attempting to describe it. The fossils attributed to *H. habilis* vary widely from one another (e.g., in the shape and size of skulls), and it is uncertain whether some of the scant skeletal fossils belong to this species. Thus, there is little agreement about which fossils should be attributed to *H. habilis* and so what its essential traits were; whether it is a unique species (some attribute a portion of the fossils to *H. rudolfensis*); and, if so, whether it properly belongs in the *Homo* genus or should be included among the australopithecines. If it does merit classification as a *Homo* species, which seems to be the general view today, there is also the heavily debated question of whether it was our direct ancestor. Most scholars currently think that it was not. Dates for *H. habilis* range from 2.5 mya at the earliest to 1.4 mya at the latest. If these dates are correct, then *H. habilis* would have coexisted with the last of the *Paranthropus* species, *H. rudolfensis* (if it is indeed a separate species) and the later *H. erectus*.

Presuming *H. habilis* was a unique species, its skull shows the beginning of several trends that would become enhanced in later *Homo* species. Chief among these is the increased size of the braincase, the average of which among the skulls found is about 650 cu cm (nearly 40 cu in), or about 50 percent larger than the average australopithecine brain and almost half the size of the brain of modern humans. The braincase is also more rounded and less narrow than that of prior hominids, and its walls are thinner. Also, the bulge of *Broca's area*, which governs speech, is visible in the cast of one of the braincases, suggesting that *H. habilis* might have been capable of some form of speech, though this interpretation is contested by a number of scholars. The sagittal crest and heavy brow bones of earlier hominids is missing, though a continuous browridge is apparent. *H. habilis*' face, while varying in some important ways among the fossils, seems to have been less prognathic than that of the typical australopith, and its nose evidently projected outward and stood farther above the mouth than in the earlier hominids.

For a photo of a reconstruction of Homo habilis by Chris Wenzel at the Westfälisches Museum für Archäologie, Herne, see http://commons.wikimedia.org/wiki/File:Homo_habilis.JPG.

The jaws of *H. habilis* are comparable in robustness to those of the gracile australopithecines, though several specimens are smaller. In general, its teeth were smaller than the prior hominids, especially the premolars and molars, and their enamel was thinner. The shape of the tooth rows was parabolic, as in modern humans.

There is some debate regarding the skeletal remains attributed to *H. habilis*, since none were found associated with its skull and teeth fossils, and *P. boisei* remains were found in the same deposits. Also, whether the remains belong to males, females or both is uncertain. Clearly, additional fossils are needed for a more adequate description of this early *Homo* species. In general, though, anthropologists seem to think that *H. habilis*'s body was similar in size and proportions to that of the smaller australopiths, with relatively long upper limbs and relatively smaller lower limbs. Estimates of its size, however, vary greatly, with upper estimates for males as high as 1.57 m (5 ft 2 in) and 52 kg (115 lbs) and females as little as 1 m (39.4 in) and 32 kg (70.5 lbs).[1] Its hands were evidently wide with large thumbs like those of later humans yet had the long, curved fingers and thumb orientation seen in the australopithecines. Its femur/thighbone was relatively long, its hip joints were larger than those of the australopithecines, and the shape of its pelvis was also more human-like. The feet seem to be more modern, though scholars disagree on whether they were arched or flat-footed and whether this made them more efficient at walking than prior species.

Whether *H. habilis* made tools is a topic of significant debate. The oldest hominid-made stone tools appeared about 2.5 mya, though no hominid remains were found with them that identified their makers. *H. habilis* fossils have often been found with other such primitive stone tools, called *Olduwan,* from the find at *Olduvai Gorge* in northern Tanzania. Most see *H. habilis* as a species that made and used these tools, though the idea that this species invented the Olduwan tool culture is not as widely held. The early Olduwan tools primarily consisted of *choppers,* which were made with only one or a few blows against another stone and were likely used for breaking bones to get at the marrow inside. The small, sharp chips or *flakes* that resulted from making the choppers were useful for cutting skin, meat, and perhaps ligaments. Such tools were probably used on scavenged animals that had been killed, partially eaten, and left behind by the large cats of the savanna that also hunted *H. habilis.*

This tool culture developed over hundreds of thousands of years, finally producing, by about 1.7 mya, "*advanced Olduwan tools,*" which included flakes shaped to scrape hides and stones worked on two sides to produce "*bifaces,*" typically used as *hand-axes*. That scavenging was *H. habilis's* only means of obtaining meat from larger game is now largely accepted, though some think it might have hunted certain small game, such as antelopes. Some evidence suggests they kept piles of stone tools in several locations—almost certainly well away from where they lived—and dragged their scavenged game there for butchering. Aside from this added meat source, *H. habilis's* diet was probably much like that of the australopithecines that preceded them in these more open grassland environments. Whether *H. habilis* was ancestral to us is debated, though it seems that those who favor that identification today have become the minority.

1. Some paleoanthropologists have suggested that *H. habilis* represents the lower height and weight range and a different species, *H. rudolfensis,* the upper range. See the discussion of *H. rudolfensis* below.

Homo rudolfensis

Uncertainties about the species *H. rudolfensis* are primarily because only a single skull and jaw have been found at a single site near Lake Turkana (formerly Lake Rudolf) in northern Kenya. No skeletal remains have been found associated with the skull and jaw fragments, so nothing can be said with any certainty about stature, bipedality, and primitive or modern body traits. Some skeletal bones found elsewhere have been attributed to *H. rudolfensis*, but that opinion is by no means shared by all paleoanthropologists. Since the skull is large, with a cranial capacity of some 750 cu cm (nearly 46 cu in), some have claimed the species was larger than *H. habilis* (at the upper range of that species' height and weight numbers noted above).

H. rudolfensis's face was long and its cheek teeth seemingly larger than those of *H. habilis*, both of which are characteristic of the australopithecines, so some have claimed it was a large-brained australopith. Other features similar to the robust members of that family have led some to claim a probable *Paranthropus* tie. Still others suggest, based on certain facial similarities, that *H. rudolfensis* represents a variant within the *H. habilis* species. Some of those who see two distinct species think *H. rudolfensis* was ancestral to *H. habilis*, and others just the reverse. Dates given for the length of time the species existed typically range from 2.5 to 1.6 mya. No tools have been found associated with the *H. rudolfensis* fossils, though it seems, based on their physiology, that they might well have created and used them.

"ERECTUS GRADE" HOMINIDS

The term *grade* refers to a group of organisms that have reached approximately the same stage of evolutionary development. It does not mean that these organisms necessarily belong to the same species, though they might. The use of the term here indicates that there are a number of similar early *Homo* remains that have been divided into as many as three distinct species by certain paleoanthropologists—*Homo ergaster*, *Homo erectus*, and *Homo georgicus*—but grouped into a single species, *Homo erectus*, by others.[2] The species name *H. georgicus*, given by their discoverers to fossils found near Dmanisi, Georgia, is now repudiated by most scholars, who identify it as one of several subspecies of *H. erectus*.[3] The major and unsettled debate

2. Another general term used for these non-Asian erectus-grade hominids is *H. erectus sensu lato*.

3. Some paleoanthropologists have termed it a sub-species of *H. ergaster*: *H. ergaster georgicus*, as opposed to *H. erectus georgicus*. It has likewise been suggested that

concerns whether *H. ergaster* and the East Asian *H. erectus* (sometimes called *H. erectus sensu stricto*) constitute two distinct species or variations of a single species, *H. erectus* (with *H. ergaster* called the "African *H. erectus*"). Still others have suggested dropping the name "erectus" and classifying the hominids as "*archaic Homo sapiens*." In what follows, I use the two-species nomenclature to clarify the apparent differences of the African and Asian populations.

Homo ergaster

H. ergaster means "working human" or "workman," reflecting its innovations in stone-tool manufacture. Most scholars agree that *H. ergaster*, with fossils dated around 1.9mya, appeared earlier than the Asian *H. erectus*. The hominid from which it evolved is widely debated, with some scholars favoring *H. habilis* and others claiming some other, possibly undiscovered, hominid. *H. ergaster's* remains have been found in both East Africa and South Africa, suggesting that the species spread throughout the continent. As noted above, the very similar Dmanisi remains have led an apparent majority of paleoanthropologists to believe that *H. ergaster* migrated out of Africa (the first hominid to do so), moved rather quickly eastward, perhaps following the game it hunted, and eventually arrived in Southeast Asia, where it became *H. erectus sensu stricto*.[4] Recent genetic analysis has tended to support this "Out of Africa" model, as has the recent discovery of *H. ergaster* remains in Asia. Erectus-grade fossils appear in Europe much later, perhaps 800,000 year ago, as a result of Asian *H. erectus* migrating westward. *H. ergaster* is estimated to have continued in Africa until about 1.3 mya.

The skull of *H. ergaster* was more human-like than any prior species, though it still bore some traits of older hominids. Its face was smaller, flatter, and less prognathic than earlier *Homo* species, had narrower, less protrudent cheekbones, and for the first time featured a narrower, outward-projecting nose with a wide base. Its cranium was higher and larger in volume than any earlier species, ranging from an average of 850–900 cu cm (52–55 cu

---

ergaster be used as a sub-species name for the African variety of *H. erectus: H. erectus ergaster.*

4. A second view, seemingly held by a minority of researchers, is that *H. erectus* evolved first in Asia and then migrated westward into Georgia and eventually Africa. The reason that *H. erectus*, not *H. ergaster*, is used by many anthropologists for these hominids, even though *H. ergaster* seems to have lived earlier than *H. erectus*, is that the Asian fossils were found first and given the *H. erectus* species name. That the first name given any species is the final designation for all fossils understood eventually to be from that species has been the general practice in paleoanthropology throughout its history.

in) in earlier members of the species to perhaps 1100 cu cm (67 cu in) later in its development, with some braincases as large as 1225 cu cm (75 cu in). This last figure overlaps the lower range of skull size range in modern humans. More primitive features include a prominent browridge, a minimal forehead, and a weak chin. *H. ergaster's* jaws were more slightly built, featured smaller molars and premolars, and shorter, less pointed canines than in previous hominids, more like ours today.

H. ergaster's body was quite human-like—tall and slender, with long legs and limb proportions very much like our own. For the first time, a hominid's legs were much longer than its arms. Males are estimated to have been as much as 1.9 m (6 ft 3 in) in height, while females averaged perhaps 1.6 m (5 ft 3 in). This implies a reduced sexual dimorphism relative to all prior hominids. *H. ergaster's* novel body shape would have allowed more efficient heat dissipation, which, along with a probable paucity of body hair, was likely an adaptation to the hot and dry environment in which it lived. Some scholars also suggest that this species developed dark skin, which would have protected it from the damaging effects of ultraviolet sunlight once their bodies began to lose hair. Its narrower, more human-like pelvis, more barrel-shaped (versus prior cone-shaped) chest, leg joint orientation, and other adaptations provided better balance and so allowed *H. ergaster* to walk and run for long distances, as we can. It, along with *H. erectus*, is understood to have been the first human ancestor to walk fully upright. All adaptations for tree climbing seen in more primitive hominids were now gone.

H. ergaster's larger brain evidently fueled the major innovations in tool making that are noted over the course of the species history. Early on, *H. ergaster* used the relatively crude Olduwan tools that *H. habilis* likely also made, but by about 1.6 mya they were apparently making a more diverse and sophisticated toolkit that paleoanthropologists have termed *Acheulean* (St. Acheul, in France, was the location of the earliest discovery of this advanced tool culture).[5] Prominent among the Acheulean innovations were bifaced hand axes, pointed picks, and straight-edged cleavers worked on both sides to produce sharper and more durable tools. They were used primarily for cracking bones to get at the marrow inside, butchering large animals, working hides, woodworking, and perhaps digging. These advanced tools were in use in Israel by about 1.4 mya but in Europe do not appear until around

5. Tools found in Georgia are of the primitive Olduwan type, supporting the belief that *H. ergaster* had already migrated out of Africa before this new Acheulean tool culture had been developed. This also accounts for the absence of Acheulean tools in Asia, the endpoint of *H. ergaster's* long migration.

800,000 years ago in Atapuerca, Spain. Early Acheulean stone tools first appear in India about 680,000 years ago.

The climate in Africa dried out and became more seasonal around 1.8 mya, creating extensive savannas, while globally the temperature was dropping. *H. ergaster* may well have been the first hominid to adjust to these new conditions. In its native Africa it likely lived relatively near streams and other sources of water, as humans and animals do today in such areas. Its food probably consisted of the tubers that were common on the savannas along with fruits, nuts, insects, smaller vertebrates, and larger game when available. Most paleoanthropologists understand *H. ergaster* to have been primarily a scavenger, though some cooperative hunting might also have occurred. It likely consumed more meat than any prior species. *H. ergaster* almost certainly shared its environment to some degree with *Paranthropus*, *H. habilis*, and perhaps other contemporaneous hominids.

Whether *H. ergaster* had learned to control fire is a topic of debate. Charcoal and charred animal bones have been found together in South African deposits possibly as old as 1.5 mya, and therefore some paleoanthropologists think the species used fire at least some of the time, perhaps for cooking and warmth. Colder temperatures during glacial periods in more elevated African regions, such as the *Gadeb Plateau* in Ethiopia, may have required mastery of fire by *H. ergaster* to survive there. As the species migrated out of Africa, it encountered temperate climates in the more northern latitudes of Georgia and China, which grew yet colder during glacial periods. These cold winter environments also suggest at least some control of fire by the erectus-grade hominids that lived in those places. Many scholars seem to think this use of fire was opportunistic, with these *Homo* species using natural fires as sources but lacking the ability to create fire by technological means. The earliest clear evidence for the controlled use of fire outside Africa comes from the *Benot Ya'aqov* site in Israel, which dates to perhaps 700,000 years ago.

H. ergaster probably lived in something like modern human hunter-gatherer societies, with social groups reflecting family ties. Available evidence suggests that their young took longer to mature than modern apes but not as long as modern humans, implying a longer period of development than in earlier species. There is also evidence that *H. ergaster* cared for sick or injured members of the group but none that they buried the dead, which would have indicated a concern for their life after death.[6]

6. Editor's note: Jarvis did not explain why he believed that the practice of burying the dead "would have indicated a concern for their life after death." Anthropologists might draw that conclusion if, in addition, they found artifacts normally useful for the living (e.g., food, ceramics, tools, etc.) buried with the dead. Burial alone does not

Homo erectus sensu stricto

For a photo of a reconstruction of H. erectus by Chris Wenzel at the Westfälisches Museum für Archäologie, Herne, see http://commons.wikimedia.org/wiki/File:Homo_erectus_new.JPG.

Much of what we have said about *H. ergaster* is applicable to *H. erectus* ("upright human") *sensu stricto* as well. Its skull was similarly large, with capacities of up to 1,150 cm (70 cu in), its face flat, its browridge thick and jutting outward, its forehead small and sloped, its nose protrudent, the space between its nose and mouth long, and its chin weak. The main differences from *H. ergaster's* skull are its somewhat higher cranium, its thicker skull bones, and several other minor features. Few skeletal remains have been found, but those that have been suggest that *H. erectus sensu stricto* had more human-like body proportions, quite similar to those of *H. ergaster*, with height, weight, and sexual dimorphism estimates falling also within the range of the African species. Some bones also indicate a similar robustness to their frames.

Some *H. erectus sensu stricto* fossils from the island of Java in Southeast Asia have been dated to approximately 1.8 mya, and erectus-grade hominids persisted there until perhaps 50,000 years ago, according to some recent estimates of certain Javanese deposits. They may even have existed as recently as 12,000 years ago, if *H. floresiensis* is determined to be a later form of *H. erectus sensu stricto*.

Tools found associated with *H. erectus* are of the older Olduwan type, including choppers, unworked flakes, and cores, although some primitive bifaces have also been found. This is not surprising if *H. ergaster* began its eastward migration by 1.8 mya, or some 200,000 years before Acheulean tool-making developed in Africa. There is evidence that suggests *H. erectus sensu stricto* also made tools from other materials, especially bone, wood, and bamboo. These tool materials reflect the largely tropical environment, with high rainfall, in which this hominid typically lived. Lush jungles produced the fruit and other vegetation that probably constituted *H. erectus sensu stricto's* main diet throughout the year. It might also have eaten some insects and small vertebrates at times.

---

support that conclusion. The lack of evidence for the practice of burying the dead—an argument from silence—might preclude belief in a life after death. However, as counter evidence, some ancient and modern people who believe in a version of life after death or reincarnation practice cremation, instead of burying the dead.

Homo erectus (or ergaster) georgicus

This erectus-grade subspecies refers to the fossils found in Dmanisi, Georgia. The remains consist of several fossil skulls, jaws, and fragments of several skeletons, all dating to approximately 1.8 mya. The skulls are primitive in form, with brain-case volumes estimated to be about 600 cu cm (36.6 cu in)—significantly smaller than that of *H. ergaster,* or *H. erectus sensu stricto,* or any other non-African hominid save for the pygmy *H. erectus floresiensis.* Its upper body is likewise primitive, though its spine and lower limbs are more advanced, indicating greater mobility. Males stood some 1.3 m (4 ft 3 in) and were appreciably larger than females. *H. erectus georgicus* was clearly bipedal.

Tooth wear and associated animal bones indicates it was omnivorous, though there is no evidence it cooked its food. It also used stone tools of the Olduwan culture.

Homo erectus pekinensis

Some paleoanthropologists believe that *H. erectus pekinensis* (or "Peking man," named for the city of Peking, now Beijing, which is near the two caves in a single hill in which the fossils were found in the 1920s and 1930s) should be classified as its own species, *Homo pekinensis.* The current consensus, however, seems to favor classifying it as a subspecies of *Homo erectus* that developed from the original tropical Southeast Asian population after moving north into the more temperate Chinese climate. Additional fossil finds would help settle this debate, though none have been found in more than seventy years. Dates for the fossils have recently been estimated at between 680,000 and 780,000 years ago.

The numerous *H. erectus pekinensis* skull and teeth fossils come from an estimated thirty-two to forty persons, a significant portion of which were adolescents. Adult brain cases were low, long, and thick, with a volume was typically about 1,050 cu cm (64 cu in). The face was vertical, though its upper and lower jaws protruded a fair bit from the plane of the face. It had a browridge that stretched across both eyes, curving downward beyond each eye and with a depression in the middle. The teeth rows of *H. erectus pekinensis* were parabolical, as in modern humans, though its cheek teeth were typically larger and its front teeth smaller than ours. Tooth size differences between males and females appear to have been greater than in modern humans as well.

Very few skeletal remains and no complete limb bones of *H. erectus pekinensis* have been found, which makes reconstructing its body difficult. Its wrist bones have joint surfaces that curve and are oriented differently from those in modern humans. Its limb bones are relatively thick, suggesting that its body may have been relatively heavy for its height, which has been estimated at 146 to 156 cm (57.5 to 61.4 in). Living in cooler climes that included snowy winters, *H. erectus pekinensis* would have done well with a stockier body (relative to that of *H. erectus sensu stricto*), since it would have retained heat better and so made survival easier.

Fire also would have helped it through the freezing weather, as would have clothes and some form of shelter, and it is likely that all of these were common features of the *H. erectus pekinensis* lifestyle. Its tools were the simpler Olduwan type, mostly flakes and simple choppers, and few indicated work on more than one side. In this they were more like tolls of *H. erectus sensu stricto* to the south than the erectus grade species to the west with their Acheulean tool culture. Over time these tools grew smaller and the flint used to manufacture them finer.

H. erectus pekinensis was likely omnivorous, eating fruits and berries when available during the warmer seasons, along with nuts, seeds, roots, and tubers, with meat likely becoming more predominant during the winter months. Associated animal remains in the two caves indicate that these proto-humans ate mostly venison, though this was supplemented with buffalo, rhino, horse, and boar meat. At least one anthropologist has suggested that *H. erectus pekinensis* was a scavenger rather than a hunter, but there is currently no consensus on this point.

Homo erectus soloensis

The remains of this *H. erectus* subspecies, found along the Bengawan Solo River on the island of Java, in Indonesia, are also known as "Solo Man" or the "Ngandong fossils," after the nearby village of that name. Once thought to be a subspecies of *H. sapiens*, they have since been reclassified to fit within the earlier hominid species. Dates for the fossils are estimated to be between 550,000 and 143,000 years ago.

Based upon analysis of the eighteen fossil crania, scholars have determined that this gracile hominid had a brain capacity ranging from 1,013 to 1,251 cu cm (61.8 to 76.3 cu in), well within the *H. erectus* range. In their other features they are likewise typical of *H. erectus*, though their material culture was more advanced. While no stone hand-axes or biface tools have been found associated with *soloensis* fossils, and only a few stone flakes and scrapers, tools made from bone, antler, and ray spines have been found. Associated animal fossil remains indicate that *Soloensis* ate a significant amount of meat, including hippos, elephants, water buffalo, deer, and cattle. Evidence also suggests that they practiced cannibalism and perhaps even headhunting.

THE "ARCHAIC HOMO SAPIENS"

"Archaic *Homo sapiens*" is a disputed category used by some paleoanthropologists to refer to a group of *Homo* species that lived from about 500,000 years ago (or more, depending on what hominids are included)[7] down to the appearance of modern humans, termed *Homo sapiens sapiens*, about 200,000 years ago. Other specialists reject the lumping of all these older *Homo* species under the single species *Homo sapiens*. For example, hominids that these latter scientists call *Homo heidelbergensis* are called by the former scientists *Homo sapiens heidelbergensis* (*heidelbergensis* designating the hominid as a sub-species of *H. sapiens*). This kind of nomenclatural confusion exists because there is currently no generally accepted definition of *Homo sapiens* as a species. In what follows I will refer to each hominid as a distinct species per the more traditional practice.

Homo gautengensis

In 2010, South-African researcher Darren Curnoe proposed the existence of this species, basing the proposal on his reinterpretation of a partial skull that for decades had been assigned to *H. habilis*. Most paleoanthropologists, however, reject his proposal. They assign to this new species other bones, from various locations, previously identified with *H. habilis*, *H. ergaster*, and even a species of *Australopithecus*. Dates suggested for the species are approximately 1.9 mya to 600,000 years ago.

According to Curnoe's analysis, *H. gautengensis* had a small brain and a small stature, estimated at some 90 cm (3 ft) in height and 50 kg (110 lbs) in weight. Its large teeth suggest that it was primarily a plant-eater. It probably spent most of its time in the trees but was bipedal on the ground. It appears to have made and used stone tools and fire, though it seemingly did not communicate by means of language.

7. Typically *Homo heidelbergensis*, *Homo rhodesiensis*, and *Homo neanderthalensis* have been included in the archaic *Homo sapiens* category, while some have also added the earlier *Homo antecessor*. A number of more recently unearthed *Homo* species, detailed below, must now also be added.

Homo antecessor

H. antecessor ("human explorer/pioneer"), so named because it appears to be the earliest hominid to have migrated into Europe, is currently understood to have lived from about 1.2 mya to 500,000 years ago (or perhaps even later). Fossils of *H. antecessor* have been found in Spain and England, suggesting that they ranged at least throughout Western Europe. The status of this species is greatly debated, owing in part to the fragmentary nature of the fossils and the fact that the main specimen comes from a juvenile. Some see *H. antecessor* as having descended from *H. ergaster* and the ancestor of *H. sapiens*; others see it as ancestral to the Neanderthals and not modern humans, but most seem to understand it as the same species as *H. heidelbergensis* with different adaptations due to the colder temperate climates in which it lived.

Skull fossils consist of a number of cranial fragments but no complete skulls, making estimates of braincase volume uncertain. Virtually all researchers accept that its brain size was upwards of 1,000 cu cm (61 cu in), with some estimates as high as 1,150 cu cm (70.2 cu in). Its mid-face was modern, being relatively flat with a protruding nose. Its brow-ridge featured an arch above each eye, its forehead was low, and its chin was weak, as in more archaic *Homo* species. It also had a protrusion at the back of the skull, known as an "*occipital bulge*." *H. antecessor's* lower jaw was thinner than *H. ergaster's*, though its molars were similarly small (as were its canines) and triple-rooted. Tooth eruption patterns appear to be similar to those of modern humans, though perhaps occurring at a more rapid rate.

H. antecessor's body seems to have been similar in size and shape to ours but somewhat more robust, though judgments about it are based only on fragmentary skeletal remains. Estimates of its height typically fall in the 1.6–1.8 m (5 ft 3 in to 5 ft 11 in) range, with the average weight of males perhaps around 90 kg (198 lbs). It had a broad chest and long forearms, and its hand, wrist, finger, and toe bones are in most respects quite like humans. *H. antecessor* apparently stood and walked as we do.

Tools found with *H. antecessor* remains were all of the more primitive Olduwan type rather than the more advanced Acheulean forms. The *hammer stones*, flakes struck on a single side, and cores indicate that their tools were used primarily for butchering animals and cracking bones to get at the marrow inside (though they may also have been used for scraping hides and woodworking). *H. antecessor* apparently lived among forests and grassy lowlands, with temperatures that ranged from seasonable but relatively mild in Southern Europe to colder winter climates farther north in England. Its diet seems to have consisted of substantial quantities of meat, probably more during cold winters, when plant foods such as herbs, nuts, fruits, seeds, tubers, and roots would have been less available. They probably scavenged some of their meat but evidently hunted larger animals such as deer and horses as well.

It seems that, while living in caves part of the time, *H. antecessor* also had a nomadic lifestyle in which they probably migrated according to the availability of their regular plant and animal foods. Cut marks similar to those found on animal bones are also found on human bones, implying that *H. antecessor* defleshed the bones of its own kind. While this might have been related to funerary rituals, the absence of evidence for such rituals until hundreds of thousands of years later makes cannibalism the more likely explanation.

Homo cepranensis

This is a proposed *Homo* species based on a single crushed skull that was unearthed during a road-construction project near the town of Ceprano in central Italy in 1994. No other fossils purporting to have come from this hominid have been found that might help sort out whether it is a new species intermediate between *H. erectus* or *H. antecessor* and *H. heidelbergensis* or simply another *H. erectus* fossil—all of which have been proposed and the last of which seems to be the current consensus. It has been suggested that it might represent a *H. erectus* form that originated in Asia and migrated into Europe. Dates for the fossil have ranged from 900,000–800,000 to 500,000–350,000 years ago and so have further complicated the interpretation of it. All that can be discerned from the reconstructed skullcap are some features like *H. erectus*, including very thick skull bones, a small brain (estimated to be 1057 cu cm or 64.5 cu in), a large browridge, and a significantly angled occipital area (at the lower back of the skull).

Homo heidelbergensis

Classification of these fossils, called *H. heidelbergensis* by those who found the first ones near Heidelberg, Germany, is plagued by the same interpretative issues we have seen regarding remains of *H. ergaster*, *H. erectus*, and *H. antecessor.* Dating of the remains from various sites has been necessarily inexact, with 600,000–400,000 years ago generally accepted as most probable, though estimates range from 800,000 to 150,000 years ago, or even up to 1.3 mya, depending on how the species is understood. The earliest dates reflect the opinion of some that *H. heidelbergensis* and *H. antecessor* are in fact the same species, or even that *H. heidelbergensis* is simply a late form of *H. ergaster*.

Its range is commonly given as pan-European, with sites found in Germany, Spain, France, England, Greece, Hungary, and Italy, as well as in Morocco and Israel. Many claim its range reached all the way down to South Africa, reflecting their belief that the African *H. rhodesiensis* and European *H. heldelbergensis* were the same species. Many believe the European *H. heidelbergensis* evolved into *H. neanderthalensis* while the African *H. heidelbergensis* (or *H. rhodesiensis*) eventually evolved into *H. sapiens*.

Skull remains found to date are incomplete, consisting of fragments of greater or lesser size. Braincase volumes range from about 1,100 to 1,450 cu cm (67.1 to 88.5 cu in) with an average of about 1,275 cu cm (77.8 cu in)—within 10 percent of the average size of modern human brains. *H. heidelbergensis* typically had a single brow-ridge that protruded from its low, relatively narrow forehead and dominated its wide-set eyes. Its nose was likewise protrudent from its relatively flat face, with a wide nasal opening. Its jaws were large and its teeth displayed distinctive characteristics that separated them from those of modern humans.

H. heidelbergensis was tall yet relatively stout, as would be expected for hominids living in colder climates. Height estimates range around 1.8 m (nearly 6 ft) for males and 1.55 m (around 5 ft 2 in) for females, with males weight perhaps as much as 90–100 kg (198–220 lbs). *H. heidelbergensis* had long bones, with limb, hand, and foot proportions very much like those of the Neanderthals that were probably their descendants.

Tools used by *H. heidelbergensis* were similar to the Acheulean types used by *H. ergaster*, though with some innovations seen also in later Neanderthals. They seem to have used some tools for butchering meat and others for preparing hides and working wood. Some evidence suggests that they used their teeth to hold materials while they cut them and their front teeth to prepare hides, among other things. They certainly hunted large game, as the bones of elephants, hippos, rhinos, horses, and deer with cut marks on them have all been found associated with *H. heidelbergensis* fossils. Their main food, however, was probably vegetable matter, as is common in modern hunter-gatherer societies.

The climate in which they lived varied with the latitude and time of year, with cooler weather predominating the farther north they lived, though the occasional glacial periods would have created colder conditions over long stretches of time. In these times, and during winters in northern areas, they likely used caves for shelter, wore skins for clothing, and used fire to help them survive. A site in France shows that they built simple shelters of rock and wood, another in Israel indicates that they were capable of controlling fire and building early forms of fireplaces, and another in Germany shows that they made and used wooden spears.

No art has been found associated with *H. heidelbergensis,* though some *red ochre,* a mineral used to create red pigment, has been found at a site in France. Some researchers believe they had even acquired the capability for some primitive form of speech. Also, a cave in Spain contained a shaft in which were found remains of some twenty-eight *H. heidelbergensis* individuals, possibly indicating that they buried their dead and perhaps even did so with some form of ritual.[8] There is no evidence they practiced cannibalism. We seem to see in *H. heidelbergensis* the beginnings of a more sophisticated culture that would continue to develop in subsequent *Homo* species.

Homo rhodesiensis

A fossil skull was found in 1921 in a lead and zinc mine in the Broken Hill (or Kabwe) region of the southern African country of Rhodesia (now Zambia). It became known, not surprisingly, as the Broken Hill Skull or Kabwe cranium and led to the claim of a new species discovery, "Rhodesian Man." Since then, comparable remains have been found at other sites not only in southern Africa but in East and North Africa as well. The remains all date from about 300,000 to 120,000 years ago. As already noted, most paleoanthropologists believe that *H. rhodesiensis* belongs to the same species as *H. heidelbergensis,* though other scholars have suggested it should be considered a member of the *H. erectus, H. neanderthalensis,* or *H. sapiens* (as *H. sapiens rhodesiensis*) species.

The three skulls found to date have many things in common with the skulls of *H. heidelbergensis,* though not in every way. The skulls are very robust. Braincase volume estimates range from as little as 1,100 cu cm (67.1 cu in) to as much as 1325 cu cm (80.9 cu in), with an average volume of perhaps 1250 cu cm (76.3 cu in). *H. rhodesiensis* featured a very prominent browridge, a broad face, a large, protrudent nose, and a parabolic palate

8. Editor's note: On September 10, 2015, the Rising Star Expedition published a report of their discovery (dating back to October 2013–March 2014) of a burial site in a South African cave containing bone fossils for at least fifteen individuals of a new species of hominin. They named it *Homo naledi,* which comes from the word for "star" in Sotho (or Sesotho), one of the languages spoken in South Africa. Although still undated, the range of dates for these bone fossils is from 2 mya (or more) to 1 mya (or less). With australopith-like features, *Homo naledi* shares *Homo erectus* as a common ancestor with *Homo sapiens.* See Berger, "*Homo naledi*"; Howley, "Homo naledi"; and Shreeve, "This Face Changes the Human Story."

like ours. No jawbones and only a few teeth in poor condition have been found, giving researchers few elements for comparison with other, roughly contemporaneous *Homo* species. The few skeletal bones likely belonging to *H. rhodesiensis* suggest a robust frame similar to that of *H. heidelbergensis,* though perhaps somewhat pigeon-toed in its lower limbs. Size estimates are uncertain but seem congruent with those of *H. heidelberegensis*. One South African scholar has suggested that some populations there were much taller, typically more than 2.13 m, or some 7 ft, in height.

Acheulean tools found in Africa at this time show advances from earlier such tools, with the earlier being thicker and less symmetrical and the later evidencing more trimming work. This may indicate the arrival of *H. rhodesiensis* there (from Europe, assuming the *H. heidelbergensis* connection). These stone tools were worked on both sides and included the usual hand axes along with a rich variety of other specialized tools. Some carefully worked symmetrical hand axes suggest an aesthetic sense, though this is by no means established fact. Horns, ivory, and bone were also worked into tools. The climate and habitat in southern Africa were essentially the same then as today. Tooth wear indicates a diet similar to that of *H. heidelbergensis*, composed of some meat but large quantities of grains, roots, and tubers. One *H. rhodesiensis* skull indicates cut marks that imply that its flesh was removed after death, indicating either cannibalism or some type of death ritual.

Homo neanderthalensis

Remains of *H. neanderthalensis* have been found from Uzbekistan and Iraq in the east, through Israel and Lebanon in the Middle East, and in most of Europe, reaching to France, Spain, and Gibraltar in the west. None has been found in Africa. Dates for the species have been given as some 230,000 years ago at the earliest to about 22,000 years ago at the latest, which implies that *H. neanderthalensis* survived, at least in Europe, well after the arrival of *Homo sapiens* there.

There is debate regarding whether Neanderthals should be classified as their own distinct species or as a subspecies of *H. sapiens, H. sapiens neanderthalensis* (as distinct from *H. sapiens sapiens*, or modern humans). Confusing the matter is the fact that two rather different Neanderthal-like forms have been found in the eastern sites, but only one of them in Europe (usually called the "classic Neanderthal" type, one well adapted to the typically colder climate there). Some researchers identify both as being Neanderthals, one type being "classic" Neanderthals and the other being "progressive"

Neanderthals, while others see them as representing two distinct species, *H. neanderthalensis* and *H. sapiens*. Since most paleoanthropologists seem to accept the latter view, Neanderthals are here presented as a distinct species.

For a photo of a reconstruction of a Neanderthal skinning its prey (in the Neanderthal Museum, Mettmann, Germany), see https://commons.wikimedia.org/wiki/File:Neandertaler-im-Museum.jpg.[9]

Many Neanderthal skulls have been discovered. They are generally large, with a mean braincase volume of about 1420 cu cm (86.7 cu in), or slightly larger than that of modern humans. Some cranial specimens reach as much as 1720 cu cm (105 cu in), and all feature a pronounced bulge at the back of the skull. Its face was long and large but less vertical than ours, with a receding forehead and chin but a more protruding middle facial area than found in either erectus-grade hominids or modern humans. It also had a very large nose. Its browridges, one above each eye, protruded significantly, as did both jaws. Its front teeth were larger and its molars somewhat narrower than ours, with a tendency to have an additional, central cusp. Neanderthals were probably lighter-skinned and had lighter-colored eyes than previous hominids, and typically blond or reddish hair and beards, all adaptations to the colder environments in which they typically lived.

Neanderthal bodies, especially those found in Europe, tended to be shorter and more robust than ours—another adaptation to the cold, similar to the body form seen in modern humans living in colder environments. Males averaged some 168 cm (66 in) in height and weighed much the same as modern Europeans do. Their chests were large and conically shaped, their vertebral columns short and massive, and their pelvises wide and thick. In general, their limb bones were thicker and shorter, with attachment areas that imply powerful muscles, meaning that they would have been on average much stronger than we are.

The craft of tool-making seemingly reached new levels of sophistication with the rise of the Neanderthals. By about 100,000 years ago Acheulean tools disappeared and the more sophisticated *Mousterian tool culture*

9. Editor's note: The following are comments by the Wiki author ("Ökologix"): "Skulptur im Neanderthal-Museum. Stellt vermutlich NICHT einen Neanderthaler dar [wie vom Dateinamen suggeriert], sondern einen Cro-Magnon [Teil der Ausstellung „Weltenwanderer": 40.000 bis 10.000 Jahre vor heute]; sollte überprüft werden, bevor das Bild verwendet wird" (http://commons.wikimedia.org/wiki/File:Neandertaler-im-Museum.jpg). My translation follows: "[The] sculpture in the Neanderthal Museum is probably NOT a Neanderthal (as the file name suggests), but a Cro-Magnon . . .; it should be checked before the image is used."

appeared. It has been suggested, based on artifacts from eastern sites, that this culture had been learned from early *H. sapiens*, but the time and location of these tools largely coincide with those of the Neanderthals, hence the general identification of Neanderthals with Mousterian tool culture. The novelty of this culture lies in the abundance of forms and the way in which the tools were created. Stone cores were first worked until they had achieved the desired shape. Then flakes of precise form desired for a given use were struck from the core, after which they were further worked to provide the various tools and weapons of this culture. This technique not only required great skill and forethought, but was also a very efficient use of the stones. Neanderthals created as many as sixty different types of points, scrapers, and other tools used for hunting, preparation of meat and skins, and woodworking. Despite this varied tool kit, no projectile weapons such as arrowheads have been found, which implies that Neanderthals hunted large animals using spears at close quarters rather than projectile weapons from a distance. Some pendants and other decorative objects, such as carved and perforated bear claws and teeth—likely strung together—have also been found associated with Neanderthals. This suggests that they had an aesthetic sensibility, but that is debated among Neanderthal experts.

The Neanderthal diet seems to have consisted mainly of meat, though this was likely supplemented by fruits and vegetables as these were available seasonally. This would have fit with the cold climates in which they often lived, where plant products were unavailable during the snowy winters, especially during the several glaciations they lived through. Some locations may have had year-round snow and ice, meaning that meat would have been the only food option; yet even warm Israeli sites indicate that Neanderthals preferred meat. Some sites provide evidence that Neanderthals were butchered and eaten, suggesting at least some cannibalistic practices, though whether they actively killed one another for food or merely scavenged the already-dead bodies of their kind is unknown.

Culturally, several important advances are in evidence at Neanderthal sites. While they often lived in caves, some also lived in the plains, where they built hearths (demonstrating that they controlled fire) and shelters made, for example, of mammoth bones with skins likely stretched over them. The great debate among paleoanthropologists is whether or to what extent Neanderthals were symbolic thinkers. This is particularly true with regard to their intentional burials. They are the earliest hominids with confirmed burials, their oldest grave thus far discovered dated at some 100,000 years ago.

Such burial of the dead may have been only an occasional practice by some Neanderthals or may have been widespread. They may have buried only loved ones, perhaps to protect their bodies from scavenging, or for hygienic or religious reasons. The main support for the last interpretation comes from what may have been ritual practices associated with the burial: especially the fact that the graves had been dug, not found; the seemingly careful positioning of the bodies in the grave; the apparent use of burial markers (at least at some sites); and the inclusion of things such as stone tools, animal bones, and perhaps also flowers, pigments, and other natural objects. Some skulls have been found with the muscle intentionally cut off, which may have been a ritual religious practice. *Animal worship* or *totemism*[10] has also been suggested as a Neanderthal religious innovation, though the evidence for this is controversial. The question of religiosity, like the question of aesthetic sensibility and symbolic thought generally, remains debated due to the lack of conclusive evidence to date for any of the relevant viewpoints. The question of whether and to what extent Neanderthals may have used language in their communication is similarly undetermined.

Also unclear are the relationships between *H. neanderthalensis*, previous *Homo* species, and *H. sapiens*. As already indicated, experts have yet to achieve consensus on whether Neanderthals should be classified as their own species or as a subspecies of *H. sapiens*. How one answers this question determines to a great extent how one understands the ancestral relationship between Neanderthals and us. If they constitute a separate species, then it is likely they were not ancestral to *H. sapiens* but instead another of the many dead-end branches in our family tree; if they are a subspecies of *H. sapiens*, then we likely descended from the Neanderthals, or they at least contributed to our genetic makeup by interbreeding. There is significant evidence that such interbreeding took place, which tends to support the single-species theory. Yet others suggest that Neanderthals belong to *H. heidelbergensis*, and still others believe that they descended from that earlier species. Clearly, then, there is no consensus regarding from whom Neanderthals descended or whether we descended from them, though the majority view seems to consider them a separate species and so a branch on the *Homo* family tree. The reason for their extinction some 22,000 years ago is similarly unclear, though competition from *H. sapiens* in a changing, warming environment for which they were physically less suited appears to be the favored hypothesis.

10. Totemism is the religious practice in which humans understand themselves to be connected with a spiritual beings (usually one per individual or social group) found in an animal or plant, which serves as the symbol or emblem of the person or group.

Homo floresiensis

The fossil remains of "Flores Man" were discovered in 2003 on Flores Island, near Bali, in Indonesia. To date the remains of nine individuals have been found, including the nearly complete skeleton and skull of a thirty-year-old woman who has been nicknamed the "Little Lady of Flores," or simply "Flo." Dates for the fossils are 74,000 to about 13,000 years ago, with some paleoanthropologists suggesting they may have occupied the island for 400,000 years or more.

The most striking thing about *H. floresiensis* was its size: it was little more than a meter (39.4 in) in height. When Flo was first discovered, some researchers thought she suffered from a growth-stunting disease, but finding the other remains, all of similar scale, refuted this view. *H. floresiensis* is now understood to be a pygmy species whose small size fits the similarly reduced size of certain other animals on the relatively small island, such as *Stegodon*, a dwarf member of the elephant family. There seems little doubt among paleoanthropologists that *H. floresiensis* belongs to a species different from *H. sapiens*; most put it in the genus *Homo*, though it has been suggested that its unique combination of both primitive and modern features justify assigning it to a different genus.

The *H. floresiensis* head was small but proportional to its overall body size, its thick skull case indicative of a brain size roughly 380 to 417 cu cm (23.2 to 25.4 cu in), comparable to that of the australopithecines and modern chimps but only a quarter of that of modern humans. Its face was very small but vertical and with a relatively high forehead. It had small browridges above each eye that were rounded to match approximately the curvature of its eyes, and the space between its nose and mouth was much smaller than ours. Its upper and lower tooth rows were narrow, its front teeth and canines were small and non-protruding, its premolars rather unusual, and its molars typically Homo-like in their relative sizes.

The skeleton featured stouter limb bones and different joint surfaces and muscle attachment areas from those found in *H. sapiens*, with armbone to thighbone ratios quite similar to those of *A. afarensis* (Lucy). Its wrist bones were similarly Lucy-like. It was bipedal, though it may have walked on all fours part of the time. Estimates of its height and weight, based on comparisons with modern African pygmies, range from 97.5 to 109 cm (3 ft 2 in to 3 ft 7 in) and from 16 to as much as 36 kg (36 to 80 lbs).

The diet of *H. floresiensis* was likely omnivorous. Associated bones indicate that they ate stegadons, large rats that developed on the island, lizards—including some very large forms similar to the "Komodo dragons" found today on nearby Komodo Island—frogs, and turtles, as well as insects, tubers, seeds, herbs, and fruits. The tropical climate of Flores Island then was essentially the same as it is today.

Their small stature and brain did not prevent them from manufacturing and using sophisticated stone tools, though there is some question whether they produced all the tools found on the island or whether *H. sapiens*, which arrived between 55,000 and 35,000 years ago, was responsible for the more advanced forms. Most of the tools found are flakes worked on both sides, blades both large and small, points and punches, and perhaps choppers as well. Many of these tools were used in hunting and butchering animals, and it is likely that they cooperated with one another in killing larger animals, such as stegadons. Evidence suggests that they used fire for cooking, and they probably had acquired language.

The current scholarly consensus is that *H. floresiensis* descended from *H. erectus* after having migrated to the island between 1,000,000 and 100,000 years ago and after which the dwarfing trend began. A volcanic eruption on the island some 12,000 years ago may have been responsible for the extinction of the species, as it seemingly was also for stegadons and other animals on the island. It may also have wiped out the *H. sapiens* on the island at that time.

Homo denisovanensis (?)

While this species of archaic *H. sapiens* has not yet been officially named—a privilege given the finder of each species of fungi, plants, and animals—the eventual name will probably be *H. denisovanensis* or something very similar. Currently the species is called the Denisova hominins or Denisovans. Alternately, as we saw in the case of Neanderthals, it might also be named *H. sapiens denisovanensis*, depending on how its relationship to our species is construed. The fossils from this new hominid were found in 2010 in Denisova Cave in Siberia. They include a finger-bone fragment, a toe bone, and a single tooth, which have been dated to around 41,000 years ago. Related artifacts, including a bracelet, have been similarly dated to around 40,000 years ago. Neanderthals and modern humans likely also lived in this region at the time.

Given such limited remains, and how little about the Denisovans can be learned from them, researchers have had to depend on the findings of their analysis of *nuclear DNA* (DNA from the nucleus of the cell, typically denoted simply as DNA) and *mitochondrial DNA* (DNA found in the mitochondria, termed *mDNA*, which is received only from one's mother) to learn more about this species. These analyses suggest that the species shared a common ancestor with Neanderthals about 640,000 years, and that this common ancestor appeared after this Neanderthal-Denisovan branch of the human family diverged from the ancestor they shared with modern humans some 804,000 years ago. At present little else can be conjectured regarding this species.

Homo sapiens idaltu

Fossil remains of this subspecies of *H. sapiens* (*idaltu* means "elder" or "first-born" in the Saho-Afar language) were discovered in 1997 in East Africa and consist of three relatively complete skulls, one from a child probably aged 6 or 7, as well as fragments of several other skulls and some teeth, all dated to between 160,000 and 154,000 years ago. This is around the date (about 150,000 years ago) and the place (Africa) that genetic analyses of both mDNA and the male Y chromosome had predicted for the origins of *H. sapiens*.

Based on this and the nature of the skulls, its discoverers have claimed that *H. sapiens idaltu* was the originating *H. sapiens* form and our immediate ancestor. This find is also claimed to provide significant support for the "*out-of-Africa theory*," which proposes that modern humans originated in Africa and spread thereafter to Europe and Asia and ultimately throughout the world (as opposed to the "*multi-regional theory*," which claims that modern humans arose simultaneously in Europe, Africa, and Asia from hominids that had earlier emigrated from Africa). It also rules out the claim that Neanderthals were our direct ancestors, given that *H. sapiens idaltu* lived earlier than *H. neanderthalensis*.

All of the three skulls found had been crushed into hundreds of pieces and had to be painstakingly reconstructed over a period of several years. The largest and most complete skull, probably from an adult male in his late twenties or early thirties, lacks a jawbone but otherwise suggests a recognizably human head with several archaic features. These latter include a more robust and elongated (front to back) skull and certain upper jaw features. Characteristics it shares with us modern humans based solely on the skull—as no skeletal remains have been found—include a higher cranium, less prominent browridges, and a very similar brain size, with the more complete adult specimen having a volume estimated at 1,450 cu cm (88.5 cu in).

More than 640 sophisticated stone tools, transitional between the Acheulean and later *Mousterian* and *Aterian* toolkits, have been found. They include cores, hand axes, flakes, flake tools, and blades. Some of these were evidently used in butchering the large mammals, such as hippos and buffaloes, which formed a significant part of their diet in addition to the plant foods they almost certainly consumed. Whether the large animals were hunted or scavenged cannot be determined. They seem to have lived along the lakeshore formed when the Awash River was temporarily dammed some quarter of a million years ago.

Some of the *H. sapiens idaltu* tool kit was also evidently used to cut away muscles from the skull bases of the deceased, after which the rear of the skull base was broken off and its edges polished. The smoothness of the skull suggests that the head was handled often. All this, plus the fact that no skeletal remains have been found with the skulls, indicates that *idaltu* practiced death rituals similar to those of some New Guinean societies today in which the skulls of ancestors are similarly treated and subsequently worshipped. Whether this also indicates that *H. sapiens idaltu* practiced ritual cannibalism cannot be determined.

The Red Deer Cave People

This hominid, to which no species name has been attached, derives its name from the Red Deer Cave (or Maludong, Malu Cave) in southern China where the first fossils were unearthed, along with many remains of the extinct red deer that were evidently a dietary staple. Additional fossils were subsequently found at a second site, Longlin Cave, also in southern China. The first fossils were discovered in 1979, but analysis of them began only in 2009. Scientists who have analyzed the fossils have offered various interpretations. Some believe that the Red Deer Cave people represent a new, archaic *Homo* species. Others believe that they belong to a *H. sapiens* population that migrated out of Africa in one of the many emigrations that evidently took place after 100,000 years ago and retained some archaic traits, as did some other *H. sapiens* populations found in Africa at about the same time.

The four partial skulls, two lower jaws, and several teeth represent a mix of archaic and modern human features. Though the archaic features are within the *H. sapiens* variability range, they lie often at the edge of those boundaries, and, according to those who analyzed the fossils, constitute a unique combination of features, especially for Eurasian humans. The skull bones are thick, more similar to those of *H. erectus* than *H. sapiens*, but well rounded. The brain, at 1,327 cu cm (81 cu in), was somewhat smaller than the *H. sapiens* average though well within the variability range of humans today. Its frontal lobe was relatively long but its parietal lobe relatively short by modern-human standards. Its face was flat yet broader and shorter than is typical in *H. sapiens*, with a prominent browridge and a quite broad nose. Its jaw protruded yet had a very weak chin. Its molars were larger than is typical for modern humans. Its diet included a significant amount of meat, particularly red deer, which the evidence suggests they cooked. A few rib and limb bones were also found with the skull remains but have not yet been described in the scientific literature. It seems that no tools have been found, since none has been mentioned to date.

What this find does indicate is that a number of relatively recent *Homo* species (or subspecies) lived together in East Asia around 50,000 years ago and likely somewhat afterward. The Denisovans were apparently the first to die out (somewhere around 40,000 years ago), followed by the Neanderthals (about 22,000 years ago), the Flores pygmies (about 12,000 years ago), and the Red Deer Cave People (about 11,500 years ago). After around 11,500 years ago, modern humans were evidently the only *Homo* species extant in East Asia or anywhere else. Whether the Red Deer Cave People interbred with any of these other human types is unknown, as attempts to recover and analyze DNA samples from the fossils have to date failed.

Homo sapiens sapiens

At long last we arrive at the final millennia of our study and the rise of our own species, *H. sapiens sapiens* ("wise, wise humans").[11] Paleoanthropologists disagree about when these first "modern humans" appeared, with dates ranging from as early as 270,000 to about 100,000 years ago, with dates around 200,000 years ago apparently the most-often cited. The dates chosen depend on what characteristics are understood by the researcher to define *H. sapiens sapiens*. Most hold to the "out-of-Africa" view that modern humans originated in East Africa and subsequently, somewhere between 125,000 and 60,000 years ago, began to migrate into the Middle East, Europe, and beyond, eventually inhabiting every continent (except Antarctica). There they encountered other *Homo* species, such as the Neanderthals and Denisovans, with whom, according to genetic evidence, some at least interbred. If species are defined by the inability of their members to mate with those of other species and bear fertile offspring—a commonly used definition—then classification of Neanderthals and Denisovans as subspecies of *H. sapiens*, as has been suggested by some paleoanthropologists, seems appropriate.

11. They might more aptly be called *Homo scientius*, or "*knowledgeable humans*," given that we today have hardly demonstrated wisdom in the way we live in relation to nature and among the various groups that constitute our human family.

The human population today has exceeded seven billion, with the majority living in temperate climates, near rivers and continental coasts, and at elevations below 600 m (roughly 2,000 ft). A bit more than half live in urban areas, and the percentage is growing. Approximately 61 percent of us live in Asia, 14 percent in Africa, 14 percent in North and South America, 11 percent in Europe, and about 0.5 percent in Australia and on islands in the world's oceans.

HUMAN NATURE

Modern humans have relatively small, flat faces with high, vertical foreheads, two small browridges separated by a slight depression, protruding chins, and unusually large braincases for the size of their bodies. Ranging between about 1310 and 1475 cu cm (80–90 cu in), and with an average size of about 1350 cu cm (82.4 cu in), the modern human brain is somewhat larger than that of *H. heidelbergensis* and somewhat smaller than that of *H. neanderthalensis*. Particularly well developed are parts of the cerebral cortex and the temporal lobes, which together are involved in sensory perception and interpretation, motor commands, spatial reasoning, short- and long-term memory, language, goal planning, decision making, personality expression, the control of social behavior, and conscious thought. It is these abilities collectively that most scholars believe are the defining mental characteristics of *H. sapiens sapiens*, enabling us to do things such as hold moral, aesthetic, and spiritual values, which, to our best knowledge, no other species can do.

The teeth of modern humans feature four incisors, two canines, four premolars and six molars arranged in a parabolic shape in each jaw. The incisors are relatively short and vertical, the canines short, essentially flush with the incisors and premolars, the same in both males and females, though the molars are relatively smaller in females. Interestingly, the rear molars, or wisdom teeth, appear to be in the process of gradually disappearing, for some people today are born without them altogether. In addition, the general robustness of the head and teeth also seem to be decreasing. The fossil record indicates that in humans living some 30,000 years ago these were some 20–30 percent more robust, and in humans living 10,000 years ago about 10 percent more robust.

In fact, the entire modern human body, generally considered, has become quite gracile compared with previous *Homo* species, though variations in height and weight relating to geography, climate, and nutrition are common. In fact, both height and weight have been trending upward for people in recent times, the former probably largely the result of better nutrition and the latter the result not only of greater availability of nutritious food but also of obesity becoming a modern epidemic. The average height for men is in the 1.75 m (5 ft 9 in) range and for women around 1.61 m (5 ft 3½ in), while weights average about 58 kg (128 lbs) and 49 kg (108 lbs) respectively, though with significant variations from these averages. Considered as a whole, nearly all the skeletal differences in modern humans compared with earlier *Homo* species, such as longer lower limb bones, large ankles, and short toes, have to do with our distinctive and exclusively bipedal locomotion and the mechanical factors required for it to operate effectively and efficiently.

The appearance of modern humans in terms of skin color, facial features, and hair type and color varies considerably among human populations and typically reflects the geographical location in which these populations evolved. Darker skin evolved in equatorial regions, where protection from the damaging ultraviolet radiation of the Sun was essential for survival, while lighter skin evolved among people living in higher and lower latitudes with less intense exposure to the Sun, probably to maximize vitamin D synthesis once meat became a lesser part of their diet.[12] Historically, the appearance of certain groupings of these features were understood to constitute distinct races; today it is recognized that such neat distinctions do not exist, given the interbreeding that has occurred among all human groups for many millennia. The term "race" has fallen out of favor among both natural and social scientists, if not among the general public. Interestingly, though humans appear less hairy than their forebears, and certainly less hairy than modern chimpanzees, they actually have more hair, though it is generally shorter, finer, and less darkly pigmented than in chimps.

The modern human diet also varies widely, from vegans and strict vegetarians who eat nothing but fruits, grains, and vegetables to some, such as Eskimos, whose diet is almost exclusively meat. Most modern humans have been and are omnivorous, consuming a variety of plants and animals, though precisely which plants and animals are consumed has historically depended on which of each were available in any given geographic area. More recently, cultural, moral, and religious beliefs have also shaped what people will eat and what they will not. Cannibalism has also been practiced among different modern human groups historically, though today it seems the practice has essentially died out.

12. *Tanning* evolved among lighter-skinned people who were exposed to dangerous amounts of sunlight as a way to block ultraviolet rays.

Homo sapiens sapiens is also known as a highly symbolic creature and the creator of increasingly sophisticated *cultures*[13] throughout its history—cultures that have sought most fundamentally to alter the environment, both physical and social, in which people live to ensure survival. The evolution of culture has already been observed in the progression of prior *Homo* species, as indicated by such things as advances in tool culture and the beginnings of ritual burial of the dead. But even in the earliest modern humans these cultural beginnings advanced at an unprecedented rate, bringing with them an explosion of cultural innovations. These advances were cumulative, culminating in the current array of unique cultures that constitute our world today. Humans are the only species that has mastered the use of fire, cooks its food, clothes itself, and creates advanced technologies, art, moral codes, and religions.

We are so deeply constituted by the symbolic systems within which we live that we will even die for symbols, such as a flag or a cross. We all seek happiness—not simply pleasurable experiences, but personal fulfillment—but this happiness, this fulfillment, means different things to different people in their cultural worlds. Happiness may be defined largely in physical terms, in the sense of possessing all the material items that are needed to survive (and, for some, far more), physical health, and other such things, or in terms of the richness and number of loving relationships one has, or in living up to some standard of belief and/or conduct according to which the value of one's life is determined. Some are also driven to understand themselves and the universe around them and/or fashion the new technologies needed to do so, while still others seek above all to understand the spiritual side of their nature, the nature of that ultimate reality that is claimed to be the true center of all reality, and how ideally to relate to that reality. For many, probably most people, happiness entails finding fulfillment in several of these ways. Happiness also depends on the way humans deal with their mortality, which is their greatest psychological burden.[14]

What has become increasingly clear in recent years is that the evolutionary path to modern humans was not the straightforward linear progression that paleoanthropologists once thought it was, but was instead messy, more shrub-like than tree-like. At various times and places these hominid cousins lived together in the same geographic regions, with some being evolutionary dead ends while others stood in the line of our remote ancestors. Exactly who our direct ancestors were remains unknown and will

13. Culture is defined as all patterns of behavior that must be learned from others, including technology, symbolic communication (e.g., language), social organization, rituals, and value systems, such as ethics and religion.

14. For more on this problem see, e.g., Streeter, *Human Nature*.

likely remain so for many decades to come; clarifying the picture will largely depend on continuing hominid finds.

EARLY HUMAN HISTORY

The history of *H. sapiens sapiens* is largely the story of the development of culture, both material and intellectual. We have seen that our subspecies came into being probably around 200,000 years ago in Africa and emigrated out of Africa and into Europe and Asia beginning sometime between 125,000 and 60,000 years ago.[15] By about 40,000 years ago, modern humans had reached Australia, by about 15,000 years ago the Americas, and the relatively remote islands in the Atlantic and Pacific between 300 and 1280 CE. Throughout this period, and down to about 10,000 years ago or 8000 BCE, humans were organized into small hunter-gatherer groups ranging from twenty to thirty to as many as one hundred members. These groups were composed of several families and lived an essentially nomadic life as they followed the wild game they hunted and the other things they ate, including fruits, grains, tubers, insects, and mollusks. Evidence for long-distance trade between *H. sapiens* groups to obtain rare items such as ochre[16] or other raw materials suggests that this practice reaches back to 100,000 years ago or earlier.

The earliest evidence of symbolic behavior comes from several sites in Africa dating to somewhat after 100,000 years ago. Tools took on a more decorative aspect, and art, chiefly in the form of body ornaments, began to appear. This material culture, termed *Aterian* from a site near the Atlas Mountains, derived from the Mousterian culture. Examples are found all across North Africa. Decorative bone harpoon tips dating to between 90,000 and 85,000 years ago have been found there (and also in the Congo) as well as small projectile points which would have been hafted on to wooden shafts to form spears and, eventually, arrows. The earliest examples of these technologies were developed to kill fish and game at a distance. These small projectile points, as well as sophisticated Aterian stone scrapers, date from around 70,000 to 65,000 years ago. Pierced and ochred shell beads used for personal adornment dated to around 82,000 years ago have also been found, as well as carved ostrich eggshells (found also in Namibia

15. Most scholars seem to favor a date no earlier than 100,000 years ago, and many, if not most, see the first major wave of *H. sapiens sapiens* leaving Africa between 70,000 and 60,000 years ago.

16. Ochre is formed by combining certain iron oxides, usually red or yellow, with clay, and is used especially for bodily decoration.

from at least 60,000 years ago), perforated bones, and ornamental stones. Elsewhere, geometrically inscribed stones, decorated bone tools, bracelets, beads, and rock art have been found in South Africa dating to around 70,000–60,000 years ago. Ochre used as body paint likely also appeared by this time, probably for decorative purposes and/or ritual practices.

Burials attributed to modern humans have been found in Europe and the Middle East dating from around 130,000 to 100,000 years ago. Some have interpreted them as evidence of religious beliefs of some sort. Evidence for religion becomes much stronger, however, in the *Upper Paleolithic Period*, beginning around 50,000 years ago.[17] The origin of *music* is also thought by some scholars to have arisen prior to 50,000 years ago, along with the rise of early forms of art, language, and other such innovations. The earliest musical forms likely employed either voice or natural objects such as sticks or stones hitting other natural objects. Many think that music originated in rhythmic patterns accompanying everyday life, such as opening nuts by hitting them with stones, where a rhythmic pattern of working might have increased the efficiency of the process. Some have suggested that it arose as an aid to mating, others think it was done purely for pleasure, and still others think it may have arisen in the context of rituals, especially religious ones. Flute-like bone pipes may have been invented prior to 50,000 years ago, though the first direct evidence appears between 40,000 and 35,000 years ago.

The *Late Paleolithic Period* witnessed the appearance of what have been claimed to be the first fully/anatomically modern humans (*H. sapiens sapiens*), typically called *Cro-Magnons* (named for the cave in southern France where the first remains were found). They are thought to have arisen in Africa around 200,000 years ago, as we have seen, and, according to most paleoanthropologists, to have migrated out of Africa somewhere around 70,000–60,000 years ago. The earliest Cro-Magnon fossil remains found are dated to around 45,000 years ago and come from a cave in Italy and a rock shelter in Lebanon.

Cro-Magnons were cultural innovators, refining and extending all aspects of prior *Homo* cultures. Like earlier humans, they were hunter-gatherers, moving around to exploit the food resources available to them in different places at different times of the year. With Cro-Magnons, however, we see the first organized campsites, some with storage pits, which seem

17. Scholars have typically divided the "*Stone Age*," the time during which humans used primarily stones as tools, into three periods: the *Lower or Early Paleolithic* (around 3.4 mya to 280,000 years ago), the *Middle Paleolithic* (about 280,000 to 50,000 years ago), and the *Upper or Late Paleolithic* (around 50,000 to between 10,000 and 6,000 years ago, depending on the region of the world being discussed).

mostly to have been occupied seasonally. Their favorite animals appear to have been reindeer. Cro-Magnons are typically associated with the much-advanced *Aurignacian culture*, though from this time forward the designation of cultures gets complicated. More than two dozen such prehistoric cultures have been identified according to date and geographical regions of the world.[18] Rather than complicating the history of this period by assigning innovations to this or that specific culture and attempting anything like a comprehensive global survey, I will simply note some of the cultural advancements of the Late Paleolithic in the West, with the approximate dates, where available, for these innovations.

In tool manufacture, Cro-Magnons skillfully produced finely-worked implements of considerable diversity, using not only flint but also bone, antlers, and ivory to make tools that had never been seen before. Each tool seems to have had a specific function, and over time these tools became standardized. Flint tools were based on the creation of fine blades and bladelets rather than simpler flakes and included chisel-shaped *burins*, likely used for carving and engraving bone, antlers, wood, and maybe even ivory, and large scrapers, and *lacloirs*, used to scrape hides, whittle down wood, and perhaps function as knives as well. Flint blades were also used to make improved harpoons, arrowheads, knives, and tools for piercing and drilling softer materials, including hides. Bone and antler tools included fishhooks and needles with eyes for making clothes and other things from animal hides. Other tools invented by around 30,000 BCE seem to have included bolas, thrown spears, bows and arrows, and nets to aid in food procurement. Tool kits continued to be expanded and refined in subsequent millennia, and more beautiful materials, such as colored quartz, jasper, and beautiful flints, began to be used even in toolmaking. By the end of the Magdalenian Period, bone was used extensively to produce barbed points, harpoons, spear heads,

18. In the West, the *Aurignacian culture* extended throughout Europe and the Middle East and is dated by different researchers from as early as 45,000 to as recently as 21,000 BCE. The probably contemporary *Chatelperronian culture* existed from around 35,000 to 27,000 BCE, ranged from Northern Spain to central France, and seems to have been developed from the earlier Mousterian culture. It has been suggested that Neanderthals were the ones who invented this culture, though this is controversial. The *Gravettian culture* has been found in France and in the southern Ukraine from as early as 30,000 to as late as 18,000 BCE and has sometimes been combined with the similar Chatelperronian culture under the *Perigordian culture* umbrella—a practice that seems to be becoming less accepted. At any rate, the Gravettian culture has been claimed by some to have been developed from the earlier Chatelperronian culture and by others from the Aurignacian. The *Solutrean culture* appeared in Spain, eastern France, and England for the relatively brief time of 20,000–15,000 BCE and was succeeded by the *Magdalenian culture* that appeared around 16,000 BCE and continued until about 10,000–8,000 BCE, by which time it extended throughout Europe.

hammers, *adzes* (tools that resemble beefed-up garden hoes and are used to carve out or smooth large-scale pieces of wood, such as logs), and wedges, many of which were engraved with animal figures.

Art also took a major step forward during this period; where earlier art seemed relatively primitive, limited in form and only occasionally produced, art after about 40,000 BCE became a common feature of human life around the globe. Even everyday functional objects were beautified with geometrical and figurative art, and such works became especially common in the Magdalenian culture at the end of the Late Paleolithic Period, around 10,000 BCE.

Cro-Magnon art came in a variety of forms, including bodily ornaments, animal and human figurines, and paintings and engravings on stone plaques and cave walls. *Bodily ornaments* included bracelets, bead necklaces, and pendants made from soft stones, such as soapstone, shells, ivory, and human and animal teeth. Bone pins were also worn, and skin- painting with a variety of colors also appears to have been practiced.

Figurines were also made from mammoth ivory, bone, antlers, and formed clay that had been strengthened by firing at high temperatures in ovens—some of the very earliest examples of ceramics, dating to around 30,000 BCE. These figurines were primarily of animals and female humans, though some materials, such as ostrich eggs, were engraved with geometrical patterns. Animal figurines included representations of mammoths, rhinoceroses, wild horses, and other now-extinct species from the environments in which they lived. Female human forms are typically called "*Venus figurines*." Dating to at least 35,000 BCE and perhaps as early as 40,000 BCE, Venus figurines range in size from 4 to 25 cm (1½ in to almost 10 in), with most having no face details, some having been painted, and many having exaggerated breasts, buttocks, hips, thighs, and/or vulvas. They reach their artistic zenith during the Magdalenian Period. Interpretations of their meaning and function vary greatly, with some researchers suggesting they represent the mother goddess or other goddesses, others that they were related to the fertility cult and the hoped-for plenty that was its object, and still others still think that they reflect a matriarchal culture, are self-portraits of individual women, or were pornographic in nature. Venus figurines have been found throughout Europe, including Russia and the Ukraine and, in the east, Siberia, China, and Japan.

Cave art was another significant innovation of Cro-Magnon people. Some caves have been engraved with images of wild animals, including horses, cows, rhinoceroses, lions, and bears, that date to as far back as 30,000 BCE or even earlier. Three caves in France show the first examples of *bas-relief* engraving, in which figures are made to stand out from the rock by chipping away at the material surrounding the figure. Each cave has a different bas-relief figure: bison in one, fish in another, and a 1.5 foot-tall female figure named the *Venus of Laussel* in the third.

Far more prevalent was cave painting,[19] a practice that goes back to perhaps 40,000 BCE, reached its full maturation in the Magdalenian Period, and died out quickly thereafter.[20] The black charcoal drawings by early Cro-Magnons evolved later into multicolored and highly realistic portrayals of significant variety. Animals were the dominant subject of the painted walls (and, occasionally, ceilings) of the caves. Horses were the most favored animals, followed by bison, rhinos, bears, panthers, saber-toothed cats, hyenas, owls, and *aurochs*, the ancestors of modern cattle. These likely reflected the fauna that surrounded the Cro-Magnons in the various particular environments in which they lived. Human images also are portrayed in these caves, ranging from simple early stick figures to the quite realistic representations in the Magdalenian Period. Interestingly, some such figures appear to be half human and half animal. Geometrical objects, such as dots, lines, and patterns of these, were also painted in the caves, as were the outlines of human hands. In some cases, earlier cave paintings were painted over by later artists.

As with the Venus figurines, so also are the meanings of these paintings and engravings debated. Some paleontologists interpret them as a form of sympathetic ritual intended to ensure a successful hunt or make the animals return when, at the end of the Magdalenian (about 10,000 BCE), the herds began to disappear as the ice ages came to an end and the climate warmed. Others think that they reflect shamanistic practices similar to those practiced by some hunter-gatherer societies today. Others consider them simply aids in teaching the young. In any case, the creation of all these different art forms clearly indicates that the Cro-Magnons had an appreciation for beauty comparable to ours today.

Other cultural innovations arose during this period. The first unambiguous evidence for Cro-Magnon music is a *flute* made from the hollow wing bone of a giant vulture and several flute fragments made of mammoth ivory dating from around 35,000 BCE. By around 30,000 BCE, Cro-Magnons

19. It should also be noted that, in addition to caves, painting was also done on stone plaques.

20. The great artistic achievements in both tools and art of the Magdalenian Period have been interpreted to mean that there was significant leisure at the time due to the abundant food sources available. It may also reflect the fact that humans in this period evidently experienced a significant growth in population and, as a result, began to live in larger communities of up to 400 to 600 people. The social organization in such large communities might have permitted specialists in art and tool-making to make their living by these means.

seem to have developed a lunar *calendar* to follow the phases of the Moon. It may have served as an aid for hunting in knowing when game, such as horses and deer, would be migrating. Pit-fired pottery has been dated as far back as 29,000 BCE. Sometime during this period, perhaps by 30,000 BCE or even earlier, *dogs* were domesticated, possibly to aid in hunting. Shelters made of rock, clay, branches, and animal hides begin to appear late in the period, as do cords made of spun, dyed, and knotted fiber that likely functioned to make braided ropes, baskets, and clothing. Primitive oil lamps also seem to have been invented by around 10,000 BCE.

Language almost certainly developed during this period, if not earlier. Many anthropologists believe that modern human behaviors appeared around 50,000 BCE, and such behaviors are understood to include complex forms of language. Language would have helped coordinate social life and enable more efficient and effective planning in the increasingly complex societies that developed later in the period. These societies may have begun to develop a systematic division of labor between the sexes and, it is suggested, might have been more egalitarian than subsequent societies.

It was certainly during this period, if not earlier, that *religion* became a significant part of human lives. As we have already noted, religion may have originated with the Neanderthals. Intentional burials, some containing stone tools and animal bones, have been found, and the removal of flesh from skulls may also have been associated with religious practices, as it is in some cultures today. Animal worship or totemism may also have been practiced, but it is with humans of the modern type that we see clear evidence for the existence of religion and religious rituals.

One of these practices is *ritual burial*. The oldest modern human burial dates to some 90,000 BCE in Israel, suggesting that this practice was brought with *H. sapiens sapiens* when it migrated out of Africa. It is a double burial, containing what are thought to be a mother and her child, and their bones have been sprinkled with red ochre. A variety of *grave goods* (material things that were buried along with the skeletons) were also found, including a wild boar mandible and, it seems, food and flowers. The practice of using red (and occasionally yellow) ochre on the bodies of the deceased became more common in the succeeding millennia, until it became widespread in the Late Paleolithic, beginning some 50,000 years or so ago.

The variety and amount of grave goods also increased in the Late Paleolithic. *Bodily ornaments*, such as necklaces, pendants, leg bracelets, caps, and even skirts made from pierced shells, bone, ivory (including deer teeth), and/or stone beads, are typically found. There are also *tools*, including stone blades, chisels, scrapers, bone-piercing points, cores, and flakes, as well as awls, batons, and other objects made from bone. Remains of the bones, teeth, and even antlers of various mammals ranging from horses, aurachs, and elk to beavers and squirrels, and even the skeletons of birds, have also been found in graves. In addition, smooth pebbles, some stained with ochre, resin and, wax have been found in some graves, as well as, it appears, food and flowers. Some graves even had decorated stone slabs placed on top of them. These grave goods were surely included in order to have the deceased person well prepared and adorned for a life after death.

There are other manifestations of what are probably best interpreted as religious behaviors related to hunting. One is the existence of a *bear cult*, in which a bear was killed and then its head and body were ritually buried separately.[21] The cave paintings of animals may also have served as part of a ritual associated with the hunt. Such rituals may have been communal practices, perhaps were led by a *shaman*, one who was understood to have access to the spirit world. The earliest burial of a shaman dates back to around 30,000 BCE.

Venus figurines also may have had religious significance, such as insuring the fertility of women and the land and perhaps even success in hunting. It has also been suggested that *ancestor cults* began at this time and that belief in and worship of gods or other supernatural beings, perhaps even a pantheon of them, may similarly have arisen.

As we have already noted, the Magdalenian culture constituted the high point of early modern human achievement but came to an end around 10,000 BCE. At that time temperatures began to warm globally as the last of the ice sheets retreated, eventually assuming their present forms near both poles. The gradual melting of these ice sheets raised sea levels significantly, destroying the evidence of seacoast cultures worldwide. These changes in temperatures and sea level changed the environments in which humans had lived and apparently caused the herds of game animals which had supplied much of the food supply during the Late Paleolithic Period to dwindle. The various material cultures that ensued, such as the *Azilian* in Western Europe, became simpler, with relatively fewer tools of cruder manufacture and little in the way of art.

Somewhere around 8000 BCE, the so-called *Neolithic Revolution* (also called the *Agricultural Revolution*) began. This was the beginning of the worldwide transition from hunter-gatherer societies to settled agricultural communities. It seems to have started in the *Fertile Crescent*, which included Mesopotamia, the eastern Mediterranean lands, and Egypt, with East Asia

21. Editor's note: Ina Wunn rejects this view (see Wunn,"Cave Bear Worship," "Bärenkult," and "Beginning of Religion").

not far behind. By around 5000 BCE, most inhabited regions of the world had transitioned to the new agricultural lifestyle.

This shift to an agricultural existence probably had its origins in people finding and protecting areas where certain edible plants naturally grew naturally. This likely led to the ability to expand these areas by careful tending and to establish new areas for such plants to grow. This may have been due to the warming climate and concomitant disappearance of the herds of game animals that had previously been a staple of their diet. This necessitated developing other stable food sources, particularly the domestication of certain types of grains, fruits, vegetables, and animals, and, as a result, the shift to permanent settled living.

The kinds of grains and other foods that people cultivated depended on what grew naturally in the area and provided the greatest yields and most calories. In the Jordan Valley, these initially included figs, barley, and probably oats. In China, rice, foxtail millet, and broomcorn millet were evidently among the first domesticated grains, while in Southeast Asia, bananas and plantains were probably among the first managed crops. In Mesoamerica, maize and squash became staple crops very early on, with beans added several millennia later, while in South America potatoes and manioc and in eastern North America sunflowers, sumpweed, and goosefoot were domesticated, though only by around 2500 BCE. In Mexico, maize, bottle gourd, and pepo squash were evidently cultivated as early as 8000 BCE. Over the centuries, many plants native to one area of the globe appeared in other areas, evidently brought by the traders who became common during this period.

Certain animals were domesticated as well, particularly those that were useful not only for providing meat but also for producing a steady supply of milk, for doing work, or for providing wool, leather, hides, and fertilizer. Among the earliest animals domesticated, apart from the dogs that were tamed very early on, were cows, sheep, goats, and pigs. These, in addition to the dromedary camel, were likely first domesticated in the Middle East. While most people evidently settled down to farm, others became nomadic herders, guiding their flocks from place to place in search of good grazing land and water.

The turn to agricultural life brought with it a series of other cultural revolutions. As farming and the food supply increased, the human population grew both worldwide and locally, creating communities that grew larger, eventually giving rise to sedentary populations living in villages and towns. The surplus food created agriculturally meant that some people could work at other trades and barter the goods they made for food. Thus, a division of labor and increased efficiencies arose. Agricultural surpluses and excess goods gave rise to trading economies, which encouraged more efficient forms of transportation and food storage, such as new types of pottery. The new pottery was the result of the invention of the potter's wheel in Mesopotamia, perhaps as early as 6000 BCE.

Some of these towns and villages became trading centers and grew to become the first cities. In cities, labor specialization became even more complex and culture developed rapidly, giving rise to more complex forms of political and economic administration, non-portable art, including architecture, and tool manufacture, seen in the invention of ground and polished stone implements such as axes and adzes. These techniques for working stone were time-consuming, but polished axes increased their mechanical strength and provided sharper cutting edges. This was particularly important as axes were increasingly used to clear land in order to make room for new agricultural fields.

Certain of these cities became the centers of the first city-states that arose at Sumer in Mesopotamia about 3500 BCE, followed around 3300 BCE by Egyptian city-states in the upper and lower Nile Valley and the early Harappan city-states along the Indus River in India. The rise of early Chinese cities along the Yellow and Yangtze Rivers may have occurred around this same time or perhaps somewhat later; the archaeological evidence is inconclusive.

The *Sumerians* developed *cuneiform* writing, which provides an interesting look at the development of early writing. Cuneiform arose around 3000 BCE as a way of keeping economic accounts of various commodities. A round reed stylus was impressed into soft clay at different angles to indicate specific numbers, to which *pictographs* (symbols resembling the things they represent) were later added to identify the commodity. The round stylus was gradually replaced by a triangularly shaped one, which by around 2600 BCE became used to make marks representing spoken syllables in addition to commodities and numbers. Writing greatly enhanced the ability of officials to administer the larger areas included in these city-states. In Egypt, writing arose somewhat later, taking the form of *hieroglyphics* (pictorial and symbolic representations of words or alphabetical elements); in the early Harappan culture, the still undeciphered, largely pictorial, Indus script was in use by at least 2500 BCE; and in China, written language consisted of *logographs* (symbols representing words or smaller linguistic units), though the date of its origins remains debated.

The first city-states also developed government bureaucracies for the administration of the state and military forces for protection or aggression, depending on the decisions of the ruler. Complex religions also arose at this time, featuring worship of deified natural elements such as the Sun, Moon,

Earth, sky, and sea, typically conceived as anthropomorphic beings. Priests, priestesses, and other functionaries arose to mediate their worship and take care of the shrines and temples dedicated to these gods. With the development of writing came religious texts, or *scriptures*, which spoke of the nature of the gods, their powers, their relationships with humankind, and what humans were required to do to please or appease them.

The larger of these early city-states soon became the centers of the first great empires, the first of which seems to have been formed by the unification of the Lower and Upper Egyptian city-states around 3100 BCE. By about 2400 BCE the *Akkadian Empire* had arisen in the Tigris-Euphrates River valley in Mesopotamia; around 2500 BCE the *Kingdom of Kerma* had developed in the Sudan; and by around 2200–2070 BCE the *Xia Dynasty* had arisen in China along the Yellow River. The *Hittite Empire*, centered in what is today northern Turkey, originated just after 1600 BCE, about the same time as the *Mycenaean civilization* began its rise in Greece. The earliest civilization in Mesoamerica was the *Mayan*, which arose around 1800 BCE.

By the sixth century BCE, the first of the very large empires appeared, the *Persian Empire*, which by around 500 BCE extended from Egypt and parts of Greece in the west all the way to the Indus River in present-day Pakistan. The Persians fell to the Greeks under Alexander the Great, and at his death in 322 BCE the *Macedonian* or *Greek Empire* ruled over all the territory of the Persian Empire and more, including rule of the unified Greek city-states. After Alexander's death his empire splintered, and the *Maurya Empire* arose in India, which united the subcontinent as well as Persia and much of what today is Afghanistan and continued until about 185 BCE. In the West, the *Roman Empire* succeeded the Macedonian as the Roman Republic began to be ruled by a series of emperors in the first century. At its greatest extent, under Trajan (ca. 117 CE), the empire stretched across much of Europe, from Britain into Germany in the north to North Africa in the south, and from Spain and Morocco in the west to Mesopotamia in the east. The first great Chinese empire began with the *Qin Dynasty*, beginning in 221 BCE, and was broken up and reunited a number of times under succeeding dynasties before assuming its current form today as the People's Republic of China.

The seventh century BCE has come to be called the "Axial Age" because of the mostly independent arising of a number of important religious developments world-wide. In China, the *Confucian* and *Taoist* traditions arose; in India, *Buddhism* and *Jainism* appeared; in Persia, *Zoroastrianism* arose; and in Palestine and Mesopotamia, *Judaism* became fully monotheistic. At about the dawn of the sixth century, philosophy began in the West with the Presocratic thinker *Thales* and would subsequently produce its most influential thinkers: Socrates (ca. 469–399 BCE), Plato (ca. 424/423–348/347 BCE), and Aristotle (ca. 384–322 BCE). Around 5/4 BCE in Nazareth of Galilee[22] a Jewish prophet, wisdom teacher, and religious and social reformer named Jesus[23] was born, carried out a ministry centered on love, and was tortured to death in his mid-30s by the local representative of the Roman Empire as a political threat (with the support of the Jewish leaders in Jerusalem, who saw him as a threat to their religious and limited political power). When, after his death, Jesus's teachings reached the larger, Gentile, Roman world, he became known as the founder of a new religion, Christianity.

We need not recount the subsequent history of the world in the Common Era, with its developing philosophies and religions, its technological advances, its wars and the suffering they have inflicted. We still live in such a world, and despite our philosophies, religions, and moral codes, there appears to be no end of the hunger for conquest and the wealth, power, and fame it brings. No one can tell what our future will be, as it depends on us and what we will choose to do collectively and individually with our freedom. It is the hope of many that our species will finally grow out of its current, embarrassing adolescence and become what it has the potential to be—a collection of diverse peoples who care as much for the welfare of the world and all its inhabitants as they do for themselves.

THE FUTURE OF HUMANKIND AND THE UNIVERSE

As for our universe, we can predict, based on current knowledge and theory, what will become of it in the future. On Earth, the continental plates will continue their inexorable movements around the globe, tearing apart and stitching together new continents as they have from the early years of the planet. For example, the Arabian Plate is currently splitting off from the African Plate, with the Red Sea and the inland low or below-sea-level regions in eastern Africa the result. Eventually the eastern part of the African Plate, the Somali Plate, will separate completely from the rest of Africa, or the Nubian Plate, with the latter continuing its current northward movement

22. Editor's note: According to Matthew and Luke, Jesus was born in Bethlehem of Judea; however, Jarvis agreed with most critical scholars that Nazareth was Jesus' birthplace, and that, in Matthew and Luke, the identification of Bethlehem as the birthplace reflects their theologies, which identify Jesus with King David.

23. The name Jesus is based on *Iēsous,* the Latin transliteration of the Greek *Iēsous.* In the NT, Jesus's full name is "Jesus son of Joseph" (Luke 3:23; John 1:45; 6:42), "Jesus from Nazareth" (Matt 21:11; Acts 10:38), and "Jesus the Nazarene" or "Jesus of Nazareth" (Mark 1:24; 10:47; 16:6; Matt 26:71; Luke 4:34; 18:37; 24:19; John 18:5, 7; 19:19; Acts 2:22; 6:14; 22:8; 26:9).

until it collides with Europe and the Mediterranean Sea exists no more. In about 50 million years my Southern California home will be in Alaska, and in some 250 million years all the continents will have joined together once again—assuming their current trajectories continue—to become another supercontinent.

Biologically, we are living in the midst of the fifth great extinction event in Earth's history. This one, however, has not been caused by massive volcanic eruptions or large objects from space colliding with our planet. It is being caused by us. It is also the case that the world is warming, and no knowledgeable and rational person believes any longer that humans are not contributing to it with our massive burning of fossil fuels. No one knows when the tipping point may be, when what has been gradual warming suddenly accelerates, bringing the net loss of large tracts of productive farmland and so famine it its wake, with the very real possibility of social chaos, even war, to secure precious food supplies. It remains to be seen whether we will continue to think that we can live well in the future without making some significant changes in our current behavior or will come to realize that all creatures are of intrinsic value and that without a healthy planet filled with millions of diverse creatures we all will suffer.

Another event that seems already to be occurring is the reversal of the polarity of Earth's magnetic field, meaning that the magnetic north and south poles flip so that a compass needle would point south instead of north. This has happened many times in our planet's history—roughly once or twice every million years—and it will continue to do so as far into the future as we can predict. One of the signs that the reversal in polarity is beginning is that the magnetic field is becoming weaker and more erratic; after this, it should take about a thousand years to complete the reversal. As the field weakens, it will permit more radiation from space to reach Earth's surface, and this radiation will endanger living things exposed to it, such as ourselves. Cancer incidence will likely soar, among other things. That this will happen is a certainty; precisely when it will occur is the only unknown.

Another cause for concern is the very real possibility that some large object from space will once again collide with Earth, bringing mass extinction as it has done before, including perhaps that of our own species. There have been some relatively near misses lately, with objects passing inside the orbit of the Moon and missing us by only several tens of thousands of miles—objects we weren't even aware of until they had passed. To insure ourselves against extinction we would do well to chart every asteroid, meteor, and comet within our cosmic neighborhood and their orbits—a process that is now underway—plan strategies to deal with any of these that seem likely to collide with Earth, and, most importantly, plant self-sustaining colonies on the Moon and Mars as quickly as we can. In this way a large impact with Earth, even should it wipe out all the humans on our planet, would not cause our species to go extinct.

On the other hand, our hunt for life elsewhere in our Solar System and, ultimately, our galaxy will continue, and, I expect, be successful. Life may well first be found in the form of microbial life on Mars, or perhaps in some novel creatures on one of the moons of Jupiter or Saturn noted previously as possible candidates for finding life in our own solar backyard. This quest to discover other forms of life in the galaxy will likely occupy humans for as long as the species exists.

As we have also noted, our galaxy is in the process of slowly drawing its two smaller companion galaxies, the Large and Small Magellenic Clouds, toward itself, and in some 20–40 million years they will collide with us, creating havoc for many millions of years before eventually being incorporated within a new and remodeled Milky Way. In fact, a massive gas ejection from these clouds, termed the *Smith Cloud*, is already colliding with the gas cloud that surrounds the Milky Way.

Our Sun will continue to shrink slowly as more of its hydrogen supply is converted to helium. This will put greater pressure on the core, which will cause its temperature to rise and so increase the Sun's brightness. The net result of this gradual heating will be that within a billion years from now Earth will become extremely hot, with the tropics and the latitudes to the north and south of them turning into enormous deserts. The polar ice will have melted along with all other frozen water deposits, and so sea levels will rise substantially, burying the present continental coastlands and so reducing the total continental land areas. If humans are still living on Earth at that time, they will probably be living at the North and South Poles, where they might still be able to grow food out in the open, or underground.

By three billion years from now, the Sun will be 40 percent hotter than its present temperature. All water above ground, including the oceans, will have evaporated, and Earth will have become a hot, barren place. If humans continue to inhabit Earth, they will have to live and grow their food beneath the surface, though most will likely have moved off-planet by then. Mars will have heated up to the point where we could live there for some tens of millions of years in relative comfort. At such temperatures water would exist in liquid form and the atmosphere would become denser, though with Mars' lesser gravity that atmosphere would be harder to hold and would not be sufficient for us to breathe without some technological assistance.

At about this time, or perhaps somewhat later, the Milky Way and the Andromeda Galaxy will begin to collide, and for hundreds of millions to perhaps several billion years they will engage in a complex gravitational dance that will finally end once the two have been completely merged. The final result will be a new galaxy, perhaps an elliptical one, of enormous mass.

By 3.5 billion years from now, the Sun will have swelled to over twice its present size, due to its loss of mass as ever more hydrogen is fused into helium and the heat from this process is radiated out into space in the increasingly powerful solar wind. This decrease in mass will produce a proportional decrease in the Sun's gravitational force, and so the orbits of the planets in the Solar System will gradually move farther out from the Sun. Earth's orbit will gradually move out from its present 154.5 million km (96 million mi) average to some 193 million km (120 million mi), though this orbital movement will not be sufficient to keep pace with the swelling of the Sun. Thus, by 3.5 billion years from now life on Earth will no longer exist; the planet will have become a scorched cinder.

About 5 billion years from now, all of the hydrogen in the Sun will have been converted into helium. Without the outward pressure created by the fusion reactions to counter the inward pull of its gravity, the Sun's outer material will collapse inward, greatly heating and pressurizing the core. Under these conditions the helium in the core will begin to fuse into lithium and beryllium, the next two elements of the Periodic Table. Because these reactions are more efficient than hydrogen fusion, more energy will be released—more than two thousand times the current energy output of the Sun. This will cause the Sun to swell to an enormous size, such that it will engulf the planet Mercury and thereafter Venus as well. At this point it will have become a *red giant* star. The lithium and beryllium will fuse to become still heavier elements, and these in turn will fuse to become even heavier elements. The red giant will eventually begin to grow unstable, creating pulsations that will throw off shells of gas enriched with elements forged in its core, forming a new planetary nebula which might eventually form into a new star system.

About 5 billion years later, or about 10 billion years from now, all of the Sun's nuclear fuel will have been fused into iron, at which point its core reactions will cease. With no outward pressure created by fusion in the core, the star's outer matter will quickly collapse, shrinking the Sun dramatically and increasing its density until its gravitational pressure inward is balanced by the *degeneracy pressure*—the pressure created by the resistance of the Sun's atoms to compression. At this point the Sun will have become a *white dwarf star*, comparable in size to Earth but with a density so great that a single teaspoon of it would weigh several tons. For billions of years the white dwarf will glow, giving off radiation as it slowly cools, until it finally becomes a kind of frozen ball termed by astronomers a *black dwarf*.

Meanwhile, dark energy will continue to accelerate the universe's expansion—assuming it is constant, as current measurements seem to indicate—meaning that over the succeeding billions of years the voids between structures in the universe will continue to increase in size even as the filaments and sheets of matter and dark matter between then continue to condense under gravity's sway. As space-time continues to be pulled outward by dark energy, it will stretch out these large-scale structures, with the result that the galaxies and galaxy clusters will begin to draw away from each other except for those close enough together that their gravitational attraction is stronger than the force of dark energy pulling them apart. Current measurements indicate that dark energy is causing the distance between galaxies to double every ten billion years. Throughout this process, gravity will still hold sway over relatively short distances. Matter, both ordinary and dark, will be drawn slowly in toward the centers of galaxies, gradually forming ever more massive black holes.

In about 20 billion years, the expansion rate of the universe will be so great that the stretching of space-time will preclude any further large scale structure formation. Every star and galaxy will be pulling away from all others. To people living at any point in the universe, the stars and galaxies will appear to be receding until finally they are so far away that their light will be unable to keep pace with the universe's expansion and so will not reach them anymore. One by one the other galaxies will disappear from the sky. In our own vicinity, by about 100 billion years from now the other galaxies of the Virgo cluster to which we belong will be so far away that they will disappear from the sky until finally only the Milky Way will be left for us to see. From Earth, the sky will be completely black in all directions save the part that looks in toward the center of our galaxy.

After that, the stars within the Milky Way will begin to disappear from Earth's view as the universe continues to expand, with the farthest stars vanishing first. The free dust and gas within and between galaxies will become so separated that they will no longer be able to condense into stars, so star birth in the universe will cease. In about 100 trillion (10^{14}) years or so, the last of the shining (hydrogen fusing) stars, the red dwarfs, will have finally exhausted their nuclear fuel, becoming white dwarfs. At this point the universe will be composed almost entirely of white dwarfs, black dwarfs, and brown dwarfs (failed stars) in roughly equal proportions, with the remaining stellar objects consisting of neutron stars and black holes. All life in the universe will probably have gone extinct by this time.

For many trillions of years thereafter, there will very occasionally be collisions between brown dwarfs and white dwarfs, either with one another or dwarfs of their own kind, forming more massive stars that give off light.

Only a few of these new stars will exist in the universe at any given time. Otherwise the universe will be relentlessly black. Essentially all that will remain of the universe will be massive black holes, neutron stars, white, black, and brown dwarfs, dead planets, and various gas and dust particles spread through the void—and dark matter and energy.

Around 10^{31} years from now, the universe's supply of protons—mostly locked up inside white, black, and brown dwarfs—will begin to decay. Current theory says they should decay into their constituent quarks, though it is possible they might eventually decay to form positrons and pions, the latter of which are highly unstable and so would quickly decay into photons. By 10^{36} years half of the protons will have decayed, with the remainder disappearing by 10^{40} years. The universe at this point will consist solely of black holes, photons, and elementary particles. Owing to their release of Hawking radiation, the massive black holes will gradually shrink, becoming smaller and smaller over succeeding billions and trillions of years, until they too disappear, leaving only a dying burst of energy to mark their deaths.

By 10^{66} years, the universe will have grown so large that the average distance between electrons would be more than 100,000 times the radius of the visible universe today. By about 10^{100} years from now the last of the black holes will have decayed away, leaving a universe composed of only elementary particles and radiation. By around 10^{116} years all remaining elementary particles should have decayed into radiation. Then, perhaps 10^{150} years from now, the universe will have reached the end of its life, having grown inconceivably large but with the temperature of its radiation finally having dwindled to absolute zero.

20

A Christian Understanding of Humankind and the Question of Life after Death

Historically, Christian theology has asserted several fundamental beliefs about human beings. First, humans are *creatures* created by God. In the understanding we have been developing in prior chapters, this means that God is ultimately responsible for our existence by creating and holding in being the laws of nature that have permitted life to exist and evolve on this planet. Thus, we are finite beings who are not ultimately the cause of our own being and so are unlike God ontologically—that is, we are not divine. Rather, we are finite, physical beings like all other animals on Earth, and as such must eventually die.

Second, traditional Christian teaching asserts that humans are created in "the image of God" (Gen 1:26–27), which means that, while we are *like* all other animals in our creaturely being and finitude, we are *unlike* all other animals in that we are in some sense like God in a way no other animal is. What this God-likeness is has been variously defined throughout the history of the church, definitions including such things as rationality, creativity, morality, and spirituality. Today, however, and especially in the understanding developed in this book, the total absence of these things in primates and other animals must be questioned. Given that God, in God's very essence, is Love, I interpret this God-likeness to be a *potential*—the potential for agapic love such that, when people become more consistently loving in the course of their lives, they are becoming more like God. In addition, to the extent that humans accept and act on God's will for them as it is given to them in each moment of their lives, humans are cocreators with God.

Third, humans are *social* creatures, meaning that they are meant to exist in relationship with others and that they are not fully themselves if they do not. The classic text here comes from Gen 2:18, in which God asserts, "It is not good that the human being should be alone." It is significant that all three of these claims—that humans are creatures created by God, exist in the image of God, and are social—are teachings central to the two creation stories in Genesis 1–2.

When we include chapter 3 along with these first two chapters in Genesis, we encounter the fourth claim, that all humans are *sinful* (with the exception of Jesus, most Christians believe). To sin is to fail to live as God's wants you to live, and, as Jesus says in Mark 12:29–31, the essence of that life according to God's will is summarized in just two commandments from the HB/OT: to "love the Lord your God with all your heart, and with all your soul, and with all your might" (Deut 6:5) and to "love your neighbor as yourself" (Lev 19:18). For Jesus, this means to love God and all people—even your enemies—agapically, being concerned with their welfare even to the exclusion of concern for yourself. Thus, the claim that all people sin is the claim that all humans have failed consistently to love agapically.

Some people may love this way a good deal of their lives, while some others appear never to do so, seemingly living entirely self-centered lives. No one loves in this way all the time, and therefore, as Paul says in Rom 3:23, "all have sinned and fall short of the glory of God." But living sinfully/lovelessly/self-centeredly is also living stupidly, according to the Bible and Christian teaching, because it guarantees that one will *not* get that which one needs most—to love and be loved—in which true human fulfillment consists. God's essential commandment to love is neither arbitrary nor meant to confine us, it is meant to help us become fulfilled in life. This, I believe, is the true wisdom that Christianity teaches, and I believe it to be true because I have experienced it in my own life ever more clearly and intensely as my life has been approaching its end.

This teaching about sinfulness necessarily presupposes that people are fundamentally free to live their lives the way they choose—which is how we experience our lives and a precondition, as we have seen, for having relationships with God and others. And if this is so, then we can understand our lives as a process of self-creation, as various existentialists, including Christian existentialists, have taught. We are, as it were, artists who create ourselves as works of art by the choices we make each day, making ourselves into something morally and spiritually good or evil, beautiful or ugly, over the course of our lives. To the extent that we behave lovingly, we make ourselves into people who are beautiful and good, and to the extent that we behave unlovingly we make ourselves into people who are ugly and evil. We know this, as we make just these judgments about people, the way they live and the effects they have on others, all the time. At the end of our lives, who we have become is finalized, and what we have made ourselves to be will range somewhere between true goodness and beauty and abject evil and ugliness. It is the way of beauty and goodness that alone brings fulfillment.

Sin, however, is not only committed by individuals; it is also committed by groups. It is even built into the social fabric of societies and their laws whenever these treat certain groups as inferior because of their gender, color, sexual orientation, wealth, or anything else. It is these social sins that have the most devastating consequences for the most people. This why it must be a fundamental Christian concern to expose and seek to end such social sins, despite the fact that some Christians are at times looking for signs of. Those who do not seek to love *all* others are not followers of Jesus and are not living in a proper relationship with God.[1]

Our task to become loving people is made much more difficult by factors both natural and social, and it is these that have traditionally been cited as constituting "original sin." On the natural side, we have been programmed genetically to survive, to go on living and reproducing for as long as possible, and this necessarily makes us self-centered. This survival drive has served us well, having made possible our own evolution as a species. It is also the case that we view the world around us from the only perspective we can have, which is with ourselves as the center of all things. This therefore tempts us to think, without even doing so consciously, that we *are* the center of everything.

On the social side, some cultures, especially those that promote individualism, teach people to be concerned above all else with their own personal priorities, such as becoming powerful, rich, and famous. Some of these cultures, like American culture today, teach that consumerism, acquiring as many *things* as possible, is the way to true happiness. The problem is that such self-centeredness is precisely the way *not* to find fulfillment, since this is found only in loving and being loved by others. What we are being called by God to do, then, is to move beyond the self-centeredness of our biological and social past into the God- and other-centeredness by which we will attain our fullest human potential as well as our true fulfillment.

In the early years of Christianity, the church understood itself to be an alternative community, in which people could live and be encouraged to live

1. For more on the nature of sin, including what motivates people to live sinfully, see, e.g., Streeter, *Human Nature*.

such agapic lives, in contrast to their self-centered urges and the values of the societies around them. Their love took concrete form, as it must if it is to be real love. Some expressions of this are shocking to modern American (and probably most other Western) ears, and so you will rarely if ever hear sermons preached on them. For example, according to the book of Acts, the early Christian community in Jerusalem lived according to what we would call economic communalism: "And all who believed were together and had all things in common; and they sold their possessions and goods and distributed them to all, as any had need" (Acts 2:44–45). And again, "Now the company of those who believed were of one heart and soul, and no one said that any of the things which he possessed was his own, but they everything in common . . . There was not a needy person among them, for as many as were possessors of lands or houses sold them, and brought the proceeds of what was sold and laid it at the apostles' feet; and distribution was made to each as any had need" (Acts 4:32, 34–35). This was what the first Christians understood lives of agapic love to entail. Not surprisingly, virtually no Christians today are calling for a return to such a state, even though concern for others as much as or above oneself would logically entail some such economic ideal. Christianity today has strayed very far from its roots—to such a degree that I do not believe that Jesus would recognize it as reflecting many of his values and teachings were he alive today.

We have spoken in the preceding chapter not only about humans but also about the future, and I believe it is important to discuss Christian teaching on the end of both humanity and the cosmos, which is collectively called *eschatology* ("study of the end"). The final end of humans and the cosmos has been the source of great speculation and misunderstanding within Christianity, and I hope at least to bring some factual information and reason to the topic.

There is a great deal of apocalyptic fervor among conservative Christians today, many of whom search through the apocalyptic writings of the Bible—the book of Daniel in the HB/OT and the apocalyptic passages in the Gospels of Matthew, Mark, and Luke and the book of Revelation in the NT—looking for signs of end-time events occurring today. Thus, they typically believe, the end of the world and the return of Jesus is coming soon. Many of them even hope for it. What few of them know, however, is what apocalyptic literature actually is or the conditions under which it arose historically. Knowing about this is essential, however, if one is to evaluate the apocalyptic perspective properly.

At least since the Babylonian exile, a significant theme in the Hebrew Scriptures has been that God called his people to help bring all the other nations of the world into the proper relationship with God, a relationship that God intended for them and constituted their salvation. As the Jewish people understood it, God gave them laws, which were *wisdom*, a deep understanding of life that brought with it the secrets to success in this life, such that if they obeyed them, their lives would be good. If they lived by God's standards, the Jews believed, the other nations would see how good their lives were and would seek to learn their secret. The Israelites would say the secret of their success was living in a right relationship with God as specified in God's commandments. Wanting the good life, other nations would one by one also come to live in right relationship with God, until eventually all nations would be converted to worship of God and so attain salvation on Earth. This "salvation" had nothing to do with the afterlife; it was all about finding fulfillment in this life. Never once in the entire HB/OT was anyone promised life in heaven with God as a reward for living in a proper relationship with God in this life. Thus, in the promises God makes to Abraham, Sarah, and their descendants, it is long life, material wealth, and many descendants that are their reward for living as God willed.

The *apocalyptic* view emerged much later, after the beginning of the second century BCE, due to an unprecedented set of circumstances. It evidently began when one of the Greek rulers over Israel following Alexander the Great, Antiochus IV, determined to make all Jews in his territory worship him as Zeus incarnate. Some Jews acquiesced, thereby saving themselves from confrontation with their Greek overlords, but in so doing they violated the first of the Ten Commandments, which prohibited worship of any other god besides their God. Other Jews, those who remained faithful to God's commandments, refused to worship Antiochus, and so began a long guerilla campaign that finally, after some twenty-five years of fighting, brought Israel its independence from Antiochus and the Greeks.

Many faithful Jews died in the struggle against the Greeks, which created a theological crisis. Fidelity to God was supposed to bring long life, wealth, and many descendants, as the Scriptures promised, yet it was the faithless Jews who were getting these things while the faithful Jews who died were denied them. This simply could not be, the Jews believed—that the evil were rewarded in life and the good punished. God would not condone such an injustice!

Fortunately for the Jews, an answer already existed that would resolve the problem, but it was not a Jewish answer—it came from the Persians and their *Zoroastrian religion*. The Jews were well aware of Zoroastrian teaching, since the Persians had ruled over them for two centuries (ca. 539–333 BCE) until Alexander and the Greeks began their conquests. According to Zoroastrianism, there were *two* gods who battled with each other throughout history: a good god and an evil god. When things were going well on earth, that meant the good god was winning; when they were going badly, the evil god was winning. Each god had his troops: the good god had his angels, who fought for him and lived with him in his kingdom, and the evil god had his demons, who abode with and fought for him. People on earth, by the way they lived, served either the good god or the evil god and after their deaths went to live with the god they had served.

Here was the solution the faithful Jews needed. Justice was not always done on earth, since sometimes the evil god was in control and so the good people might die. Thus, the final justice must come *after death*—a wholly new "revelation" for the Jews. The evil might indeed prosper when evil ruled the world, but the good would receive a much greater reward after death, while the evil would find the evil god no friend but rather a torturer.

While this solution resolved the justice issue, it also created a serious problem for the Jews—the existence of a second god. That was impossible for the Jews, and so a substitute was found. The book of Job depicted one of God's angels whose job it was to travel the earth and bring back information to God about what was happening there, including who failed to live according to God's commandments and so deserved punishment. (God was imaged as like a human king, who would sit in his castle and send out "messengers"—which is the literal meaning of the word "angel"—so that he would know what was going on in his kingdom.) This angel was called "the *satan*," a Hebrew noun that meant "adversary" or "accuser." Since the Accuser accused Job of being faithful only because he had lived a charmed life and sought to demonstrate it by inflicting Job with every adversity possible, he seemed the best candidate for an evil angel; and so eventually "the satan" became Satan, the evil one opposed to God, the Devil.

This, then, is how the persecuted, faithful Jews—and, later, faithful Christians—came to view the world of their experience, one in which Satan temporarily held sway, causing them to be persecuted, but whom God would soon overthrow and once again rule on earth. They did not ask why God, who was good, would allow an evil angel to run the world and bring destruction and death to faithful people. Instead, they described the world as they experienced it and affirmed a faith that God would mete out justice and reward the faithful dead. The new kind of literature that appeared to reflect this new, more pessimistic view of life on earth, was called *apocalyptic literature*, from the Greek term *apocalypsis*, which means "revelation."[2]

Apocalyptic literature is a highly symbolic type of writing, employing biblical imagery from earlier times to show that evil would soon be

2. The book of Revelation is also called the Apocalypse (or the Apocalypse of John or the Revelation of John, John being the name of the persecuted Christian to whom God gives the revelation about what is going on and what its outcome will be).

overthrown by God and that those who did not give up their faith to save themselves in this life would be rewarded with life everlasting in a paradisiacal heaven with God. Apocalyptic literature is usually presented as a revelation from God given to some past hero of the faith that predicts what will happen in the future, though it is actually written in the midst of the persecution. In this way, the persecution that the faithful are experiencing appears to have been foretold, confirming that God is indeed still in control and that therefore all will turn out well in the end for those who keep their faith. Thus, the *book of Daniel* was written during the time of persecution under Antiochus IV.[3] Mark 13:3–37 is called the *little apocalypse* (used also in Matthew 24 and Luke 21). It was probably written around the time of Emperor Nero's persecution of Christians in Rome from 64 to 68 CE and the nearly simultaneous Roman suppression of a Jewish rebellion between 66 and 74 CE, resulting in the psychologically devastating destruction of the holy city of Jerusalem and God's Temple in 70 CE. The book of *Revelation* was written during the persecution of Christians under Emperor Domitian in the Roman province of Asia (now western Turkey) around 95–96 CE.

The production of apocalyptic literature by Jews and Christians had come to an end by around 200 CE, but its influence continued, causing a succession of Christian thinkers and their followers to proclaim that the end of the world and Christ's second coming were imminent. The social upheaval that such predictions caused in the church led early Christian leaders to renounce these predictions—which, of course, never proved to be accurate anyway, since this crystal-ball view of the books was not what they were or had been intended to be. When the contents of the NT were being settled in the fourth century CE, these leaders did not want the Christian apocalypses included. In the end, though, one apocalypse was included, the book of Revelation, due to its supposed tie to the author of the Gospel and three letters of John. The inclusion in the NT of this apocalypse, along with the "little apocalypse" in Mark, Matthew, and Luke, ensured that this type of literature would continue to be read and misunderstood by most people of future generations, who would take the "prophecies" in them as predictions of their own future, rather than as statements about current reality for Christians of the past who were undergoing persecution.

On the one hand, we have the traditional Jewish view that God's people are called to work with God in bringing all people into loving relationships with God, since in that way they will find their fulfillment and greatest joy in life. On the other hand, we have the apocalyptic view, in which the world is divided into good and evil people, and Christians define themselves as the

3. Daniel was written in the year 165 BCE or so.

good ones and wait for the others to be judged by God and cast into hell, where they will be punished forever. It should be obvious why the apocalyptic view is popular among at least some Christians. It allows them to hate their enemies (whoever they might be) under the cover of wanting justice and to be delighted that those others, whoever they are, will be punished forever. It is a very human attitude, but also a sinful one. There is no love in this view, only vengeance, as well as an image of God as judge who sends bad people to hell. This is, as we have seen, diametrically opposed to Jesus's teaching that God loves all people and constantly works for their happiness and fulfillment, and that we are to love everyone, even our enemies, for it is only in this way that we, as well as our enemies, will find salvation in life. Given this, it should be clear that Jesus was not the kind of apocalyptic preacher that Mark and those who followed him claimed he was.

It is noteworthy that Paul, too, adopted this apocalyptic vision, though he differed from the Synoptic Gospels and Revelation in some significant ways. He did believe that the risen Christ would return soon: he speaks of "we who are alive, who are left until the coming of the Lord" (1 Thess 4:15), implying that he and/or others would still be alive when Christ returned. Clearly he was wrong in this expectation, and so his view provides no support for the apocalyptic perspective. That many early Christians believed in this apocalyptic vision does not mean that it came from Jesus; and there are very good reasons to believe it did not, despite being put into his mouth by the Synoptic evangelists.

I have been trying in this book to develop an alternative vision of Christianity, one in which people are drawn to a God who loves them and constantly seeks to bring about the fulfillment of their lives in every minute, rather than the common, supposedly Christian understanding today, in which people fear God because, if they do not live correctly, God might condemn them to hell when they die. The common view is built on fear. The other view, which is built on love, is the true heart of Christianity.

As we have already argued, the future is open, not predetermined, and it is we ourselves and those who succeed us who will determine the future. The world will be what we make of it, for good or ill, depending on whether we heed the divine call to love or nature's and society's call to self-centeredness.

If this is what a proper Christian eschatology looks like for the species, what can we say about the other traditional focus of eschatology, the ultimate fate of the individual? A continuation of the individual's subjective existence after death has traditionally been termed "the Christian hope." The theological ground for this hope lies solely in the love of God, who may, out of love, want to continue the relationship begun in this life with those who acknowledge God or to initiate it with those who did not respond to God's offer of love while in this life. How that might be possible, no one can say. If all physical reality exists within God, then it is God who ultimately determines what is possible within that reality. If God determined in the infinite past that some creatures, perhaps those of sufficiently developed mind (and not just humans), would continue after death in a spiritual form analogous to God's form of spiritual existence, then that possibility is built into the overarching rules governing all universes actual and possible.

If subjective existence after death is what awaits us, what survives of us must be the fullness of our personal being, so that there is essential continuity between who we were in our mortal form and who we will be in our spiritual form. Paul spoke of a resurrection of the whole person, including the body, though not in its present form; rather, he claimed that a new "spiritual body" would replace the one made of flesh and blood (1 Cor 15:35–57). However we might conceive the nature of this continued existence beyond death, the key point is that it is the individual in all her or his essential individuality that continues. And key is the recognition that if such subjective immortality is indeed a reality, it is solely because of the love of God, who gives us this gift out of pure grace. It is nothing we have earned or ever could earn.

The nature of that continued life has been imaged in vastly different, mostly symbolic, ways in the NT and Christian history. While whatever we say must be speculative, it seems that at the least we would continue in our relationships with God. For those who had a loving relationship with God in this life, that would mean a continuation and deepening of that relationship into the everlasting future, bringing a bliss never possible in mortal life; for those who did not have a loving relationship with God in this life, it would mean the initiation of that relationship and its growth and deepening richness into everlastingness, along with the joy and fulfillment it will bring. This is what an agapic God would desire.

Traditional Christian doctrine, following the apocalyptic model, has postulated a final judgment in which all people are judged by God and sent either to heaven or hell, depending on how they have lived their lives in relation to God. We have rejected that notion as unworthy of the Perfectly Agapic God, though there may be some truth in it. I have been greatly influenced here by Marjorie Suchocki, who has suggested that after death, as we come into increasing intimacy with the divine Spirit, we come to view our lives and ourselves as God (and perhaps other people) viewed us, feeling God's pleasure as we have done loving things and God's disappointment with us when we have not. Since God has been present with us throughout our lives, the perfect divine memory of us is the objective truth of who we were in this life, and therefore we are confronted with the real truth of who we were, not the relatively rosy picture we fashion by reworking our memories in order to feel better about ourselves.

In this way there is a something like a final judgment, but it is not one enacted by God resulting in people being assigned to either heaven or hell. Instead, it is a judgment we make regarding who we were in this life, based on God's infallible memory of us (our "objective immortality" in God) in every moment of our life. The more that life was one of love, the more we experience the pleasure that God had in us. On the other hand, the more self-centered we were and the more unlovingly we treated others, the more we will experience God's disappointment in us. The former experience we might term various degrees of heaven, and the latter degrees of hell. However, over time, as we come more and more into intimate relationship with God, we constantly experience the great divine love for us, and it is by means this that we gradually come to forgive ourselves as God had already forgive us. In the end, it will only be the love of God that we know, and in this we will experience our greatest fulfillment, joy, and peace. This seems to me the most likely nature of life after death, given an understanding of God as Perfect Agapic Love, should continuing life after death in fact be what awaits all humans. At the very least, we will continue to exist in the mind of God as we are remembered and enjoyed for the positive values we enacted during our lives.

21

God and the History of the Universe

We have learned several things from our study of the nature and history of the universe, the science that undergirds it, and what a scientifically informed and coherent Christian theology might look like. These are particularly important for anyone attempting to construct an integrated scientific and theological worldview.

One thing we have learned has to do with *the nature of science*, including what it can and cannot do. The proper object of scientific study is the physical universe—and, if they exist and become accessible to us, the other universes. Science provides us today, and will continue to provide humanity, with the best possible knowledge of the nature of the universe and the things that exist within it and how they behave. Sometimes scientific progress is revolutionary, overthrowing prior understandings, such as when relativity theory demonstrated that Newton's understandings of the nature of space and time were incorrect. For the most part, however, scientific progress is evolutionary, moving by increments and refinements ever closer toward a final understanding of the nature of the universe and its components.

On the other hand, science cannot do some things. It cannot *prove* any of its theories. Rather, scientific theories are more or less *probably* true, with that probability depending on the strength of the evidence in support of the theory that accumulates over time. Furthermore, these scientific theories are not the final truth about the universe but only the best understanding we have at any given time. The understanding of the nature and history of our universe that we have reviewed in this book is the best that science can offer at this point in history, yet we should expect that it will almost certainly, in some if not most areas, be refined or even superseded in the future. Another major limitation of science is that, by definition, it cannot tell us anything about whether any other type of reality—specifically, spiritual reality—might exist.

In addition, science, at least thus far, has not been very good at addressing some things. For example, neither the natural sciences nor the social sciences have had much to say about subjective human experiences, especially one-time experiences, such as aesthetic experiences or unique spiritual experiences. What the sciences have had to say about these things has tended to be simplistic and reductionistic—especially when these have been considered within the context of the materialistic, reductionistic, and deterministic metaphysical beliefs of some scientists that discuss them.

It is important to recall, too, that while virtually all scientists would agree on the mathematical formulas that describe physical reality, they do not always agree on how to interpret their meaning—or, at times, even whether some kinds of scientific study are legitimate, such as research on psychic phenomena. Scientists have their own metaphysical views and interpretational proclivities, as do all people, and these influence what they understand the math to be telling them. Consequently, most scientists understand quantum mechanics to indicate that indeterminacy or freedom exists on the quantum scale, but other scientists believe that the results fit within a deterministic, multiverse worldview. Or, with regard to the history of the universe, including the evolution of life on our planet, some scientists understand that the production of novel complex entities, such as molecules from atoms or humans from primates, brings with it new, emergent qualities that did not exist before and could not have been predicted from knowledge of the prior forms from which they arose. Other scientists reject this idea, arguing that reductionism provides the only proper explanation of physical reality and that, with sufficient knowledge, the higher forms and their behaviors could have been predicted simply from a fuller knowledge of the nature of the fundamental particles that constitute them. It seems, then, that there are fundamentally two kinds of metaphysical belief: religious/spiritual and atheistic, as suggested in the first chapter, with most scientists adhering to one of these beliefs or the other.[1] Thus, some scientists identify themselves as atheists, or materialists, and many of these also believe in determinism and reductionism alone in interpreting the meaning of scientific data. They also often claim that science is the only legitimate source of knowledge. Other scientists believe in the existence of a spiritual

1. There is, of course, a third option, agnosticism, the belief neither in atheism nor in religion/spirituality and so claiming no metaphysical beliefs at all. Although this position may be held intellectually for good reasons, the functional result of this belief would likely be difficult, if not impossible, to distinguish from atheism.

reality, denying that physical reality is the only kind of reality, and these tend to affirm that indeterminism/freedom/openness exists in the universe in addition to determinism; that holistic study, including top-down causality, is a valid method in science as well as reductionism, or bottom-up causality; and that there are other valid sources of knowledge about reality in addition to science (such as private individual moral, aesthetic, and spiritual experiences). Since there is no way to prove which understanding of reality is true, each scientist, like each person, must decide for him- or herself which worldview seems to make the most sense, given the totality of his or her knowledge and experience.

For the NASA/ESA/Hubble image of the Butterfly Nebula, see http://www.nasa.gov/multimedia/imagegallery/image_feature_2526.html#.U9wSF_ldWsg.

Nonetheless, while disagreement persists among scientists in the areas noted above (and others), virtually all scientists agree upon some general features of the universe. First, it is hard to imagine a scientist who would not agree that the universe is *a wondrous place filled with stunning beauty*. From the smallest scales to the largest, we witness the order and beauty of nature created by only a handful of natural laws. We can see the ordered beauty of galaxies and nebulae at the level of the largest scales; and even destruction at that level can be beautiful, as in supernova explosions of stars. But even on the smallest scales, scientists have noted the beauty of certain molecules, such as DNA, with its rich symmetry.[2] A richness of detail exists in all of nature that no artist could ever duplicate.

For NASA, ESA, and the Hubble SM4 ERO Team's image of Stephan's Quintet galaxies, see http://www.nasa.gov/mission_pages/hubble/multimedia/ero/ero_stephan_quintet.html.

Furthermore, the universe is a *dynamic*, rather than static, place. Physical reality is constantly changing, such that not only is everything within the universe evolving, but also the space-time fabric itself is changing, expanding outward. The universe, therefore, is a *process*, one in a never-ending state of *becoming*. It is also one in which beauty continues to evolve.

2. See, for example, Kovac and Weisberg, *Roald Hoffmann*.

For the NASA, ESA, HEIC, and The Hubble Heritage Team (STScI/AURA) image of the Cat's Eye Nebula, see http://www.nasa.gov/multimedia/imagegallery/image_feature_211.html.

Another feature of our universe that is fundamental to science is that the universe has evolved from utter simplicity in the earliest fraction of a second of its history to enormous *complexity*, such as we see in the universe and in life on Earth today. Not only has the universe moved from simplicity to complexity in the course of its history, it has also moved from uniformity (such as the hydrogen and helium universe that originally existed) to *diversity* (such as the enormous variety of atoms and molecules that exist today). This dynamic universe has thus created untold *novelty* throughout its history, from the physics of its earliest beginnings to the chemistry of atoms to, eventually, geology and biology (on at least this planet). Every new form of life on Earth over the past nearly four billion years attests to novelty.

It is also the case that the universe is at heart thoroughly *relational*, so that we cannot adequately describe anything within it in isolation; rather, we can appropriately understand one of its components only in terms of its relationships to all the other things around it. Each entity contributes to changes in the things around it and is at the same time changed by those things. Consequently, the evolution of life on this planet has been, and continues to, be a thoroughly relational process.

Furthermore, scientists believe that the universe is wholly *law-governed*, even though they might offer different, and even contradictory, interpretations of the nature of those laws and of what they might mean. For example, according to the determinist, the laws of nature determine everything that happens in the universe. According to the majority of scientists, however, who see indeterminism as well as determinism in operation in the universe, natural laws determine the behaviors of certain things (such as conglomerates like rocks) while providing the limits within which other entities possessed of some degree of freedom (from quarks to humans) behave. These laws of nature are what give order to the cosmos. Without them, the cosmos would no longer be a "*uni*-verse"; instead, it would be sheer chaos.

For the NASA, ESA, S. Beckwith (STScI), and The Hubble Heritage Team (STScI/AURA) image of the Whirlpool Galaxy and companion galaxy, see http://hubblesite.org/gallery/wallpaper/pr2005012a.

Another feature of the physical universe that scientists agree upon is that the laws of nature are *mathematical* in nature. Consequently, we can adequately describe them only by mathematical equations. Mathematics is a kind of language. Because only a minority of people ever come close to mastering it, scientists attempt to translate the mathematics of any given theory into abstract conceptual language—which is why we sometimes encounter differences of opinions among scientists on the meaning of such laws in their translated formulations. So fundamental is mathematics to the description of theories that scientists have used the math to predict things about the world that other scientists have later found to be true through experiments, such as Paul Dirac's equations that predicted the existence of antimatter, which was later discovered experimentally.

This fact of the fundamentally mathematical nature of the universe has been, at least for some scientists, rather amazing. This has been called the "*unreasonable effectiveness*" of mathematics in describing nature. The problem begins with the recognition that mathematics in itself has nothing to do with the world. It is abstract, existing in the realm of ideas alone. Consequently, many kinds of mathematics can be imagined and studied, including kinds that are irrational in the commonsense view of things. We have seen one example in imaginary numbers, in which the square of a number is negative. There are also anticommuting numbers.[3] In both cases, science has found that these kinds of pure, counterintuitive mathematical constructs are necessary to describe the natural laws that govern the world. The problem of the "unreasonable effectiveness" of mathematics in science, therefore, is the question of *why* this should be the case. Why should *abstract* mathematics describe the nature of the *real* world? This is a problem that concerned Einstein and a number of physicists after him. It seems that we cannot resolve it scientifically.

Yet another issue that the mathematical nature of physical reality raises is that typically, if not always, "beautiful" or "elegant" mathematical equations describe the true nature of reality, while "ugly" equations usually, if not always, do not. Not surprisingly, then, beautiful equations are often compared to art, such as music, sculpture, or poetry. Indeed, when one studies the arts, one finds that mathematics is frequently deeply engrained within them. In music, for example, the various notes are related to one another mathematically; and in the visual arts, such as painting or sculpture, various types of symmetries are often fundamental to the work. For this reason, early-twentieth-century philosopher of mathematics Bertrand Russell claimed, "Mathematics, rightly viewed, possesses not only truth, but

3. See the definition on p. 105.

supreme beauty—a beauty cold and austere . . . yet sublimely pure, and capable of a stern perfection such as only the greatest art can show."[4]

It is rather remarkable that this mathematical beauty is made concrete in nature, and that it is equations that have mathematical beauty that describe, as we have seen, most, if not all, of the fundamental laws of the universe. To a physicist this is a stunning discovery. What it means is that, at the deepest level of physical reality, truth and beauty are intimately tied. It is the harmonious interconnection of such beautiful equations that will ultimately tell us the full truth about the universe! So powerful is this connection that the quantum physicist Paul Dirac spent a significant portion of his later career seeking beautiful equations and, when such equations were found, sought then to find what it was in nature they described. The profound question here is *why* beauty and truth should be so integrally tied in nature, when there seems to be no good reason why they should be.

The ancient Greek philosophers were concerned with a third value in addition to truth and beauty: goodness. Is it possible that the *true* descriptions of the laws of nature could be *good* as well as *beautiful*? When considering inanimate things, goodness is typically ascribed to them according to how well they do what they are intended to do. For example, we judge a watch to be good to the degree that it tells time accurately and continues to do so over the course of many years. If we apply this standard to the laws of nature, it does seem that we can consider them good. They have, to the best of our knowledge, existed continuously throughout the entire history of the universe and have performed brilliantly in fulfilling their function, which is, at a minimum, to create order in the universe. If the laws of nature also have the function of creating diverse and complex entities, including living beings, then this is another way in which we might judge those laws to be good.

Another question naturally follows from this recognition: If the laws of nature can be good, true, and beautiful, can we also speak of the universe—and the things produced within it by the operation of these laws—as good, beautiful, and true? I would suggest that we can, though in somewhat different ways according to the kinds of things we consider (as we will see below). Also important to recognize is that, while the laws of nature are *inherently* good, true, and beautiful, created things have these three characteristics as *potentialities*, so that any individual thing has them only to a certain degree. For example, regarding the *beauty* of inanimate things, we can consider, for example, the structure of different molecules and the different degrees of beauty they possess relative to each other. Our criteria for analyzing their

4. Russell, *Mysticism and Logic*, 60.

beauty would likely emphasize symmetry, where we would judge more symmetrical molecules to be more beautiful than asymmetrical ones. Some chemists, as we have noted, do actually speak in this way, referring to certain molecules, such as the intricate and symmetrical double-helix of DNA, as being exquisitely beautiful compared with other molecules.

We usually judge the beauty of a thing, however, not only by its individual form but also by how it stands in relation to the other things around it—its *harmony*, or lack thereof, with other things. For example, on the largest scales of nature, most of us see beauty in things such as nebulae, which are composed of dust and gas and the stars to which they are giving or have given birth, each more or less beautiful in itself. It is the way in which all the individual things within the cloud relate harmoniously to each other, and to the essentially empty space that surrounds them, that constitutes part of their beauty as well. Thus, we judge some nebulae to be more beautiful than others, based on the shapes and various colors of their clouds, and the relative distributions and contrasts of the gas, dust, and stars within them.

To test this idea, go to the Hubble Space Telescope website, http://hubblesite.org/ (see the QR code above left) and http://www.nasa.gov/mission_pages/hubble/main/#.U91tq_ldWsg (see the QR code above right) and view the many nebulae it has photographed, some of which appear in this book. You will almost certainly find some more beautiful than others.[5]

It seems that nature is brimming with beauty in varying degrees. But is beauty something objective in a thing or merely the result of a person's subjective judgment? We might expect, given the different types and experiences of people and their sometimes differing judgments about the beauty of a thing, that beauty is not something inherent in a thing. Some cross-cultural studies, however, suggest that this is not the case. For example, one study concerned people's judgments regarding the relative beauty of human faces,

5. In determining what constitutes a thing as beautiful, we are touching on the philosophy of art, which we will not discuss further here, but which the interested reader might want to pursue elsewhere by examining the many books on the subject, some of which are in the bibliography. See, e.g., Barrow, *Between Inner Space and Outer Space;* Coogan, "Process and Art"; and Parker, "Art, Creativity, and Process Thought."

including those of different races and regions of the world from those doing the judging. What they found was a consistent overall pattern of agreement regarding which faces were more or less beautiful, irrespective of the race or culture of the persons doing the judging. I would suggest that comparative study of astronomical entities such as nebulae would yield similar cross-cultural judgments of their relative beauty. Such studies suggest that the beauty of a thing is an objective characteristic, something inherent in it.

In this light, we might also tie the beauty of a thing to its truth. For example, with watches or clocks, we do make judgments on which are more beautiful than others and might well find consensus on which is the most beautiful of all watches in a given sample. Therefore, we might say that the more beautiful a watch is, the more truly it is (the ideal of) a watch. This example illustrates how beauty and truth can be tied together in inanimate things.

As regards the *goodness* of things, we can employ the same criteria to inanimate things that we applied to the laws of nature. With regard to things that exist in the universe, therefore, we can judge their relative goodness according to the degree to which they fulfill their purposes and do so consistently over time. For example, as we noted above, a good watch would be one that told time precisely and continued to do so over time, whereas a bad watch would be one that failed to do either or both. The degree of goodness of a thing, therefore, would be inherent within it and dependent on its relationship to the things around it. Consequently, as with beauty, goodness also would be part of the truth of that thing—what it really is. We might say that the truth of a thing is a matter of degree, so that the more perfectly good and beautiful a thing is, the more truly it is that thing. For example, the more a watch fulfills its potential to be a good and beautiful watch, the more truly it is a watch (that is, the ideal of a watch).

Up until now, we have been speaking of the goodness, truth, and beauty of inanimate things in the universe, such as molecules, nebulae, and watches. However, these three qualities take on additional meanings when talking about living things, and especially those that have reached the kind of mental development that we see in ourselves.[6] Our mental capabilities

6. As I have previously suggested, it seems that creatures exist along a continuum of lesser to greater complexity and correspondingly of lesser to greater interiority and freedom. This suggests that there might also be a continuum of meanings of beauty, goodness, and truth that are appropriate to the different kinds of creatures that do not have the same kind of interiority or subjectivity that we do. For the present, at least, knowledge of which creatures have precisely what degree of freedom and subjectivity is beyond our scientific reach. Still, most of us who have had pets know that different dogs, cats, birds, and the like have different personalities. That implies a higher degree of freedom of behavior, and so interiority, than less complex animals. For this reason,

bring with them a degree of self-awareness and freedom of behavior unknown to beings with lesser minds. This freedom and self-awareness mean that, to a great extent, we can and do, in fact, create ourselves—that is, our "inner selves," ourselves as "persons"—over the course of our lives. We do so by what we think, what we feel (for feeling is also an integral part of our interior self), and what we do from moment to moment throughout our lives. Thus, with regard to beauty, we make ourselves into unique individuals, each differing in beauty from all others, not only physically (in our faces, our body shapes, and the like), but also subjectively, in the kinds of persons we choose to be. Thus, we speak of people being beautiful on the inside, or of having beautiful spirits.

The inner beauty of a person seems to be intimately tied to that person's inner goodness, which in this human context means the person's *morality*. One's morality is properly determined by the way one treats other persons, oneself, other living creatures, and the natural world. Also, like inner beauty, one's moral goodness exists in various degrees. For the religious person, there is another dimension to personal goodness in addition to morality: what we might call spiritual goodness, or spirituality. Where moral goodness concerns one's relationships with other finite beings, spiritual goodness concerns one's relationship with ultimate reality, which for Christians and all other theists is God. For the religious person, therefore, the degrees to which a person embodies morality and spirituality constitute the degree of inner goodness of that person.

We perceive, at some level, that beauty and goodness are intimately related in people, as the way we sometimes speak indicates. For example, when something morally good occurs, such as a reconciliation between former friends who have been alienated or someone helping another person in need, we seem at least as likely to say "Was not that beautiful?" as we are to say "Was not that good?" Similarly, when we see people arguing or fighting, we are as likely to say, "That's really ugly!" as we are to say, "That's really wrong!" Such aesthetic terms can be used, and in fact are used, as equivalent to moral terms with regard to many, if not all, human interactions. It seems reasonable, then, to suggest that goodness is the form that beauty takes with regard to the inner person. Physical (external) beauty and goodness (how well a person's body functions) are largely genetically determined, not self-created (though, in some ways, human efforts can aid them) and are, therefore, far less important than the self-created beauty that is the goodness of a person. Together, therefore, a person's inner beauty and goodness must be

I speak in this section only of humans, who have a high degree of interiority and freedom, both of which we experience in ourselves and other humans.

added to the physical goodness and beauty in determining the truth of that person, who that person really is, and how well that person fits the ideal truth of what a human being could be. Accordingly, we might say, in light of our discussion of inanimate things, that the more fully a person exemplifies such beauty and goodness, the more truly human that person is.

These reflections show that the values in the ancient Greek philosophical triad—goodness, truth, and beauty—are deeply interconnected, not only in the laws of nature but also in the universe and the entities that populate it. But here yet another profound question arises: Why should these three values—goodness, truth, and beauty—be so integrally related in the laws of nature and in every entity in the universe, including the unique ways in which these values exist in intelligent, largely self-created creatures such as ourselves? This question follows on our prior queries of why the laws of nature should exist and be mathematical, why it is beautiful equations that usually (if not always) describe these natural laws, why the universe itself is so filled with beauty on all scales, and why beauty and truth should be so closely related in the natural laws. It is here, I believe, that a theistic, and particularly Christian, understanding of reality can offer coherent and rational answers to these questions—answers that flow naturally from the Christian theological understanding—but that no atheistic worldview that I know of provides or perhaps can provide.

The key to the Christian answers to these questions lies in a consideration of the essential nature of God as understood in the Christian tradition. As we have seen, the central, nonnegotiable Christian claim about God is that God is Perfect Agapic Love. In the Christian tradition, moreover, various past and present theologians who have reflected on the ancient Greek triad have also asserted that God is the perfection of Goodness, Truth, and Beauty. As such, God is the ultimate standard against which all finite goodness, truth, and beauty are judged.[7]

If this is true, as I think Christian theologians must assert, then these three must be intimately united within God in a way that is fundamental to God's nature. Since God's essential nature is Agapic Love, then that must mean that God's Truth, Goodness, and Beauty are fundamentally related, not only to each other, but also to God's Love in some way. I would suggest that the way these three attributes are essentially united in God is that Goodness, Truth, and Beauty reflect the different perspectives from which we can view or experience God's Love. That is, God's perfect Goodness is God's Love considered from the moral perspective; God's perfect Beauty

7. This claim can be seen, for example, in Thomas Aquinas's fourth argument for the existence of God (*Summa Theologica*, I, etc.).

is God's Love considered from the aesthetic perspective; and God's perfect Truth is God's Love considered from a cognitive perspective (that is, the essential Truth about God is that God is Love). It is this Goodness, Beauty, and Truth of God's Love that constitute the attractive power of God, by which the Infinite Agapic Spirit works to promote the realization of these values in the universe as fully as possible.

It has been commonplace in the history of Christian thought to say that, just as a work of art reflects the person who created it, so the universe would bear signs of the one who created it. It is from this perspective that Christian theology offers answers to the questions we posed above (as well as a host of others). Why the universe is rationally understandable, why it has laws to govern it, and why these laws are mathematical in nature is because they are creations of a Rational Mind—God—who chooses to create an ordered cosmos and does so by choosing rational, mathematical laws to govern it and then holds them in being. Reflective of this line of thought was *Srinivasa Ramanujan* (1887–1920), the great twentieth-century mathematician, who claimed, "An equation is uninteresting to me unless it expresses a thought of God."[8]

The reason we can describe the laws that govern the universe with beautiful equations is that Beauty and Truth (and Goodness) are one in God, and, therefore, the laws that God has established to govern the universe reflect this unity within God. So also, the reason beauty, truth, and goodness are united in each creature to some degree according to its nature is that they are united in the God who is ultimately responsible for their creation via the good, true, and beautiful laws of nature. A Christian, in fact, should *expect* that the universe would reflect these characteristics, since they are present in the God who has created and continues in each moment to create our universe by means of the laws of nature and the divine influence. Thus, the Christian understanding completes the understanding of reality that science offers us, going beyond what science can tell us to offer answers to the questions science itself raises but cannot answer. In addition, it incorporates the value-experiences of morality, aesthetics, and spirituality in a way that science has never succeeded in doing, while also making room for other ways of apprehending truth than by means of the senses alone. Finally, in the Christian view, the universe and its history have meaning and are not some senseless accidents.

In this perspective, then, the evolution of the universe as a whole and of life on this planet becomes more than just the product of the random set of natural laws that happen to govern the universe, as some have postulated.

8. Quoted by Ranganathan, *Ramanujan*, 88.

Rather, the laws that gave rise to our universe and its life, and that govern the evolutionary process, are the products of a Loving God who has had more than the one goal of giving life and seeking to help each creature that has ever evolved to experience fulfillment in whatever way is possible for it. God has had a second goal—that of eventually evolving beings possessed of intelligence and freedom who are therefore able to create themselves in significant ways and who will hopefully use this ability to fulfill their potentials for the higher forms of goodness, truth, and beauty by becoming people of agapic love toward other creatures, themselves, and God. In this way, they help the universe move toward the maximum possible realization of these values, which is the divine goal, and so increasingly mirror the nature of the God who made it.

In this universe, the one we inhabit and the only one we *know* to exist, God has established the laws of nature such that life will be produced on every planet throughout the universe that has the conditions necessary for it to arise. These natural laws also make it likely that life will eventually evolve highly complex and therefore intelligent and significantly free creatures. Such creatures will be able to perceive and consciously conceive the existence and nature of God and so, hopefully, choose to live in an agapic relationship with God, self, and all creatures. In this way, they find their ultimate fulfillment as well as help to further the realization of the divine values of goodness, truth, beauty, and love in the world around them.

This understanding is expressed in the traditional Christian teaching that humans are created in the image of God, but are not themselves God. We are like God in that we possess (to some degree) certain traits that God possesses, such as intelligence, will, creativity, freedom, beauty, goodness, and love. However, whereas these traits are supremely present in God, because they are inherent in the divine nature, for us they are potentials that we might or might not attain. The attainment of them is up to us individually, depending on how we choose to use our freedom to create ourselves as persons over the course of our lives. In this potential lie our greatest goodness and beauty, and the truth of who God intended us to be—loving persons—both for our own good and for the good of others. To the degree that we attain these values, God is pleased with us and experiences joy and fulfillment in our achievement. This agape, this goodness, truth, and beauty, is the moral and spiritual goal of our lives, both as individuals and as a species.

Our own experiences as persons support this understanding of goodness, truth, beauty, and love as inherent values that we all carry within us. We all hunger for goodness, truth, beauty, and love in our lives, even if we are not aware of it. We want to know the *truth* about things—think about the endless string of "why" questions young children often ask—and we hate it when people lie to us. We also want to be truthful people, knowing and speaking the truth, even though we may not be consciously aware of it or consistently do so. We want *goodness* in our lives and hate evil and the suffering it brings us and others about whom we care. We also want to be good people, even though sometimes we get lost in our own self-centeredness and are not all that good. We want to have *beauty* in our lives and to avoid ugliness. We also want to be beautiful ourselves, even though we sometimes are not physically and/or personally beautiful. Most of all, we want and, indeed need, *love* in our lives—to give it and to receive it. The more we experience this giving and receiving of love, the better our lives are, and the more fulfilled we as persons will be. I think that any person who reflects thoughtfully upon his or her life will find these things to be true.

The understanding of the nature of God that I have been emphasizing explains why people have this hunger for goodness, truth, beauty, and love in their lives. We have these hungers because they are realities in the God who is ultimately responsible for our existence, and at some level we realize they are potential realities in our lives, too, that we can make actual if we choose to do so. In addition, I have suggested that all of us experience the constant presence of God with us in the deeper, subconscious parts of our minds, and so have actually experienced something of God's perfect Goodness, Truth, Beauty, and love ourselves. Having experienced them and the kind of personal fulfillment they bring, our minds, both subconsciously and consciously, hunger for more of this goodness, truth, beauty, and love, and do so throughout our lives.

The problem is that goodness, truth, beauty, and love are never found in such purity in the world around us; they exist only in relative degrees, due to the imperfection of the world and the things, such as ourselves, that inhabit it. This is why, no matter how much goodness, truth, beauty, and love we experience in this life, we always yearn for more—because we have experienced these values more purely and fully, albeit, for most of us at least, subconsciously. These hungers for goodness, truth, beauty, and love are thus ultimately a hunger for the God who is the perfection of them all. Perhaps it was in part some such recognition that motivated Augustine to say to God in the prayer that opens his *Confessions*, ". . . you made us for (literally, toward) yourself, and our hearts find no peace until they rest in you."[9] The ultimate fulfillment of our needs for goodness, truth, beauty, and love, then, cannot happen in this life. It will occur only if and when we experience God more fully and directly in the spiritual life that most Christians believe

9. Augustine, *Confessions*, 21.

lies beyond this one. The best we can do here and now is to experience goodness, truth, beauty, and love in our relationships with the world, its creatures, and preeminently, other people. It is in this that we will find as much fulfillment as we can possibly have on earth.

This might all seem highly theoretical. However, I hope I have shown that it is also eminently practical, offering something that humans throughout history have desired: *wisdom*. Wisdom is knowledge about the nature of life, its deeper structures, and the way things really work in the world. If one has this wisdom and lives according to it, he or she will have the best life possible. Having such a life is everyone's goal, but not all life-paths will get us to that goal. According to Christian teaching, the things highly prized in the world that many people think will bring them fulfillment, such as wealth, power, and fame, will inevitably fail to do so. My own experience in life, and the experiences of others that I know, tell me that this is true. Counterintuitively, Christian wisdom claims that the only way to attain maximum joy and fulfillment in life is to live an agapic life, one in which you *do not* seek your own fulfillment but instead seek the fulfillment of others and in so doing attain your own. This is because agapic love shown toward others will almost always be returned, sooner or later, for this kind of love is finally irresistible—being, as it is, what we all crave most deeply. Thus, the selfless love you give will return to you in the love others show you, because you have loved and so helped them to be fulfilled. Agapic love breeds agapic love, even in the less-than-perfect forms in which we less-than-perfect humans show it.

It is the further Christian claim that, if all people were to live this way, with everyone loving and being loved agapically, then everyone would be fulfilled in the deepest sense. The biblical term for this state is the *reign* (usually translated, less appropriately, as *kingdom*) *of God*, and it has been the hope of Jews and Christians alike for millennia. In such a world, everyone would win and none would lose—unlike most situations in our competitive culture, where winners can be winners only if there are losers. This ought to tell us that something is wrong about the way we have chosen to live together as a society. Moreover, although standard human morality might agree that love is a noble virtue, I think that only a religious person, such as a Christian, would affirm the view that this love should be agapic—totally self-giving and forgetful of the self.

CONCLUSION

In this book, I have sought to:

- inform readers about scientific and Christian thought, including dispelling various common and popular but incorrect ideas about them, especially the idea that science and theology (and/or faith) are necessarily incompatible,
- present the whole sweep of the history of the universe and the place of humans in that history, and to suggest where and how God has been active, and still is active,
- show that a rational Christian theology is fully compatible with contemporary science and can even offer some potentially useful perspectives on matters of concern to science that science itself has not been successful in explaining, and
- offer a worldview that integrates science and Christian theology.

In this worldview, the ultimate reason for the existence of the universe is the Agapic Love of God, who desires to bring into being creatures that can be the recipients of this love. God creates primarily by establishing the laws of nature that bring order to the universe and determine its workings, either directly or indirectly. The ordered cosmos and the laws that govern it exist through time because they are grounded in, or held in being by, God, who is the ultimate source of order in the universe and so is properly understood to be its Creator. As God is everlasting, without beginning or end, so too, it seems likely, is the universe.[10] Throughout its history, God has been establishing universes filled with creatures upon which God has showered love.

In our universe, God created the laws of nature and has continued to hold them in being. God has worked within these laws to bring forth a whole host of different and ever more complex creatures: from subatomic particles to atoms to molecules to stars. Stars create in their cores all the elements necessary to make planets and then send these elements and more out into space at their explosive deaths. Over time, gravity has caused these elements to unite around some nascent, later-generation stars and form

10. As I have already noted, this is true whether our universe is the only one that exists or whether other universes have existed, exist now, and/or will exist in the future. If the multiverse does exist, and if the fundamental laws of each universe are somewhat different from others and are in some way self-generated, as some scientists have suggested, the existence of these varieties of laws would be possible only by God's establishment of some overarching set of meta-laws that determines how these various natural laws for each universe have been produced and the possible forms these sets of laws can take. Natural laws do not arise spontaneously out of nothing; they need some prior laws that permit them to arise.

planets, some of which have had conditions necessary to promote the existence and flourishing of life. On these special planets, such as ours, the natural laws that God established caused the most primitive life-forms to come into existence. Single-celled life then evolved into simple multicellular life, and from there into progressively more complex forms of the plant, animal, and fungus kingdoms, including all the different phyla within each of them.

Within the animal kingdom, evolution, influenced by God at the level of individual creatures in their freedom, created a vast variety of more complex creatures, moving from one kind of being's dominance to the next in the long history of Earth. In the waters, the earlier domination of plants and simple animals eventually made way for the rise of fish. These were followed by amphibians, which made the move to land and were the first dominant animal species there. Then reptiles came to dominate the land, seas, and air, and after them the mammals. Eventually primates arose among the mammals, some of which eventually evolved into human beings.

Throughout all of this process, God has been active. For all of the creatures that have ever lived, God has sought to work with each according to whatever freedom it possessed to find its greatest fulfillment, its "salvation," as well as that of all others. God has also had another goal in the evolutionary process: the creation of intelligent, self-aware creatures capable of consciously discerning the existence of the Infinite Agapic Spirit, engaging in a loving relationship with God, and seeking to help fulfill God's loving intention for all other creatures. The natural laws that God established, either directly or indirectly, were set in such a way that the attainment of this goal would be the inevitable outcome—though one not wholly predictable in terms of the specific forms of such creatures—in numerous places in the universe. This is why the history of our universe has seen a movement from simplicity to complexity in its creatures, entailing a movement from lesser toward greater interiority, creativity, and freedom, which has permitted increasingly profound forms of relationship between the Infinite Agapic Spirit and the creatures. God has worked within this general process to bring about these desired ends by the power of God's Love, which is the attractive force of the beauty, goodness, and truth of love, seeking to call forth a similar response of love from every creature, each according to its ability to do so, not only for God but also for the other creatures in their lives.

In human beings, the goal of creating beings capable of conscious awareness of God, and of entering into an intentional agapic relationship with this Infinite, Loving Spirit, has been attained. Such relationships are the richest possible that God can have with finite creatures—at least on this planet. Those who do so engage in a conscious relationship with God are able then to align their purposes in the world with God's purposes. As we have seen, God's purpose for us is to live an agapic life and so to attain our full human potential, our full humanity, which is our highest goodness, truth, and beauty. Living such an agapic life means that our preeminent goal is to love all others and to do so in each moment of every day, making other people's lives—and, ideally, other creatures' lives, too—better in whatever ways we can.

This is how Christians understand Jesus to have lived. It is no easy task! Perhaps it is even finally impossible for us to do it fully, but we can at least do our best to live this way, and in so doing make the world and ourselves better. The proper function of the Christian and other theistic religions, then, is to aid in the process of raising our typically subconscious experience of God's Love into conscious awareness and then making explicit the nature of God as Agape and God's agapic intention for the world and for the life of each person. This is our highest calling as individuals and as a species: to partner with God in trying to make the world a place where there are no losers but only winners.

To me, this is a beautiful worldview, one that satisfies the demands of reason, what I know about science and theology, and my own experiences in life. It is a worldview that draws people to God, not out of fear or some self-centered desire, but simply as a response to God's Perfect Agapic Love, a love that is good, true, and beautiful and seeks nothing for itself but everything for each of us. It calls for a world in which all people are so loved by all others that they feel valued, have their basic needs met, and find the personal fulfillment that loving and being loved in this way brings. Somewhat like the physicist who believes that the beauty of the mathematical equations means they must express a truth of nature, I find the Christian worldview I have offered here to be deeply satisfying and, finally, too beautiful (and good) not to be true!

Appendix

Geologic Time Scale[1]

Eon/Era/Period/Epoch	**Time** (millions of years ago)
Precambrian Eon:	
Hadean Eon	4,600–4,000
Archaean Eon[A]	4,000–2,500
Proterozoic Eon	2,500–541
Phanerozoic Eon:	
Paleozoic Era:	
Cambrian period	541–485
Ordovician period	485–444
Silurian period	444–416
Devonian period	416–359
Carboniferous period	359–299
Permian period	299–252
Mesozoic Era:	
Triassic period	252–201
Jurassic period	201–145
Cretaceous period	145–66
Cenozoic Era:	
Paleogene period	66–23
Neogene period	23–2.58
Quaternary period	2.58–present

A. Formerly Archaeozoic.

1. Editor's note: Jarvis used a 2009 time scale. This time scale is a simplified adaptation of the time scale at http://www.stratigraphy.org/ICSchart/ChronostratChart2015-01.jpg.

Bibliography

RESOURCES IN PRINT AND ONLINE

Aczel, Amir D. *God's Equation: Einstein, Relativity and the Expanding Universe*. London: Piatkus, 2000.

Adler, Ralph A., and Robert Horman, *Genesis of the Big Bang*. Oxford: Oxford University Press, 2001.

Alston, William P. *Perceiving God: The Epistemology of Religious Experience*. Ithaca, NY: Cornell University Press, 1991.

Armstrong, Karen. *The Case for God*. New York: Knopf, 2009.

Augustine. *Concerning the City of God against the Pagans*. New York: Penguin, 1984.

———. *Confessions*. New York: Penguin, 1961

———. *The Enchiridion on Faith, Hope and Love*. Chicago: Regnery, 1961.

Ayala, Francisco J. *Darwin and Intelligent Design*. Minneapolis: Fortress, 2006.

———. *Darwin's Gift to Science and Religion*. Washington, DC: Joseph Henry, 2007.

Baker, Catherine. *The Evolution Dialogues: Science, Christianity, and the Quest for Understanding*. 2nd ed. Washington, DC: American Association for the Advancement of Science, 2007.

Barbour, Ian G. *Nature, Human Nature, and God*. Theology and the Sciences. Minneapolis: Fortress, 2002.

Barnes, Eric. "Inference to the Loveliest Explanation." *Synthese* 103 (1995) 251–77.

Barrow, John D. *Between Inner Space and Outer Space: Essays on Science, Art and Philosophy*. Oxford: Oxford University Press, 2000.

———. *The Book of Nothing: Vacuums, Voids, and the Latest Ideas about the Origin of the Universe*. New York: Vintage, 2002.

———. *The Constants of Nature: The Numbers that Encode the Deepest Secrets of the Universe*. New York: Random House, 2004.

———. *The Origin of the Universe*. New York: Basic Books, 1994.

Barrow, John D., and Frank J. Tipler. *The Anthropic Cosmological Principle*. Oxford: Oxford University Press, 1986.

Barrow, John D. et al., eds. *Fitness of the Cosmos for Life: Biochemistry and Fine-Tuning*. Cambridge Astrobiology 2. Cambridge: Cambridge University Press, 2007.

Barton, Nicholas H., et al. *Evolution*. Cold Spring Harbor, NY: Cold Spring Harbor Laboratory, 2007.

Beard, K. Christopher. *Hunt for the Dawn Monkey: Unearthing the Origins of Monkeys, Apes, and Humans.* Berkeley: University of California Press, 2004.

Becker, Ernest. *The Denial of Death.* New York: Free Press, 1973.

Berger, Lee R., et al. "*Homo naledi*, a New Species of the Genus Homo from the Dinaledi Chamber, South Africa." *eLife* (September 10, 2015). http://lens.elifesciences.org/09560/index.html.

Bjornerud, Marcia. *Reading the Rocks: The Autobiography of the Earth.* New York: Westview, 2005.

Bojowald, Martin. *Once before Time: A Whole Story of the Universe.* New York: Knopf, 2010.

Bonner, John T. *The Evolution of Complexity by Means of Natural Selection.* Princeton: Princeton University Press, 1988.

Brasier, Martin, and Jonathan Antcliffe. "Decoding the Ediacaran Enigma." *Science* 305 (August 20, 2004) 1115–17.

Brocks, Jochen J., and Nicholas J. Butterfield "Early Animals Out in the Cold." *Nature* 457 (February 5, 2009) 672–73.

Brooke, John Hedley. *Science and Religion: Some Historical Perspectives.* Cambridge History of Science. Cambridge: Cambridge University Press, 1991.

Bryson, Bill. *A Short History of Nearly Everything.* New York: Broadway, 2004.

Buick, Roger. "Early Life: Ancient Acritarchs." *Nature* 463 (Feb. 18, 2010) 885–86.

Cairns-Smith, A. G. *Seven Clues to the Origin of Life: A Scientific Detective Story.* Cambridge: Cambridge University Press, 1985.

Carr, Bernard, ed. *Universe or Multiverse?* Cambridge: Cambridge University Press, 2007.

Carroll, Sean. *Dark Matter, Dark Energy: The Dark Side of the Universe.* Chantilly, VA: Teaching Company, 2007.

Castelvecchi, Davide. "Half-life (More or Less)." *Science News* 174 (November 22, 2008) 20–23.

Catling, D. C., et al. "Why O_2 Is Required by Complex Life on Habitable Planets and the Concept of Planetary 'Oxygenation Time.'" *Astrobiology* 5 (2005) 415–38.

Cavalier-Smith, Thomas. "Deep Phylogeny, Ancestral Groups and the Four Ages of Life." *Philosophical Transactions of the Royal Academy B (Biological Sciences).* http://rstb.royalsocietypublishing.org/content/365/1537/111.full?sid=aa3c3afc-c967-4769-aeca-5a318836bfb4.

Cela-Conde, Camilo J., and Francisco J. Ayala. *Human Evolution: Trails From the Past.* Oxford: Oxford University Press, 2007.

Chaisson, Eric J. *Cosmic Evolution: The Rise of Complexity in Nature.* Cambridge: Harvard University Press, 2001.

Cheetham, Nicolas. *Universe: A Journey from Earth to the Edge of the Cosmos.* London: Smith-Davies, 2005.

Chen, Xiang. "A Different Kind of Revolutionary Change: Transformation from Object to Process Concepts." *Studies in History and Philosophy of Science* 41 (2010) 182–91.

Clark, Mary E. *In Search of Human Nature.* London: Routledge, 2002.

Clayton, Philip. *Adventures in the Spirit: God, World, Divine Action.* Minneapolis: Fortress, 2008.

———. *God and Contemporary Science.* Edinburgh: Edinburgh University Press, 1997.

———. *Mind and Emergence: From Quantum to Consciousness*. Oxford: Oxford University Press, 2004.

Clayton, Philip, and Arthur Peacocke, eds. *In Whom We Live and Move and Have Our Being: Panentheistic Reflections on God's Presence in a Scientific World.* Grand Rapids: Eerdmans, 2004.

Clayton, Philip, and Paul Davies. *The Re-Emergence of Emergence: The Emergence Hypothesis from Science to Religion.* Oxford: Oxford University Press, 2006.

Close, Frank, et al. *The Particle Explosion.* 1986. Reprinted, Oxford: Oxford University Press, 1994.

Cobb, John B., Jr. *A Christian Natural Theology.* Philadelphia: Westminster, 1965.

———, ed. *Back to Darwin: A Richer Account of Evolution.* Grand Rapids: Eerdmans, 2008.

Cobb, John B., Jr., and David Ray Griffin. *Process Theology: An Introductory Exposition.* Philadelphia: Westminster, 1976.

Collins, Francis S. *The Language of God: A Scientist Presents Evidence for Belief.* New York: Free Press, 2006.

Coogan, Jack. "Process and Art: Form as Correlative Principle." Unpublished paper for the conference "Process Theology and the Arts," Center for Process Studies, Claremont Graduate School, Claremont, CA, 1987.

Corey, Michael A. *The God Hypothesis: Discovering Design in Our "Just Right" Goldilocks Universe.* Lanham, MD: Rowman & Littlefield, 2001.

Cornwell, John, ed. *Explanations: Styles of Explanation in Science.* Oxford: Oxford University Press, 2004.

Courtillot, Vincent. *Evolutionary Catastrophes: The Science of Mass Extinction.* Translated by Joe McClinton. Cambridge: Cambridge University Press, 1999.

Cowen, Ron. "Exoplanets Make Pictorial Debut." *Science News* 174:12 (Dec. 6, 2008) 5–6.

———. "The Hunt for Habitable Planets." *Science News* 174:13 (Dec. 20, 2008) 16–20.

———. "Icy Rock Takes Long Way Home." *Science News* 174:6 (Sept. 13, 2008) 12.

———. "Mars Conceals Buried Icy Treasure." *Science News* 174:13 (Dec. 20, 2008) 10.

——— "Mars May Host Salty Reservoirs." *Science News* 175:8 (Apr. 11, 2009) 12.

———. "Moons Are Icy All the Way Through." *Science News* 175:9 (Apr. 25, 2009) 13.

———. "Multiverse Hosts Stellar Diversity." *Science News* 174:6 (Sept. 13, 2008) 12.

———. "Plumes of Martian Methane Hint at Possible Underground Microbial Life." *Science News* 175:4 (Feb. 14, 2009) 10.

———. "The Solar System's Big Bang." *Science News* 175:4 (Feb. 14, 2009) 26–29.

———. "Superstorms Swirl at Saturn's Poles." *Science News* 174:10 (Nov. 8, 2008) 9.

Craig, William Lane, and Quentin Smith. *Theism, Atheism, and Big Bang Cosmology.* Oxford: Clarendon, 1993.

Creager, Kenneth C. "Solid Evidence in the Inner Core." *Nature* 454 (Aug. 14, 2008) 833–35.

Curnoe, Darren. "Defining 'Human'—New Fossils Provide More Questions than Answers." *The Conversation* (Mar. 15, 2012). http://theconversation.com/defining-human-new-fossils-provide-more-questions-than-answers-5847.

———. "A Review of Early Homo in Southern Africa Focusing on Cranial, Mandibular and Dental Remains, with the Description of a New Species (Homo gautengensis sp. nov.)." *HOMO—Journal of Comparative Human Biology* 61 (2010) 151–77.

———, et al. "Human Remains from the Pleistocene-Holocene Transition of Southwest China Suggest a Complex Evolutionary History for East Asians." *PloS One* 7:3 (2012) e31918.

Damasio, Antonio. *Descartes' Error: Emotion, Reason and the Human Brain.* New York: Putnam, 1994.

d'Aquili, Eugene, and Andrew B. Newberg. *The Mystical Mind: Probing the Biology of Religious Experience.* Theology and the Sciences. Minneapolis: Fortress, 1999.

Darwin, Charles. *The Descent of Man.* New York: Penguin Classics, 2004

———. *The Origin of Species.* New York: Gramercy, 1979.

Davies, Paul. *Cosmic Jackpot: Why Our Universe Is Just Right for Life.* New York: Houghton Mifflin, 2007.

———. *The Fifth Miracle: The Search for the Origin and Meaning of Life.* New York: Simon & Schuster, 1999.

———. *The Goldilocks Enigma: Why Is the Universe Just Right for Life?* New York: Houghton Mifflin, 2006

———. *The Mind of God: The Scientific Basis for a Rational World.* New York: Simon & Schuster, 1992.

———. *Superforce: The Search for a Grand Unified Theory of Nature.* New York: Simon & Schuster, 1984.

Dawkins, Richard. *The Blind Watchmaker: Why the Evidence of Evolution Reveals a Universe without Design.* New York: Norton, 1986.

———. *The Selfish Gene.* Oxford: Oxford University Press, 2006.

Deacon, Terrence William. *The Symbolic Species: The Co-Evolution of Language and the Human Brain.* New York: Norton, 1997.

Deamer, David W., and Gail R. Fleischaker, eds. *Origins of Life: The Central Concepts.* Boston: Jones & Bartlett, 1994.

Dean, William D. *Coming To: A Theology of Beauty.* Philadelphia: Westminster, 1972.

Denton, Michael J. *Nature's Destiny: How the Laws of Biology Reveal Purpose in the Universe.* New York: Free Press, 1998.

Dirac, Paul A. M. *The Principles of Quantum Mechanics.* Oxford: Clarendon, 1930.

———. "The Quantum Theory of the Electron." *Proceedings of the Royal Society of London, Series A, Containing Papers of a Mathematical and Physical Character* 117 (Feb. 1928) 610–24.

Dixon, Thomas. "Scientific Atheism as a Faith Tradition." *Studies in History and Philosophy of Science* 33:2 (July 2002) 337–59.

Dobzhansky, Theodosius. *Mankind Evolving: The Evolution of the Human Species.* New Haven: Yale University Press, 1962.

Domning, Daryl P., and Monika Helwig. *Original Sinfulness: Original Sin and Evil in the Light of Evolution.* Aldershot, UK: Ashgate, 2006.

Drake, Nadia. "NASA Finds 'Definitive' Liquid Water on Mars." *National Geographic* (September 28, 2015). http://news.nationalgeographic.com/2015/09/150928-mars-liquid-water-confirmed-surface-streaks-space-astronomy/.

———. "Water on Mars: What Does It Really Mean?" *National Geographic* (Sept. 29, 2015). http://news.nationalgeographic.com/2015/09/150928-mars-liquid-water-life-space-astronomy/?utm_source=NatGeocom&utm_medium=Email&utm_content=wild_science_20151005_A&utm_campaign=Content&utm_rd=835721024/.

Draper, John William. *A History of the Conflict between Religion and Science.* New York: Appleton, 1874.

Dubay, Thomas. *The Evidential Power of Beauty: Science and Theology Meet.* San Francisco: Ignatius, 1999.

Dyson, Freeman. *Origins of Life.* Rev. ed. Cambridge: Cambridge University Press, 1999.

Eastman, Timothy E., and Hank Keeton, eds. *Physics and Whitehead: Quantum, Process, and Experience.* SUNY Series in Constructive Postmodern Thought. Albany: State University of New York Press, 2004.

Edwards, Rem B. *What Caused the Big Bang?* Amsterdam: Rodopi, 2001.

Ehrenberg, Rachel. "Animal Ancestor probably Survived Ancient Ice Age: Chemical Fossils Date Back to at Least 635 Million Years Ago." *Science News* 175:5 (Feb. 28, 2009) 13.

———. "Fossil Stokes Bird-Origin Feud." *Science News* 183:13 (June 29, 2013) 10.

Eigen, Manfred. *Steps towards Life: A Perspective on Evolution.* Oxford: Oxford University Press, 1992.

Emiliani, Cesare. *Planet Earth: Cosmology, Geology, and the Evolution of Life and Environment.* Cambridge: Cambridge University Press, 2007.

Ernest Becker Foundation. "A Four-part Sketch of Ernest Becker and His Work." http://www.ernestbecker.org/a-five-part-introdution-to-ernest-becker.html.

Fenchel, T. M. *The Origin and Early Evolution of Life.* Oxford: Oxford University Press, 2002.

Ferngren, Gary B. *Science and Religion: A Historical Introduction.* Baltimore: Johns Hopkins University Press, 2002.

Ferris, Timothy. *The Whole Shebang: A State-of-the-Universe Report.* New York: Simon & Schuster, 1997.

Flew, Anthony, and Roy Abraham Varghese. *There Is a God: How the World's Most Notorious Atheist Changed His Mind.* New York: HarperOne, 2007.

Fortey, Richard. *Earth: An Intimate History.* New York: Knopf, 2004.

———. *Life: A Natural History of the First Four Billion Years of Life on Earth.* New York: Knopf, 1998.

Freeman, Ken, and Geoff McNamara. *In Search of Dark Matter.* Berlin: Springer, 2006.

Fretheim, Terence E. *God and World in the Old Testament: A Relational Theology of Creation.* Nashville: Abingdon, 2005.

Fry, Iris. *The Emergence of Life on Earth: A Historical and Scientific Overview.* New Brunswick, NJ: Rutgers University Press, 2000.

Garrison, Tom. *Oceanography: An Invitation to Marine Science.* 6th ed. Belmont, CA: Thomson, 2007.

Gates, S. James, Jr. "The Fine Art of Not Knowing." http://www.youtube.com/watch?v=-nQDNNYyj4. 2:59–3:04 min.

———. "Who Is Afraid of Music?" Lecture 2 in *Superstring Theory: The DNA of Reality.* Chantilly, VA: Teaching Company, 2006. http://www.thegreatcourses.com/tgc/courses/course_detail.aspx?cid=1284.

Geggel, Laura. "'Bizarre,' Human-Size Sea Scorpion Found in Ancient Meteorite Crater" (Sept. 01, 2015). http://www.livescience.com/52052-ancient-sea-scorpion-fossils.html.

Gell-Mann, Murray. *The Quark and the Jaguar.* New York: Freeman, 1994.

Giacobini, Giacomo. "Richness and Diversity of Burial Rituals in the Upper Paleolithic." *Diogenes* 214 (2007) 19–39.

Gibbons, Ann. *The First Human: The Race to Discover Our Earliest Ancestors*. New York: Doubleday, 2006.

Giberson, Karl. "The Anthropic Principle: A Postmodern Creation Myth?" *Journal of Interdisciplinary Studies* 9 (1997) 63–90.

———. *Saving Darwin: How to Be a Christian and Believe in Evolution*. New York: HarperCollins, 2008.

Glendenning, Norman K. *After the Beginning: A Cosmic Journey through Space and Time*. London: Imperial College Press, 2004.

Goldsmith, Donald. *Einstein's Greatest Blunder? The Cosmological Constant and Other Fudge Factors in the Physics of the Universe*. Cambridge: Harvard University Press, 1995.

Gould, Stephen Jay. *Ever since Darwin: Reflections in Natural History*. New York: Norton, 1977.

———. *Leonardo's Mountain of Clams and the Diet of Worms: Essays on Natural History*. London: Cape, 1998.

———. "Nonoverlapping Magisteria." *Natural History* 106 (Mar. 1997) 16–22.

———. *Rocks of Ages: Science and Religion in the Fullness of Life*. Library of Contemporary Thought. New York: Ballantine, 1999.

———. *The Structure of Evolutionary Theory*. Cambridge, MA: Belknap, 2002.

———. *Wonderful Life: The Burgess Shale and the Nature of History*. New York: Norton, 1989.

Grant, Andrew. "Kepler Honored, not Mourned." *Science News* 183:12 (June 15, 2013) 10–11.

Grant, Edward. *Science and Religion 400 B.C. to A.D. 1550: From Aristotle to Copernicus*. Baltimore: Johns Hopkins University Press, 2006.

Greene, Brian. *The Elegant Universe: Superstrings, Hidden Dimensions, and the Quest for the Ultimate Theory*. New York: Norton, 1999.

———. *The Fabric of the Cosmos: Space, Time, and the Texture of Reality*. New York: Random House, 2004.

Grene, Marjorie, and David Depew. *The Philosophy of Biology: An Episodic History*. Evolution of Modern Philosophy. Cambridge: Cambridge University Press, 2004.

Gribbin, John. *Deep Simplicity: Bringing Order to Chaos and Complexity*. New York: Random House, 2004.

———. *In Search of the Big Bang: Quantum Physics and Cosmology*. New York: Bantam, 1986.

———, and Martin J. Rees. *Cosmic Coincidences: Dark Matter, Mankind, and Anthropic Cosmology*. New York: Bantam, 1989.

Gundling, Tom. *First in Line: Tracing Our Ape Ancestry*. New Haven: Yale University Press, 2005.

Gurnis, Michael. "Sculpting the Earth from Inside Out." *Scientific American* 284:3 (Mar. 2001) 40–47.

Guth, Alan. *The Inflationary Universe: The Quest for a New Theory of Cosmic Origins*. Reading, MA: Addison-Wesley, 1997.

Habermas, Gary. "My Pilgrimage from Atheism to Theism: A Discussion between Anthony Flew and Gary Habermas." *Philosophia Christi* 6:2 (2004) 197–211.

Harrison, Edward Robert. *Cosmology: The Science of the Universe.* 2nd ed. Cambridge: Cambridge University Press, 2000.

Hartshorne, Charles. *Creative Synthesis and Philosophic Method.* Lanham, MD: University Press of America, 1983.

———. *Man's Vision of God and the Logic of Theism.* Hamden, CT: Archon, 1964.

———. *Omnipotence and Other Theological Mistakes.* Albany: State University of New York Press, 1984.

Haught, John F. "Darwin and contemporary Theology." *Worldviews* 11 (2007) 44–57.

———. *God after Darwin: A Theology of Evolution.* 2nd ed. Philadelphia: Westview, 2008.

Hawking, Stephen W. *Black Holes and Baby Universes.* New York: Bantam, 1994.

———. *A Brief History of Time.* New York: Bantam, 1988.

———. *A Brief History of Time, Updated and Expanded Tenth Anniversary Edition.* New York: Bantam, 1996.

———. *The Theory of Everything: The Origin and Fate of the Universe.* New ed. Beverly Hills: Phoenix, 2007.

———. *The Universe in a Nutshell.* New York: Bantam, 2001.

——— et al. *The Large-Scale Structure of Space-Time.* Cambridge Monographs on Mathematical Physics. Cambridge: Cambridge University Press, 1975.

———, and Roger Penrose. *The Nature of Space and Time.* Princeton: Princeton University Press, 1996.

Hawley, John F., and Katherine A. Holcomb. *Foundations of Modern Cosmology.* 2nd ed. Oxford: Oxford University Press, 2005.

Hays, Richard B. *The Moral Vision of the New Testament: Community, Cross, New Creation: A Contemporary Introduction to New Testament Ethics.* San Francisco: HarperSanFrancisco, 1996.

Hazelton, Roger. *A Theological Approach to Art.* Nashville: Abingdon, 1967.

Hazen, Robert M. *Genesis: The Scientific Quest for Life's Origins.* Washington, DC: Joseph Henry, 2005

———. *Origins of Life.* Chantilly, VA: Teaching Company, 2005.

Hoffman, Paul F., and Daniel P. Schrag. "Snowball Earth." *Scientific American* 282:1 (Jan. 2000) 68–75.

Hofstadter, Albert, and Richard Kuhns, eds. *Philosophies of Art and Beauty: Selected Readings in Aesthetics from Plato to Heidegger.* Chicago: University of Chicago Press, 1964.

Hofstadter, Douglas R. *Gödel, Escher, Bach: An Eternal Golden Braid.* 20th anniversary ed. New York: Basic Books, 1999.

Hogan, Craig. *The Little Book of the Big Bang: A Cosmic Primer.* New York: Springer, 1998.

Holder, Rodney D. *God, the Multiverse, and Everything: Modern Cosmology and the Argument from Design.* Aldershot, UK: Ashgate, 2004.

Holland, John. *Emergence: From Chaos to Order.* New York: Basic Books, 1999.

Horgon, John. *The End of Science: Facing the Limits of Knowledge in the Twilight of the Scientific Age.* Reading, MA: Addison-Wesley, 1996.

Howley, Andrew. "Homo naledi: 1,500 Fossils Revolutionize Human Family Tree." Rising Star Expedition, National Geographic Society (Sept. 10, 2015). http://voices.nationalgeographic.com/2015/09/10/homo-naledi-1500-fossils-cave-revolutionize-human-evolution-family-tree/.

Hoyle, Fred, and N. C. Wickramasinghe. *Astronomical Origins of Life: Steps towards Panspermia.* Dordrecht: Kluwer Academic, 2000.

Jablonka, Eva, and Marion J. Lamb. *Evolution in Four Dimensions: Genetic, Epigenetic, Behavioral, and Symbolic Variation in the History of Life.* Cambridge: MIT Press, 2005.

James, William. *The Will to Believe and Other Essays in Popular Philosophy.* New York: Longmans, Green, 1897.

Javaux, Emmanuelle J., et al. "Organic-walled Microfossils in 3.2-Billion-Year-Old Shallow Marine Siliciclastic Deposits." *Nature* 463 (Feb. 18, 2010) 934–38.

Johnson, George. *Fire in the Mind: Science, Faith, and the Search for Order.* New York: Vintage: 1996.

Jones, Mark, and Robert Lambourne. *An Introduction to Galaxies and Cosmology.* Cambridge: Open University Press, 2005.

Jungerman, John A. *World in Process: Creativity and Interconnection in the New Physics.* Albany: State University of New York Press, 2000.

Kalajdzievski, Sasho, and R. Padmanabhan. *Math and Art: An Introduction to Visual Mathematics.* Boca Raton: CRC, 2008.

Kasting, James F. "The Origins of Water on Earth." *Scientific American* 13:3 (2003) 28–33.

Kauffman, Stuart A. *At Home in the Universe: The Search for the Laws of Self-Organization and Complexity.* Oxford: Oxford University Press, 1995.

———. *Investigations.* New York: Oxford University Press, 2000.

———. *Reinventing the Sacred: A New View of Science, Reason, and Religion.* New York: Basic Books, 2008.

Keel, William C. *The Road to Galaxy Formation.* 2nd ed. Berlin: Springer, 2007.

Knoll, Andrew H. *Life on a Young Planet: The First Three Billion Years of Evolution on Earth.* Princeton: Princeton University Press, 2003.

Koenigswald, G. H. R. von. *The Evolution of Man.* Translated by Arnold J. Pomerans. Rev. ed. Ann Arbor Science Library. Ann Arbor: University of Michigan Press, 1976.

Kolb, Edward, and Michael Turner. *The Early Universe: Frontiers in Physics.* Boulder: Westview, 1994.

Kovac, Jeffrey, and Michael Weisberg, eds. *Roald Hoffmann on the Philosophy, Art and Science of Chemistry.* Oxford: Oxford University Press, 2012.

Kragh, Helge. *Conceptions of Cosmos: From Myths to the Accelerating Universe: A History of Cosmology.* Oxford: Oxford University Press, 2007.

Krauss, Lawrence M. *Hiding in the Mirror: The Mysterious Allure of Extra Dimensions, From Plato to String Theory and Beyond.* New York: Viking, 2005.

Kuhn, Thomas. *The Structure of Scientific Revolutions.* Chicago: University of Chicago Press, 1962

Lahav, Noam. *Biogenesis: Theories of Life's Origin.* Oxford: Oxford University Press, 1999.

Lamsdell, James C., et al. "The oldest described eurypterid: a giant Middle Ordovician (Darriwilian) megalograptid from the Winneshiek Lagerstätte of Iowa." *BMC Evolutionary Biology* 15:169 (Sept. 1, 2015). http://www.biomedcentral.com/1471-2148/15/169.

Lederman, Leon, and Christopher T. Hill. *Symmetry and the Beautiful Universe.* Amherst, NY: Prometheus, 2004.

Lemonick, Michael. *Echo of the Big Bang.* Princeton: Princeton University Press, 2003.

Leslie, John. *Universes.* London: Routledge, 1989.

Lincoln, Don. *Understanding the Universe: From Quarks to the Cosmos.* Hackensack, NJ: World Scientific Publishing Co., 2004.

Lindberg, David C. *The Beginnings of Western Science: The European Scientific Tradition in Philosophy, Religion and Institutional Context, Pre-History to A.D. 1450.* 2nd ed. Chicago: University of Chicago Press, 2008.

Lindberg, David C., and R. L. Numbers, eds. *God and Nature: Historical Essays on the Encounter between Christianity and Natural Science.* Berkeley: University of California Press, 1986.

Lipton, Peter. *Inference to the Best Explanation.* 2nd ed. London: Routledge, 2004.

Livingstone, David N. *Darwin's Forgotten Defenders: The Encounter Between Evangelical Theology and Evolutionary Thought.* Grand Rapids: Eerdmans, 1987.

Livio, Mario. *The Accelerating Universe: Infinite Expansion, the Cosmological Constant, and the Beauty of the Cosmos.* New York: Wiley, 2000.

Longair, Malcom. *The Cosmic Century: A History of Astrophysics and Cosmology.* Cambridge: Cambridge University Press, 2006.

Luisi, Pier Luigi. "Some Open Questions about the Origin of Life." In *Fundamentals of Life,* edited by Gyula Palyi, et al., 287–301. Paris: Elsevier, 2002.

Macdougall, Douglas. *Frozen Earth: The Once and Future Story of Ice Ages.* San Diego: University of California Press, 2004.

Manson, Neil A. *God and Design: The Teleological Argument and Modern Science.* London: Routledge, 2003.

Marques-Bonet, Tomas, et al. "A Burst of Segmental Duplication in the Genome of the African Great Ape Ancestor." *Nature* 457 (Feb. 12, 2009) 877–81.

Masci, David. "Scientists and Belief." Religion and Science in the United States, Pew Research Religion & Public Life Project (Nov. 5, 2009). http://www.pewforum.org/2009/11/05/scientists-and-belief/.

May, Brian, Patrick Moore, and Chris Lintott. *Bang: The Complete History of the Universe.* London: Carlton, 2006.

May, Gerhard. *Creatio ex Nihilo: The Doctrine of "Creation out of Nothing" in Early Christian Thought.* Edinburgh: T. & T. Clark, 1995.

Mayr, Ernst. *What Evolution Is.* New York: Basic Books, 2001.

McGrath, Alister E. *Dawkins' God: Genes, Memes and the Meaning of Life.* Oxford: Blackwell, 2004.

———. *A Fine-Tuned Universe: The Quest for God in Science and Theology.* Louisville: Westminster John Knox, 2009.

———. *The Open Secret: A New Vision for Natural Theology.* Oxford: Blackwell, 2008.

———, ed. *The Order of Things: Explorations in Scientific Theology.* Oxford: Blackwell, 2006.

———. *Nature. A Scientific Theology,* Vol. 1. London: Continuum, 2001.

———. *Reality. A Scientific Theology,* Vol. 2. London: Continuum, 2002.

———. *Theory. A Scientific Theology,* Vol. 3. London: Continuum, 2003.

McMullin, Ernan. "Cosmic Purpose and the Contingency of Human Evolution." *Theology Today* 55 (1998) 389–414.

Medawar, Peter B. *The Limits of Science.* Oxford: Oxford University Press, 1985.

Miller, Kenneth R. *Finding Darwin's God: A Scientist's Search for Common Ground between God and Evolution.* New York: HarperCollins, 1999.

———. *Only a Theory: Evolution and the Battle for America's Soul.* New York: Penguin, 2008.

Mithen, Steven. *After the Ice: A Global Human History 20,000–5,000 B.C.* Cambridge: Harvard University Press, 2006.

Mlodinow, Leonard. *Feynman's Rainbow: A Search for Beauty in Physics and in Life.* New York: Vintage, 2011.

Moltmann, Jürgen. *God in Creation: An Ecological Doctrine of Creation. The Gifford Lectures 1984–1985.* London: SCM, 1985.

Morowitz, Harold J. *The Emergence of Everything: How the World Became Complex.* New York: Oxford University Press, 2002.

Morris, Simon Conway. *Life's Solution: Inevitable Humans in a Lonely Universe.* Cambridge: Cambridge University Press, 2003.

———, ed. *The Deep Structure of Biology: Is Convergence Sufficiently Ubiquitous to Give a Directional Signal?* West Conshohocken, PA: Templeton Foundation, 2008.

Morris, Simon Conway, and J. S. Peel. "The Earliest Annelids: Lower Cambrian Polychaetes from the Sirius Passet Lagerstätte, Peary Land, North Greenland." *Acta Palaeontologica Polonica* 53:1 (2008) 137–48.

Morrison, David, and Tobias Owen. *The Planetary System.* San Francisco: Benjamin Cummings, 2002.

Motz, Lloyd, and Jefferson Hane Weaver. *The Story of Physics.* New York: Avon, 1989.

Mukhanov, Viatcheslav. *Physical Foundations of Cosmology.* Cambridge: Cambridge University Press, 2005.

Murphy, Nancey, and William R. Stoeger, eds. *Evolution and Emergence: Systems, Organisms, Persons.* Oxford: Oxford University Press, 2007.

NASA. "NASA Spacecraft Returns New Images of Pluto En Route to Historic Encounter." (Feb. 4, 2015). https://www.nasa.gov/press/2015/february/nasa-spacecraft-returns-new-images-of-pluto-en-route-to-historic-encounter/.

NASA, Jet Propulsion Laboratory, California Institute of Technology. "Finding Another Earth" (July 23, 2015). http://www.jpl.nasa.gov/news/news.php?feature=4666/.

NASA, Jet Propulsion Laboratory, California Institute of Technology. "NASA's Kepler Mission Discovers Bigger, Older Cousin to Earth" (July 23, 2015). http://www.jpl.nasa.gov/news/news.php?feature=4665/.

National Radio Astronomy Observatory. "Massive Cloud Speeding toward Collision with Milky Way." *Science Daily* (1/13/08). www.sciencedaily.com/releases/2008/10/080112153747.htm/.

Nelson, Paul A. "The Role of Theology in Current Evolutionary Reasoning." *Biology and Philosophy* 11 (1996) 493–517.

Newberg, Andrew B. *Principles of Neurotheology.* Ashgate Science and Religion Series. Farnham, UK: Ashgate, 2010.

Newberg, Andrew B., and Mark Robert Waldman. *Born to Believe: God, Science, and the Origin of Ordinary and Extraordinary Beliefs.* New York: Free Press, 2006.

———. *Why We Believe What We Believe: Uncovering Our Biological Need for Meaning.* New York: Free Press, 2006

Newberg, Andrew B. et al. *Why God Will Not Go Away: Brain Science and the Biology of Belief.* New York: Ballantine, 2001.

Newman, John Henry. *An Essay in Aid of a Grammar of Assent.* Notre Dame: University of Notre Dame Press, 1979.

Nicolson, Ian. *The Dark Side of the Universe: Dark Matter, Dark Energy, and the Fate of the Cosmos.* Baltimore: Johns Hopkins University Press, 2007.

Nietzsche, Friedrich. *Beyond Good and Evil.* Translated by Marianne Cowan. Chicago: Regnery, 1955.

Oerter, Robert. *The Theory of Almost Everything: The Standard Model, The Unsung Triumph of Modern Physics.* New York: Pi, 2005.

Ogden, Schubert M. "Love Unbounded: The Doctrine of God." *Perkins School of Theology Journal* 19:3 (1966) 5–17.

———. *The Reality of God and Other Essays.* San Francisco: Harper & Row, 1963.

———. *The Understanding of Christian Faith.* Eugene, OR: Cascade Books, 2010.

Olson, Richard G. *Science and Religion 1450–1900: From Copernicus to Darwin.* Baltimore: Johns Hopkins University Press, 2006.

O'Neil, Jonathan, et al. "Neodymium-142 Evidence for Hadean Mafic Crust." *Science* 321 (Sept. 26, 2008) 1828–31.

Outler, Albert C. "The Wesleyan Quadrilateral in Wesley." *Wesleyan Theological Journal* 20 (1985) 7–18.

Pagel, Bernard. *Nucleosynthesis and Chemical Evolution of Galaxies.* Cambridge: Cambridge University Press, 1997.

Pappalardo, Robert T., et al. "The Hidden Ocean of Europa." *Scientific American* 281:4 (1999) 54–63.

Parker, Becky. "Art, Creativity, and Process Thought." Unpublished manuscript for the conference "Process Theology and the Arts," Center for Process Studies, Claremont Graduate School, Claremont, CA, 1987.

Paley, William. *Natural Theology: Or Evidences of the Existence and Attributes of the Deity Collected from the Appearances of Nature.* London: Faulder, 1802.

Peacocke, Arthur. *Paths from Science towards God: The End of All Our Exploring.* Oxford: Oneworld, 2001.

———, ed. *The Sciences and Theology in the Twentieth Century.* Notre Dame: University of Notre Dame Press, 1981.

———. *Theology for a Scientific Age: Being and Becoming—Natural, Divine and Human.* Minneapolis: Fortress, 1993.

Pelikan, Jaroslav. *The Emergence of the Catholic Tradition (100–600). The Christian Tradition: A History of the Development of Doctrine*, vol. 1. Chicago: University of Chicago Press, 1971.

———. *The Growth of Medieval Theology (600–1300). The Christian Tradition: A History of the Development of Doctrine*, vol. 3. Chicago: University of Chicago Press, 1971.

———. *The Spirit of Eastern Christendom (600–1700). The Christian Tradition: A History of the Development of Doctrine*, vol. 2. Chicago: University of Chicago Press, 1971.

Penrose, Roger. *The Road to Reality: A Complete Guide to the Laws of the Universe.* New York: Random House, 2004.

———, et al. *The Large, the Small and the Human Mind.* Cambridge: Cambridge University Press, 1997.

Perkins, Sid. "Dates in Doubt for Oldest Evidence for Complex Life: New Analyses of Ancient Rocks Hint Biomarkers Came Later." *Science News* 174:11 (Nov. 22, 2008) 5–6.

———. "Scientists Tap into Deep Magma." *Science News* 175:2 (Jan. 17, 2009) 13.

———. "Step-by-step Evolution." *Science News* 175:3 (Jan. 31, 2009) 30–33.

———. "Tiny Crystals in Australian Rocks Suggest Earlier Debut for Oxygen." *Science News* 175:8 (Apr. 11, 2009) 9.

Petit, Charles. "UltraMassive: As Big as It Gets." *Science News* 174:9 (Oct. 25, 2008) 18–23.

Plummer, Charles C. et al. *Physical Geology.* New York: McGraw-Hill, 2007.

Polanyi, Michael. *Personal Knowledge: Towards a Post-Critical Philosophy.* Chicago: University of Chicago Press, 1958.

Polkinghorne, John C. *Belief in God in an Age of Science.* New Haven: Yale University Press, 1998.

———. *Exploring Reality: The Intertwining of Science and Religion.* New Haven: Yale University Press, 2005.

———. *The God of Hope and the End of the World.* New Haven: Yale University Press, 2002.

———. *Quarks, Chaos & Christianity: Questions to Science and Religion.* Rev. ed. New York: Crossroad, 2005.

———. *Traffic in Truth: Exchanges between Science and Theology.* Minneapolis: Fortress, 2002.

Popa, Radu. *Between Necessity and Probability: Searching for the Definition and Origin of Life.* Berlin: Springer, 2004.

Prevost, Robert. *Probability and Theistic Explanation.* Oxford: Clarendon, 1990.

Principe, Lawrence M. *Science and Religion.* Chantilly, VA: Teaching Company, 2006.

Prothero, Donald R. *Bringing Fossils to Life: An Introduction to Paleobiology.* 2nd ed. New York: McGraw-Hill, 2004.

Prothero, Donald R., and Robert H. Dott, Jr. *Evolution of the Earth.* 8th ed. New York: McGraw-Hill, 2010.

Quammen, David. "Was Darwin Wrong?" *National Geographic* 206:5 (Nov. 2004) 2–35.

Ranganathan, Shiyali Ramamrita. *Ramanujan, the Man and the Mathematician.* Great Thinkers of India. Bombay: Asia Publishing House, 1967.

Reasoner, Mark. *Romans in Full Circle: A History of Interpretation.* Louisville: Westminster John Knox Press, 2005.

Rees, Martin. *Before the Beginning: Our Universe and Others.* Cambridge: Perseus, 1998.

———. *Just Six Numbers: The Deep Forces that Shape the Universe.* New York: Basic Books, 2000.

———. *New Perspectives in Astrophysical Cosmology.* 2nd ed. Cambridge: Cambridge University Press, 2000.

———. *Our Cosmic Habitat.* Princeton: Princeton University Press, 2001.

Regal, Brian. *Human Evolution: A Guide to the Debates.* Santa Barbara: ABC-CLIO, 2004.

Reid, Robert G. B. *Biological Emergences: Evolution by Natural Experiment.* Cambridge: MIT Press, 2007.

Rosenblum, Bruce, and Fred Kuttner. *Quantum Enigma: Physics Encounters Consciousness.* Oxford: Oxford University Press, 2008

Rudwick, Martin J. S. *The Meaning of Fossils: Episodes in the History of Paleontology.* 2nd ed. Chicago: University of Chicago, 1976.

Ruse, Michael, and Joseph Travis, eds. *Evolution: The First Four Billion Years.* Cambridge: Harvard University Press, 2009.

Russell, Bertrand. *Mysticism and Logic and Other Essays.* London: Longman, 1919.

Russell, Robert John. *Cosmology: From Alpha to Omega.* Minneapolis: Fortress, 2008.

Russell, Robert John et al., eds. *Physics, Philosophy, and Theology.* Rome: Vatican Observatory, 1988.

———, eds. *Evolutionary and Molecular Biology: Scientific Perspectives on Divine Action.* Rome: Vatican Observatory, 1999.

Saey, Tina Hesman. "Dissing a Loaded Label for Some Unicellular Life." *Science News* 175:8 (Apr. 11, 2009) 5–6.

———. "Shared Differences: The Architecture of our Genomes is Anything but Basic." *Science News* 175:9 (Apr. 25, 2009) 16–20.

———. "Team Decodes Neandertal DNA." *Science News* 175:6 (Mar. 14, 2009) 5–7.

Sample, Ian. "'Red Deer Cave People' Could Be New Species of Human." *The Guardian* (Mar. 14, 2012). http//guardian.co.uk/science/2012/mar/14/red-deer-cave-people-species-human.

Sanders, Laura. "Forget Bird-Brained." *Science News* 174:9 (Oct. 25, 2008) 14.

Sawyer, G. J., and Viktor Deak, creators; text by Esteban Sarmiento, et al. *The Last Human: A Guide to Twenty-Two Species of Extinct Humans.* New Haven: Yale University Press, 2007.

Schneider, Michael S. *A Beginner's Guide to Constructing the Universe: Mathematical Archetypes of Nature, Art and Science.* New York: Harper Perennial, 1995.

Schneider, Peter. *Extragalactic Astronomy and Cosmology: An Introduction.* Berlin: Springer, 2006.

Schopf, J. William. *Cradle of Life: The Discovery of Earth's Earliest Fossils.* Princeton: Princeton University Press, 2001.

Sheppard, Anne. *Aesthetics: An Introduction to the Philosophy of Art.* Oxford: Oxford University Press, 1987.

Shreeve, Jamie. "This Face Changes the Human Story. But How?" *National Geographic* (Sept. 10, 2015). http://news.nationalgeographic.com/2015/09/150910-human-evolution-change/.

Siegfried, Tom. "It Is Likely That Times Are Changing." *Science News* 174:6 (Sept. 13, 2008) 26–28.

———. "Light in the Dark." *Science News* 183:1 (Jan. 12, 2013) 18–20.

———. "Strings Link the Ultracold with the Superhot." *Science News* 175:9 (Apr. 25, 2009) 26–28.

Silk, Joseph. *The Big Bang.* 3rd ed. New York: Holt, 2002.

——— *The Infinite Cosmos: Questions from the Frontiers of Cosmology.* Oxford: Oxford University Press, 2008.

———. *On the Shores of the Unknown: A Short History of the Universe.* Cambridge: Cambridge University Press, 2005.

Sinclair, Nathalie, et al., eds. *Mathematics and the Aesthetic: New Approaches to an Ancient Affinity.* New York: Springer, 2006.

Singh, Simon. *Big Bang: The Origin of the Universe.* New York: Harper Perennial, 2005.

Smith, Cameron M., and Charles Sullivan. *The Top 10 Myths about Evolution.* Amherst, NY: Prometheus, 2007.

Smith, John Maynard, and Eors Szathmary. *The Major Transitions in Evolution.* Oxford: Freeman, 1995.

———. *The Origins of Life: From the Birth of Life to the Origin of Language.* New York: Oxford University Press, 1999.

Smith, Michael D. *The Origin of Stars.* London: Imperial College Press, 2004.

Smolin, Lee. *The Life of the Cosmos*. New York: Oxford University Press, 1997.

———. *Three Roads to Quantum Gravity*. New York: Basic Books, 2001.

Smoot, George, and Keay Davidson. *Wrinkles in Time: Witness to the Birth of the Universe*. New York: HarperCollins, 1994.

Sobel, Dava. *The Planets*. New York: Penguin, 2006.

Sponheim, Paul R. *The Pulse of Creation: God and the Transformation of the World*. Minneapolis: Fortress, 1999.

Srivastava, Mansi, et al. "The Amphimedon queenslandica Genome and the Evolution of Animal Complexity." *Nature* 466 (Aug. 5, 2010) 720–26.

Stanley, Steven M. *Earth System History*. 2nd ed. New York: Freeman, 2005.

Stern, Alan. "Debates Over Definition of Planet Continue and Inspire." *Science News* 174:12 (Dec. 6, 2008) 32.

Streeter, Jarvis. *Human Nature, Human Evil, and Religion: Ernest Becker and Christian Theology*. Lanham, MD: University Press of America, 2009.

Strogatz, Steven. *Chaos*. Chantilly, VA: Teaching Company, 2008.

———. *Sync: How Order Emerges from Chaos in the Universe, Nature, and Daily Life*. New York: Hyperion, 2003.

Suchocki, Marjorie Hewitt. *God, Christ, Church: A Practical Guide to Process Theology*. New revised ed. New York: Crossroad, 1992.

Susskind, Leonard. *The Cosmic Landscape: String Theory and the Illusion of Intelligent Design*. New York: Little, Brown, 2005.

Tarbuck, Edward J., and Fred Lutgens. *Earth: An Introduction to Physical Geology*. 9th ed. Upper Saddle River, NJ: Prentice Hall, 2008.

Tegmark, Max. "On the Dimensionality of Spacetime." *Classical and Quantum Gravity* 14:4 (1997) L69–L75.

Thirring, Walter. *Cosmic Impressions: Traces of God in the Laws of Nature*. Philadelphia: Templeton Foundation, 2008.

Thistlethwaite, Susan Brooks, ed. *Adam, Eve, and the Genome: The Human Genome Project and Theology*. Theology and the Sciences. Minneapolis: Fortress, 2003.

Thomas Aquinas, Saint. *Summa Contra Gentiles*, Book 1. Translated by Anton Pegis. Notre Dame: University of Notre Dame Press, 1975.

———. *Summa Theologica. First Complete American Edition*. Translated by the Fathers of the English Dominican Province. Vol. 1. New York: Benziger: 1947.

Thomas, Paul J., et al., eds. *Comets and the Origin and Evolution of Life*. Berlin: Springer, 1997.

Tipler, Frank J. *The Physics of Immortality: Modern Cosmology, God and the Resurrection of the Dead*. New York: Doubleday, 1994.

Tyson, Neil DeGrasse, and Donald Goldsmith. *Origins: Fourteen Billion Years of Cosmic Evolution*. New York: Norton, 2004.

———, et al. *One Universe: At Home in the Cosmos*. Washington, DC: Joseph Henry, 2000.

Ulmschneider, Peter. *Intelligent Life in the Universe: From Common Origins to the Future of Humanity*. New York: Springer, 2003.

van Andel, Tjeerd H. *New Views on an Old Planet: Continental Drift and the History of the Earth*. 2nd ed. Cambridge: Cambridge University Press, 1994.

Vilenkin, Alex. *Many Worlds in One: The Search for Other Universes*. New York: Hill & Wang, 2007.

Wadhwa, Meenakshi, and Sara S. Russell. "Timescales of Accretion and Differentiation in the Early Solar System: The Meteoritic Evidence." In *Protostars and Planets IV*, edited by V. Mannings and Sara S. Russell, 995–1018. Tucson: University of Arizona Press, 2000.

Walter, Fabian, et al. "A Kiloparsec-scale Hyper-starburst in a Quasar Host Less Than 1 Gigayear after the Big Bang." *Nature* 457 (Feb. 5, 2009) 699–700.

Ward, Keith. *Divine Action: Examining God's Role in an Open and Emergent Universe.* Philadelphia: Templeton Foundation, 2007.

———. *Religion and Creation.* Oxford: Oxford University Press, 1996.

———. *Re-thinking Christianity.* Oxford: OneWorld, 2007.

Ward, Peter Douglas, and Donald Brownlee. *Rare Earth: Why Complex Life is Uncommon in the Universe.* New York: Springer, 2003.

Watts, Fraser. *Creation: Law and Probability.* Theology and the Sciences. Minneapolis: Fortress, 2008.

Weinberg, Steven. *Dreams of a Final Theory.* New York: Pantheon, 1992.

———. *The First Three Minutes: A Modern View of the Origin of the Universe.* 2nd ed. New York: Basic Books, 1993.

Weidenreich, Franz, and G. H. R. von Koenigswald. "Morphology of Solo Man." *Anthropological Papers of the American Museum of Natural History* 43:3 (1951) 205–90.

WGBH Educational Foundation. "Magnetic Storm." *NOVA* (Nov. 18, 2003).

Wharton, David A. *Life at the Limits: Organisms in Extreme Environments.* Cambridge: Cambridge University Press, 2002.

White, Andrew Dickson. *A History of the Warfare of Science with Theology in Christendom.* New York: Appleton, 1896

White, Randall. "Systems of Personal Ornamentation in the Early Upper Palaeolithic: Methodological Challenges and New Observations." In *Rethinking the Human Revolution: New Behavioural and Biological Perspectives on the Origin and Dispersal of Modern Humans*, edited by Paul Mellars, et al., 287–302. Cambridge, UK: McDonald Institute for Archaeological Research, 2007.

Whitehead, Alfred North. *Adventures of Ideas.* New York: Free Press, 1967.

———. *The Function of Reason.* Boston: Beacon, 1958.

———. *Modes of Thought.* New York: Free Press, 1968.

Whittle, Mark. *Cosmology: The History and Nature of Our Universe.* Chantilly, VA: Teaching Company, 2008.

Wigner, Eugene. "On the Unreasonable Effectiveness of Mathematics in the Natural Sciences." *Communications on Pure and Applied Mathematics* 13 (1960) 1–14.

Wiseman, James A. *Theology and Modern Science: Quest for Coherence.* New York: Continuum, 2002.

Witze, Alexandra. "At Home in the Universe." *Science News* 181:12 (June 16, 2012) 22–25.

———. "Indian Ocean Set Stage for Humans." *Science News* 181:5 (Mar. 10, 2012) 11.

Wolfson, Richard. *Einstein's Relativity and the Quantum Revolution: Modern Physics for Non-Scientists.* 2nd ed. Chantilly, VA: Teaching Company, 2000.

Wolpert, Lewis. *The Unnatural Nature of Science.* Cambridge: Harvard University Press, 1993.

Worthing, Mark William. *God, Creation, and Contemporary Physics.* Theology and the Sciences. Minneapolis: Fortress, 1996.

Wunn, Ina. "Cave Bear Worship in the Palaeolithic." *Cadernos Lab. Xeolóxico de Laxe Coruña* 26 (2001) 457–63.

———. "Bärenkult in urgeschichtlicher Zeit. Zur religiösen Deutung mittelpaläolithischer Bärenfossilien." *Zeitschrift für Religionswissenschaft* 7 (1999) 3–23.

———. "Beginning of Religion." *Numen* 47 (2000) 417–52.

Wysession, Michael E. *How the Earth Works.* Chantilly, VA: The Teaching Company, 2008.

Zee, Anthony *Fearful Symmetry: The Search for Beauty in Modern Physics.* Princeton: Princeton University Press, 1999.

RESOURCES ONLINE

www.aip.org/history/cosmology/. The American Institute of Physics site, Cosmic Journey: A History of Scientific Cosmology.

www.astro.ucla.edu/~wright/cosmolog.htm/. Edward N. (Ned) Wright's Cosmology Tutorial, with current information on developments in cosmology.

www.astro.virginia.edu/~dmw8f/BBA_web/index/frames/.html/. Mark Whittle's site on Big Bang acoustics.

www.britannica.com/. A good source for general science information.

www.cpepweb.org/. The Contemporary Physics Education Project's site on particle physics, including The Universe Adventure.

www.en.wikipedia.com/. Many excellent articles on the whole sweep on scientific topics.

www.enc.tfode.com/. Another online encyclopedia that has articles on scientific topics.

http://www.esa.int/Our_Activities/Space_Science/Rosetta/Highlights/Landing_on_a_comet/. About Europe's "comet chaser" spacecraft landing on a comet November 2014.

www.eso.org/public/. Site of the European Organization for Astronomical Research in the Southern Hemisphere.

www.firstgalaxies.ucolick.org/galaxies.html/. Lick Observatory site exploring the first galaxies in the early universe.

www.gwu.edu/~darwin/BiSc151/Eukaryotes/Eukaryotes.html/. The George Washington University site on evolution.

www.hubblesite.org/. NASA's Hubble Telescope site, with loads of information and photographs taken by Hubble.

www.humanorigins.si.edu/. The Smithsonian site regarding human evolution.

www.jwst.nasa.gov/. NASA's site for the James Webb Space Telescope, which was scheduled to launch in 2014 and replace the Hubble Space Telescope.

www.map.gsfc.nasa.gov/universe/. NASA's Universe 101 site for general audiences.

www.palaeos.org. An accessible site regarding the history of life on earth.

www.particleadventure.org/. Lawrence Berkeley National Laboratory site, The Particle Adventure, dealing with particle physics in an accessible form.

www.plato.stanford.edu. The excellent Stanford University encyclopedia of philosophy.

www.preposterousuniverse.com/writings/cosmologyprimer/. Sean Carroll's Cosmology Primer for general audiences.

www.science.nationalgeographic.com/. The website of the National Geographic Society, featuring many helpful articles on all aspects of science.

www.sciencenews.org/. The website for the Society for Science and the Public, publisher of the periodical *Science News,* containing excellent articles on major discoveries in all areas of science.

www.superstringtheory.com/. Patricia Schwarz's site on superstring theory for non-specialists.

www.talkorigins.org/faqs/astronomy/bigbang.html/. Bjorn Feuerbacher and Ryan Scranton's site, Evidence for the Big Bang, refuting creationist pseudoscience.

www.ucmp.berkeley.edu/. The website of the University of California Museum of Paleontology.

www.usgs.gov/. U.S. Geological Survey's Science for a Changing World site, dealing with biology, geology, ecology, and geography.

Index of Names and Authors

Index of Subjects

Index of Biblical Passages

1 Samuel

2 Samuel

2 Kings

1 Chronicles

2 Chronicles

Psalms

Isaiah

Daniel

Jonah

Matthew

Mark

Luke

John

Acts

Romans

1 Corinthians

Galatians

Ephesians

Philippians

1 Thessalonians

1 Timothy

2 Timothy

32500372R00279

Made in the USA
San Bernardino, CA
07 April 2016